Multiscale and Innovative Kinetic Approaches in Heterogeneous Catalysis

Multiscale and Innovative Kinetic Approaches in Heterogeneous Catalysis

Special Issue Editors

Pascal Granger
Yves Schuurman

MDPI • Basel • Beijing • Wuhan • Barcelona • Belgrade

MDPI

Special Issue Editors

Pascal Granger
CNRS
France

Yves Schuurman
Université Claude Bernard Lyon 1
France

Editorial Office
MDPI
St. Alban-Anlage 66
4052 Basel, Switzerland

This is a reprint of articles from the Special Issue published online in the open access journal *Catalysts* (ISSN 2073-4344) from 2018 to 2019 (available at: https://www.mdpi.com/journal/catalysts/special_issues/Kinetic_Heterogeneous)

For citation purposes, cite each article independently as indicated on the article page online and as indicated below:

LastName, A.A.; LastName, B.B.; LastName, C.C. Article Title. *Journal Name* **Year**, *Article Number*, Page Range.

ISBN 978-3-03921-179-1 (Pbk)
ISBN 978-3-03921-180-7 (PDF)

Contents

About the Special Issue Editors

Yves Schuurman is a Senior Researcher (Directeur de Recherche) at IRCELYON. IRCELYON is a shared laboratory between CNRS and University of Lyon that is devoted to catalysis and the environment. Schuurman's research interests lie in the area of reaction mechanisms and kinetics of heterogeneous catalyzed reactions with a special focus on transient and dynamic methods. The areas of application include syngas conversion, C1 chemistry and, more recently, thermochemical biomass conversion. He is the co-author of more than 140 papers.

Pascal Granger is Distinguished Professor at the University of Lille, member of the French Chemical Society. His research is focused on reaction mechanisms and kinetics of catalytic reactions involved in post-combustion catalysis and catalytic DeNOx and DeN2O processes for mobile and stationary sources. His research interests are also focused on the development of perovskites as potential substitutes of PGM. Granger is the co-author of more than 139 papers and co-editor of two Wiley Books dedicated to perovskite materials (2016) and one book series entitled 'Past and Present in DeNOx catalysis' (2007) published by Elsevier.

catalysts

MDPI

Editorial

Multiscale and Innovative Kinetic Approaches in Heterogeneous Catalysis

Yves Schuurman [1] and Pascal Granger [2],*

[1] Institut de Recherches sur la Catalyse et l'Environnement, Ircelyon UMR 5256, Université Claude Bernard Lyon 1, 2 Avenue Albert Einstein, 69626 Villeurbanne CEDEX, France; yves.schuurman@ircelyon.univ-lyon1.fr

[2] Univ. Lille, CNRS, Centrale Lille, ENSCL, Univ. Artois, UMR 8181—UCCS—Unité de Catalyse et Chimie du Solide, F-59000 Lille, France

* Correspondence: pascal.granger@univ-lille.fr; Tel.: +33-320-434-938

Received: 19 April 2019; Accepted: 20 May 2019; Published: 31 May 2019

Kinetics and reactor modeling for heterogeneous catalytic reactions are prominent tools for investigating, and understanding, the catalyst functionalities at nanoscale, and related rates of complex reaction networks. Prominent developments were achieved in the past three decades from steady-state to unsteady state kinetic approaches facing important issues related to the transformation of more complex feedstocks using a wide variety of reactor designs, including continuous flow reactors, fluidized reactors, recirculating solid reactors, pulse reactors, Temporal Analysis of Product (TAP) reactors with sometimes a strong gap in the operating conditions from ultra-high-vacuum to high pressure conditions. In conjunction, new methodologies have emerged giving rise to more sophisticated mathematical models, including the intrinsic reaction kinetics and the dynamics of the reactors and spanning a large range of length and time scales, from the nanoscale of the active site to the reactor scale. Recently, the development of steady-state isotopic transient kinetic analysis coupled with in situ and in operando techniques is aimed at gaining more insight into reactive intermediates.

The objective of this special issue is to provide diverse contributions that can illustrate recent advances and new methodologies for elucidating the kinetics of heterogeneous reactions and the necessary multiscale approaches for optimizing the reactor design.

Among the different contributions provided in this special issue, two articles review and summarize the use of elegant methodologies. In the frame of microkinetic approaches for catalytic reactions, the isolation of the real intermediates among various adsorbates and the calculation of more accurate kinetic and thermodynamic parameters to refine kinetic models are still challenging. In this context, the development of new analytical tools, such as adsorption equilibrium infrared spectroscopy, provides an alternative to classical surface science studies—offering the opportunity to get more accurate heat of adsorptions of co-adsorbed species, and taking the coverage dependency into account in more realistic operating conditions [1]. In general, the extrapolation of kinetic models in very different operating conditions than those applied for its development must be taken with caution leading to unrealistic deviations and over interpretations. In practice, microkinetics cannot be sufficient to get a proper description of complexity, as mentioned by Standl and Hinrichsen [2], who proposed both lumped and microkinetic approaches in catalytic olefin cracking and methanol-to-olefin over zeolites. Useful general and specific recommendations for future modeling of complex networks are given by these authors.

Full papers also proved the usefulness of kinetic approaches especially in the context of an energy transition. By way of illustration, Song [3] paid attention to dehydration of 2,3-butanediol to 1,3-butadiene and methyl ethyl ketone produced from various biomasses instead of fossil resources. It was found that 1D reactor modelling taking into account interfacial and intra particles gradients can provide important information for further development of commercial processes. Nowadays,

Catalysts **2019**, *9*, 501

computer-aided design can be essential for the prediction of reactor performances. At the macroscopic scale, the use of empirical rate equations is not rigorous and precise enough to fit boundary conditions whereas microkinetic approaches should in principle provide more robust models, but sophistication is usually synonymous with time-consuming. Gossler et al. [4] developed a relevant methodology for reactor simulation for gaining time. It is worthwhile to note that these complex approaches coexist with more conventional approaches performed in the kinetic regime. Such studies can be useful to get more relevant structure-reactivity relationship taking uniformity in the gas phase and the catalyst bed composition as shown by Urmès et al. [5], who concluded that the selective hydrogenation of acetylene on supported palladium-based catalyst involves a single active site. Temporal analysis of products is able to investigate the catalyst behavior in wide conversion range, especially at high conversion generally encountered in more realistic conditions. Because transport regimes can be modeled, those transient experiments can provide the time response of a surface exposed to ammonia, and distinguish between the ability of cobalt and iron to store a mixture of hydrogenated ad-species or predominantly N or NH [6]. Finally, two contributions report lab-scale experiments on structured catalysts, e.g., dense filamentous graphite [7,8], and illustrating the best practice at lab-scale through the comparison of catalysts in powder and tableted form to examine the impact of internal diffusion limitation on the determination of kinetic parameters.

Conflicts of Interest: The authors declare no conflict of interest

References

1. Bianchi, D. A Contribution to the Experimental Microkinetic Approach of Gas/Solid Heterogeneous Catalysis: Measurement of the Individual Heats of Adsorption of Coadsorbed Species by Using the AEIR Method. *Catalysts* **2018**, *8*, 265. [CrossRef]
2. Standl, S.; Hinrichsen, O. Kinetic Modeling of Catalytic Olefin Cracking and Methanol-to-Olefins (MTO) over Zeolites: A Review. *Catalysts* **2018**, *8*, 626. [CrossRef]
3. Song, D. Modeling of a Pilot-Scale Fixed-Bed Reactor for Dehydration of 2,3-Butanediol to 1,3-Butadiene and Methyl Ethyl Ketone. *Catalysts* **2018**, *8*, 72. [CrossRef]
4. Gossler, H.; Maier, L.; Angeli, S.; Tischer, S.; Deutschmann, O. CaRMeN: An Improved Computer-Aided Method for Developing Catalytic Reaction Mechanisms. *Catalysts* **2019**, *9*, 227. [CrossRef]
5. Urmès, C.; Schweitzer, J.-M.; Cabiac, A.; Schuurman, Y. Kinetic Study of the Selective Hydrogenation of Acetylene over Supported Palladium under Tail-End Conditions. *Catalysts* **2019**, *9*, 180. [CrossRef]
6. Wang, Y.; Kunz, M.R.; Siebers, S.; Rollins, H.; Gleaves, J.; Yablonsky, G.; Fushimi, R. Transient Kinetic Experiments within the High Conversion Domain: The Case of Ammonia Decomposition. *Catalysts* **2019**, *9*, 104. [CrossRef]
7. Xu, Z.; Duong-Viet, C.; Ba, H.; Li, B.; Truong-Huu, T.; Nguyen-Dinh, L.; Pham-Huu, C. Gaseous Nitric Acid Activated Graphite Felts as Hierarchical Metal-Free Catalyst for Selective Oxidation of H_2S. *Catalysts* **2018**, *8*, 145. [CrossRef]
8. Obalová, L.; Klegova, A.; Matějová, L.; Pacultová, K.; Fridrichová, D. Must the Best Laboratory Prepared Catalyst Also Be the Best in an Operational Application? *Catalysts* **2019**, *9*, 160.

Review

A Contribution to the Experimental Microkinetic Approach of Gas/Solid Heterogeneous Catalysis: Measurement of the Individual Heats of Adsorption of Coadsorbed Species by Using the AEIR Method

Daniel Bianchi

Institut de Recherches sur la Catalyse et l'Environnement de Lyon, IRCELYON, UMR 5256, Université Claude Bernard-Lyon 1, Bat Chevreul, 43 Bd du 11 Novembre 1918, 69622 Villeurbanne, France; daniel.bianchi@ircelyon.univ-lyon1.fr

Received: 28 May 2018; Accepted: 27 June 2018; Published: 29 June 2018

Abstract: The two first surface elementary steps of a gas/solid catalytic reaction are the adsorption/ desorption at least one of the reactants leading to its adsorption equilibrium which can be or not disturbed by the others surface elementary steps leading to the products. The variety of the sites of a conventional catalyst may lead to the formation of different coadsorbed species such as linear, bridged and threefold coordinated species for the adsorption of CO on supported metal particles. The aim of the present article is to summarize works performed in the last twenty years for the development and applications of an analytical method named Adsorption Equilibrium InfraRed spectroscopy (AEIR) for the measurement of the individual heats of adsorption of coadsorbed species and for the validation of mathematical expressions for their adsorption coefficients and adsorption models. The method uses the evolution of the IR bands characteristic of each of coadsorbed species during the increase in the adsorption temperature in isobaric conditions. The presentation shows that the versatility of AEIR leads to net advantages as compared to others conventional methods particularly in the context of the microkinetic approach of catalytic reactions.

Keywords: heats of adsorption; FTIR spectroscopy; AEIR method; Temkin model

1. Introduction

One of the aims in gas/solid heterogeneous catalysis is to correlate the global rate of the reaction (defined as the catalytic activity) including in its unit a property of the catalyst (i.e., weight, specific surface area, amount of active sites) to the elementary steps on the surface forming the detailed mechanism of the reaction. This constitutes the microkinetic approach of heterogeneous catalysis [1] which imposes the assessment of the kinetic parameters of the surface elementary steps such as the rate constants: k (pre-exponential factor and activation energy) and adsorption coefficients: K (pre-exponential factor and heat of adsorption) and the determination of the global rate of the reaction from the detailed mechanism [1]. The kinetic parameters of the surface elementary steps can be obtained either by experimental procedures on model surfaces (surface sciences approach) and conventional powdered catalysts or by theoretical methods (i.e., DFT (Density Functional Theory) calculations). From the detailed mechanism, the global rate of the reaction can be obtained by different either classical [2,3] or more recent [4–6] methods. Using, kinetic parameters from different origins may lead to ambiguities in the conclusions of a microkinetic study considering that they are dependent on the composition, morphology and structure of the catalyst (material gap) which control the type of adsorption sites (terraces, steps and corners for metal supported catalysts) and the interactions between adsorbed species. Similarly, surface sciences studies of catalytic reactions use mainly low adsorption pressures (as compared to conventional conditions of heterogeneous catalysis) which may neglect the contribution of weakly adsorbed species (pressure gap). The use of kinetic parameters from different experimental and theoretical approaches is mainly due to the large numbers of surface elementary steps of a detailed mechanism of a catalytic reaction even for bi- and tri-atomic reactants (i.e., CO/O_2, CO/H_2, CO/NO, NH_3/NO). This is due to the fact that the number of either ruptures or creations of bonds in a surface elementary step must be limited. For instance, the number of surface elementary steps in detailed mechanisms of microkinetic studies is: 12, for NH_3 synthesis from N_2/H_2 (or its decomposition) on ruthenium catalysts [7–9]; 22, for NO/H_2 on Pt catalysts [10]; 26, for the production of C1 and C2 species from CO/H_2 on cobalt catalysts [11] and 32, for the ethylene oxidation on Ag catalysts [12]. However, among the surface elementary steps of a detailed mechanism, few of them control the global rate of the reaction: they constitute the kinetic model of the reaction [2,3]. This has led our group developing the experimental microkinetic approach (abbreviation EMA) of heterogeneous gas/solid catalytic reactions such as CO/O_2 [13,14] and CH_4 production from CO/H_2 [15,16] on Pt/Al_2O_3 catalysts. The main point of the EMA is that considering a plausible kinetic model of a catalytic reaction (based either on literature data or a formal kinetic approach), all the kinetic parameters of interest are obtained by experimental procedures on the conventional dispersed solid catalyst. This prevents using kinetic parameters coming from different sources (surface sciences, conventional catalysts, DFT) and overcomes the impacts of the material and pressure gaps. Note that it is the concurrence between the experimental catalytic activity and that from the EMA which validates the procedure; otherwise the plausible kinetic model must be reconsidered.

The first surface elementary steps of any gas/solid catalytic reaction are the adsorption with a rate R_a (rate constant k_a, activation energy E_a equal to 0 for non-activated chemisorption) followed by the desorption with a rate R_d (rate constant k_d, activation energy E_d) of at least one reactant. For $R_a - R_d = 0$, these two elementary steps lead to the adsorption equilibrium of the reactant on the sites which is characterized by the adsorption coefficient $K_a = k_a/k_d$ and a heat of adsorption $E = E_d - E_a$ (K_a determines the coverage of the adsorption sites for adsorption temperature T_a and pressure P_a). During a catalytic reaction, the adsorption equilibrium of a reactant can be disturbed or not by others surface elementary steps with a rate R_s and its coverage is controlled by the reaction equilibrium $R_a - R_d - R_s = 0$ [2,3]. However, in numerous kinetic studies of gas/solid catalytic reactions, it is often assumed that the adsorption equilibrium is not disturbed by the catalytic reaction. It is well known that the adsorption of a reactant may lead to the formation of different adsorbed species. For instance, the reactant CO of the CO/O_2 and CO/H_2 reactions on supported metal particles may lead to the formation of linear, bridged and threefold coordinated CO species which are well characterized by their IR bands in distinct wavenumber ranges [17]. Similarly, for the selective catalytic reduction of NO by NH_3 in excess O_2 (NH_3-SCR) on $x\%$ $V_2O_5/y\%$ WO_3/TiO_2 catalysts the adsorption of NH_3 leads to $NH_{3ads\text{-}L}$ and NH_4^+ species on the Lewis and Brønsted sites respectively which provide distinct IR bands ([18] and references therein). The role of each coadsorbed species in the reaction is a key point of microkinetic studies. For instance, their respective coverages for a composition of the reactive gas mixture are fixed by their individual heats of adsorption and adsorption coefficients implying that the determination of their values constitutes the first stage of the EMA of the reaction. Moreover, considering that the final aim of a microkinetic study is to express the global rate of the reaction as a function of the kinetic parameters of the surface elementary steps, it is of interest that experimental studies validate mathematical expressions for the adsorption coefficients and the adsorption models for each coadsorbed species formed by the reactants and others compounds of the reactive mixtures (i.e., H_2O for NH_3-SCR to consider the impact of NH_3/H_2O co-adsorption on the catalytic activity). This is the aim of an original method named Adsorption Equilibrium InfraRed spectroscopy (AEIR) which has been particularly developed in the last twenty years. This method is based on the evolutions of the IR bands characteristic of each adsorbed species during the increase in the adsorption temperature T_a in isobaric conditions. The aim of the present article is to summarize the development and the applications of this method.

2. Context of the Development of the AEIR Method

2.1. Classical Methods for the Measurement of the Heats of Adsorption

The driving force of this development was the difficulties obtaining the data of interest for a EMA of a catalytic reaction by conventional methods such as the isosteric heats of adsorption of a gas via volumetric/gravimetric measurements, the activation energy of desorption from temperature programmed desorption experiments and the differential/integral heats of adsorption using microcalorimetry. The isosteric heat of adsorption is based on the measurement either in isothermal or isobaric conditions of the coverages of a gas [19]. This allows one determining different couples (T_a, P_a) leading to the same coverage which provide the average isosteric heat of adsorption Q_{iso} at different coverages via the Clausius-Clapeyron equation [19]:

$$\left(\frac{\partial \ln P_a}{\partial(\frac{1}{T_a})}\right)_\theta = \frac{-Q_{iso}}{R} \tag{1}$$

where R is the ideal gas constant. These measurements are tedious to perform and time-consuming, while the method is strongly affected by experimental uncertainties. Moreover the formation of several adsorbed species leads to average values of limited interest in line with the aims of the EMA. Moreover, Equation (1) imposes the use of a large number of experimental data often associated to successive pretreatments of the catalyst which may affect its properties (i.e., sintering of the supported metal particles). Microcalorimetry [20] provides the differential and integral heats of adsorption of a gas according to its coverage. However different experimental difficulties can be encountered such as: the presence of several adsorbed species leading to average values; the impact of gaseous impurities [21]; the non-equilibrium nature of the adsorption at low temperatures [22,23] and the contribution of parallel reactions to the adsorption at high temperatures. Moreover, isosteric methods and microcalorimetry do not provide mathematical expressions for the adsorption coefficient and the adsorption model. The difficulties of these two analytical methods explains the success of temperature programmed desorption (TPD) methods which reveal easy the presence of coadsorbed species having different activation energy of desorption via their rates de desorption [19,24,25]. TPD methods consist adsorbing a gas at a temperature low enough to obtain a very low rate of desorption and then increasing the temperature T_d in inert atmosphere to desorb progressively the different adsorbed species according to their activation energy of desorption. This leads to a succession of peaks during the increase in T_d characterized by the temperature T_m of their maximum [19,24,25]. Equations based on classical theories of the adsorption (i.e., the kinetic theory of the gases and the statistical thermodynamics) provide kinetic parameters of interest such as the activation energy of desorption from T_m [19,24,25]. However, the TPD method imposes a careful design of the experiment in line for instance with the criteria proposed by Gorte et al. [26,27] to prevent the contribution of mass and heat transfers and to neglect the readsorption. These criteria show that readsorption can be rarely prevented using conventional catalysts [28–30]. In these conditions mathematical formalisms may provide the heats of adsorption of the adsorbed species [28–31]. However, similarly to microcalorimetry difficulties in the exploitation of the TPD spectra come from the contribution of parallel surface processes such as surface reconstructions and reactions with impurities (i.e., O_2, H_2O). Moreover, for heterogeneous surfaces the TPD peak (without and with readsorption) of an adsorbed species is very broad [31] leading to strongly overlapped peaks for coadsorbed species restricting significantly the access to the kinetic/thermodynamic parameters of interest for an EMA.

2.2. Precursor Works Using IR Spectroscopy for the Measurement of the Heats of Adsorption

Different early works were dedicated to the use of IR spectroscopy in this field. These studies either assumed or established the validity of a linear relationship between (a) the amount of the adsorbed species and (b) the area of its characteristic IR band according to the Beer-Lambert law which is the basis of the quantitative exploitation of FTIR (Fourier-transform infrared spectroscopy) spectra. The first studies were made by surface sciences via Infrared reflection adsorption spectroscopy (IRAS). For instance, Kottle et al. [32] measured the isosteric heats of adsorption of a linear CO species on an evaporated golf film for T_a and P_a in the ranges of 300–383 K and 7–530 Pa using an IR band in the 2120–2115 cm^{-1} range according to the coverage. There was a scatter in the data from Equation (1) and the authors provided the average of the heats of adsorption in two coverage ranges: 13.4 kJ/mol and 12.4 kJ/mol in the coverage ranges 0.1–0.6 and 0.3–0.6 respectively. Similarly, Richardson et al. [33] measured the isosteric heats of adsorption of a linear CO species (IR band at 2161 cm^{-1}) adsorbed on a NaCl film: 13 ± 3 kJ/mol (via a series of isotherms) after the validation of the Beer-Lambert law. Moreover, after showing that the isotherms were consistent with the Langmuir adsorption model:

$$\theta(T_a, P_a) = \frac{K(T_a) \quad P_a}{1 \quad + \quad K(T_a) \quad P_a} \tag{2}$$

the values of the adsorption coefficient $K(T_a)$ in the temperature range of the experiments were compared to mathematical expressions from the statistical thermodynamics approach of the adsorption. This allowed the authors obtaining an estimation of the partition function of the adsorbed species. The Goodman's group [34] have used a similar procedure to study the isosteric heats of adsorption of a linear CO species adsorbed on Cu(100) characterized by an IR band in the range 2086–2064 cm^{-1} according to the coverage. They used seven isotherms for adsorption pressures in the range 10^{-3}–130 Pa showing that the isosteric heat of adsorption varied from 70 kJ/mol to 53 kJ/mol in the coverage range of 0–0.15 ML. The same group has applied the procedure to measure the isosteric heats of adsorption of a linear CO species (IR band in the range of 2096–2053 cm^{-1} according to the coverage) adsorbed on Pd film on Ta(110) using eight isobars in the range of $\approx 10^{-7}$–130 Pa [35]. The authors showed clearly that the isosteric heat of adsorption decreased roughly linearly with the increase in the coverage from ~96 kJ/mol to ~40 kJ/mol in the coverage range of 0–0.35 ML. Similar studies have been performed for the bridged CO species on Pd crystals (IR band in the range 1968–1947 cm^{-1}) on Pd(100) [36]: 121 ± 8 kJ/mol in the coverage range 0.45–0.55 ML and Pd(111) [37,38] with a linear decrease in the isosteric heat of adsorption from 145 kJ/mol to 103 kJ/mol in the coverage range of $\approx$0–0.3 ML.

In parallel to the studies using IRAS, IR spectroscopy has been applied to the measurement of the heats of adsorption of CO species on conventional catalysts. These works concerned mainly weakly adsorbed species (heats of adsorption $< \approx 100$ kJ/mol) due to the limited performances of the IR cells (working mainly in static conditions) to maintain, the catalyst at high temperatures and pressures on the IR beam. The first works were dedicated to the study of linear CO species adsorbed on the Lewis sites of metal oxides. For instance, Paukshtis et al. [39] studied the individual heats of adsorption of two coadsorbed linear CO species (IR bands in the range of 2150–2230 cm^{-1}) on different Lewis acidic sites of 16 dispersed metal oxides such as MgO, Al$_2$O$_3$, ZrO$_2$, TiO$_2$. Assuming the validity of the Beer-Lambert law for the area (named A) of the IR bands they showed that the different isotherms followed the Langmuir model and using Equation (2) they obtained the heats of adsorption of the different L CO species from the plots $[\ln(A/A_0) - 1] = f(1/T)$ for each IR bands (A_0: the area at saturation of the sites). Thus on ZrO$_2$ they determined that the individual heats of adsorption of two coadsorbed linear CO species characterized by IR bands at 2203 and 2183 cm^{-1} were of 36 and 28 kJ/mol respectively. In a following study [40], the authors have studied the heats of adsorption of a pyridine species on Al$_2$O$_3$ on ZrO$_2$ characterized by an IR band at 1445 cm^{-1}. However, this adsorbed species did not followed the Langmuir model and the authors provided the isosteric heats of adsorption in the range 100–160 kJ/mol by using high temperatures (range 400–700 K) and low adsorption pressures (range 1.3×10^{-2}–3200 Pa). Yates et al. [41,42] have performed similar measurements for the L CO species adsorbed the Lewis sites of Al_2O_3 and SiO_2. On Al_2O_3 they noted that two species were formed providing strongly overlapped IR bands at 2195 cm^{-1} (main IR band) and 2213 cm^{-1}. However, using isobaric conditions (T_a in the range of 180–350 K and P_a = 659 Pa) they provided via the Langmuir model their average heat of adsorption: 20.9 kJ/mol. Garrone et al. have studied the linear CO species on the Lewis sites of different metal oxides: TiO_2 [43], ZrO_2 [44] and Na-Z5M5 [45] using isothermal conditions at $T_a \approx 300$ K for $P_a <\approx 13$ kPa. The point of interest in these studies was that the isotherms on TiO_2 and ZrO_2 were compared to the Temkin adsorption model to take into account the heterogeneity of the adsorption sites.

Considering the characterization of the kinetic/thermodynamic parameters of adsorbed species with high heats of adsorption relevant of catalytic reactions, the first applications of the IR spectroscopy were dedicated to the measurement of the activation energy of desorption. This was linked to the limited performances of the IR cells using experimental conditions representative of heterogeneous catalytic reactions [46] and reference therein. For instance, Soma-Noto and Sachtler [47] have used this procedure to measure the activation energy of the L and B CO species adsorbed on Pd/Al_2O_3 and Pd-Ag/Al_2O_3 by studying the evolution of their characteristic IR bands with the duration of the isothermal desorption in vacuum in the range 373–540 K: 113 kJ/mol and 171 kJ/mol respectively. Similarly, some authors have developed TPDIR procedures: this consisted studying the evolution of the intensities of the IR bands of the adsorbed CO species on Pt/Al_2O_3 [48] during the linear increase of the desorption temperature T_d. This provided the evolution of the coverage θ_X of a adsorbed X species with T_d giving the curves $d\theta_X/dT_d$ which were exploited according to classical TPD procedures. This means that the same difficulties than those associated to the TPD procedure may contribute to the experimental data such as the consumption of the adsorbed species by reactions with H_2O and O_2 impurities [49]. The design of microreactor IR cell using gas flow rates at atmospheric pressure and high temperatures on the IR beam has allowed the characterization of the adsorption equilibrium of adsorbed species in experimental conditions representative of heterogeneous catalysis. The difficulties in the design of these IR cells come from the association of a small optical path length (range 2–3 mm) to limit the overlap of the IR spectra of the gaseous and adsorbed species and high temperatures due to the limited thermal stability of the IR windows and their sealing materials [46] and references therein. For instance, using a microreactor IR cell, Bell et al. [50] have determined the heat of adsorption of the linear CO species on Ru/Al_2O_3 in the coverage range 1–0.85 using x% CO/H_2 gas mixtures and three isotherms at T = 498, 523 and 548 K. In this small coverage range, the experimental data were consistent with the Langmuir model (Equation (2)) indicating a heat of adsorption of 106 kJ/mol. Using a similar IR cell, Kohler et al. [51] have measured the heats of adsorption of linear CO species on unreduced (IR band at 2132 cm^{-1}) and reduced (IR band at 2090 cm^{-1}) x% Cu/SiO_2 solids with x in the range of 2–10. On the unreduced solids three isotherms at 358, 378 and 441 K with $P_a \leq 20$ kPa showed that the L CO species followed the Langmuir model leading to a heat of adsorption of 25 kJ/mol. This was confirmed by using the isosteric method showing that the heat of adsorption was independent on the coverage: $\approx$29 kJ/mol in the coverage range of 0.1–0.9. For the reduced solids, isotherms at 358, 378, 441 and 493 K showed that the coverage of the L CO species was not consistent with the Langmuir model and the isosteric method indicated that the heat of adsorption increased with the decrease in the coverage according to a profile consistent with the Freundlich model with values at low coverages varying with the copper content for $\approx$50 kJ/mol to 28 kJ/mol for x = 9.5 to x = 2.1 [51]. Clarke et al. [52] have confirmed the value at low coverage (range of 0–0.18) for a reduced 7% Cu/SiO_2 by using as approximation the Langmuir model: 35 kJ/mol.

The AEIR method has been developed in line with these precursor works using the adsorption of CO on Pt containing catalysts as case study [26,53–55]. The first step was the design of a microreactor IR cell allowing a significant increase of the highest temperature (until 900 K) as compared to literature data [46] (this improvement was imposed by the high heats of adsorption at low coverages of the

L CO species on Pt particles). The aim of the experimental procedure of the AEIR method was to combine measurements at the adsorption equilibrium (i.e., this prevents the impacts of heat and mass transfers) and temperature programmed procedures (rapidity of the experiment). Considering our interests for the EMA of gas/solid catalytic reactions, the aims of the exploitation of the IR spectra were the measurement of the individual heats of adsorption of coadsorbed species via the validation of mathematical expressions for the adsorption coefficients and adsorption models provided by classical theories of the adsorption. Two applications of the AEIR method are used to support the presentation: the adsorptions of CO on Pt/Al_2O_3 and NH_3 on TiO_2 based catalysts considered as the first steps of the EMA of catalytic reactions such CO/H_2 and de-NO$_x$ from NH_3-SCR respectively.

3. The Adsorption Equilibrium InfraRed Spectroscopy Method

3.1. IR Cell Microreactor for the Application of the AEIR Method

The AEIR method has been developed using a homemade stainless steel microreactor IR cell in transmission mode working at atmospheric pressure [46]. It has been designed (see Figure 1 in [46]) taking into account previous models and literature data. Briefly, a short path length (≈2.2 mm) limits the contribution of the gas phase to the IR spectra of the adsorbed species allowing using adsorption pressure of CO until ≈20 kPa. The originality of the IR cell is that the two CaF$_2$ IR windows delimiting the path length in the heating part of the cell, are positioned on polished flat flanges without sealing materials (the two windows was maintained by using vacuum on one of their faces). This permits using temperatures in the range of 300–900 K with an heating rate of ≈0.1–20 K/min [46]. The powdered catalyst (weight in the 40–200 mg range) was compressed into a disk (diameter = 18 mm) positioned on the IR beam between the two CaF$_2$ windows. In a recent work, it has been shown that a DRIFT cell can be used for the AEIR method (using the pseudo absorbance mode) taking into account that according to its design, heat transfers may create some difficulties to know the exact temperature of the fraction of the solid submitted to the IR beam for T >≈ 623 K [56].

3.2. Experimental Procedure of the AEIR Method

After pretreatment of the catalyst at high temperatures (i.e., H$_2$ reduction at 713 K for 2.9% Pt/Al_2O_3 and O$_2$ oxidation for TiO$_2$ based solids) it is cooled to 300 K. Then the switches H$_2$ (or O$_2$) → He → x% G/He (i.e., G either CO or NH$_3$) lead to the adsorption of G at the adsorption pressure $P_a = x \, 10^3$ Pa. After the stabilization of the IR bands of the adsorbed species, indicating the attainment of the adsorption equilibrium, the adsorption temperature T_a is increased ($\alpha \approx 10$ K/min) in the presence of x% G/He following the changes in the IR spectra of the adsorbed species. This provides the change in the intensities (in absorbance mode) of the IR bands characteristic of each adsorbed species X_{ads} at the adsorption equilibrium as a function of T_a in isobaric condition. It has been shown that the gas/solid system evolves by a succession of adsorption equilibriums taking into account that the high adsorption pressure and the moderate heating rate lead to a fast change from an adsorption equilibrium at (T_a, P_a) to that at (T_a + dT_a, P_a) [30]. Similarly to the classical methods dedicated to the measurement of the heats of adsorption, surface processes parallel to the adsorption (i.e., surface reconstruction, CO dissociation) may contribute to the change of the intensity of the IR bands of the adsorbed species. However, the AEIR method permits to take into account these contributions by comparing the intensities of the IR bands at different adsorption temperatures during the first heating (i.e., 713 K) and cooling (i.e., 300 K) cycle in x% G/He. Often differences are observed due to reconstruction [53–55] and CO dissociation [57]. However, these processes are ended after the first heating/ cooling cycle in x% G/He as attested by the repeatability of the intensities of the IR bands of the adsorbed species during a second heating/cooling cycle in x% G/He: this means only the IR spectra of the second cycle (stabilized surface) are exploited via the AEIR method. The intensities of the IR bands of the adsorbed species can be modified by another process as observed for the adsorption of CO on supported Ag° particles [58]. After a first heating/cooling cycle, it has been observed that the IR band of the B CO species (at 1994 cm^{-1}) increases during the heating stages in parallel to the decrease in the IR band of the L CO species (at 2044 cm^{-1}). The reverse situation is observed during the cooling stages [58]. A similar process has been described by Müslehiddinoglu and Vannice [59] during the isothermal desorption at 300 K of the adsorbed CO species on Ag° particles. According to literature data [58] and references therein, this has been ascribed to an intensity transfer (in the 1/1 ratio) from the IR band of the B CO species to that of the L CO species. This transfer does not contribute significantly to the observations for different situations either if the amount of B CO species is low as compared to that of the L CO species (i.e., Pt/Al$_2$O$_3$) or if the two adsorbed species have different heats of adsorption allowing the significant decrease in the coverage of one of them without affecting that of the second species. For others situations the AEIR method does not apply.

As example of experimental data of the AEIR method, the inset of Figure 1 shows that the adsorption of 1% CO/He at 300 K on 2.9% Pt/Al$_2$O$_3$ leads to an IR spectrum with three IR bands at 2073, 1878 and 1835 cm^{-1} ascribed [53–55] to linear, bridged and three fold coordinated CO species (named L, B and 3FC CO species respectively). Figure 1 gives the evolution of the IR band of the L CO species during the second increase in T_a for 1% CO/He. Similar spectra are obtained for the B and 3FC CO species.

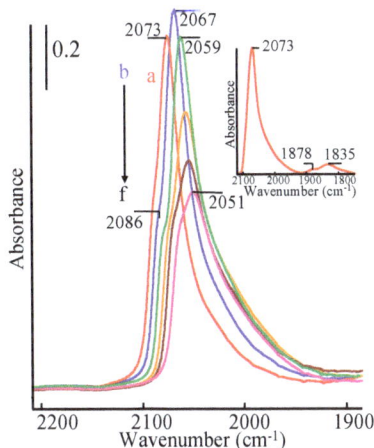

Figure 1. Evolution of the IR band of the linear CO species adsorbed on a reduced 2.9% Pt/Al$_2$O$_3$ catalyst with the adsorption temperature T_a at P_a = 1 kPa. (a–f) T_a = 378, 453, 533, 633, 693 and 783 K. **Insert:** IR bands of the different adsorbed CO species on 2.9% Pt/Al$_2$O$_3$ for T_a = 300 K and P_a = 1 kPa.

3.3. Exploitation of the IR Spectra According to the AEIR Method

Considering the Beer-Lambert law, the amount of each adsorbed species X is proportional to the intensity of its IR band. This allows one obtaining the experimental evolutions of the coverage of each adsorbed species: θX_{ex}, with T_a, in isobaric conditions from the change in its IR band as for the L CO species in Figure 1:

$$\theta X_{ex} = \frac{A_X(T_a)}{A_{XM}} \tag{3}$$

where $A_X(T_a)$ and A_{XM} are the area of the IR band (in absorbance mode with values lower than $\approx$1.1) characteristic of the X species at T_a and at saturation of the sites respectively. The value of A_{XM} is obtained by ascertaining that the area of the IR band is not modified by either the increase in T_a in isobaric conditions or the increase in P_a in isothermal conditions. This is often the situation for the adsorbed CO species on metal particles at T_a = 300 K for P_a in the range 1–10 kPa due to their high heats of adsorption at full coverage. However, for weakly adsorbed species such as the linear CO species on the Lewis sites of metal oxides: ZrO_2 [60] and TiO_2 [61] the saturation of the adsorption sites is not obtained at 300 K for the highest adsorption pressure available with the IR cell. For this situation, an estimation of A_{XM} is obtained according to the procedure of Kohler et al. [51]: assuming that the adsorption follows the Langmuir's model (Equation (2)) for non dissociative chemisorption, then the plot of $(1/A_X(300\,K))$ vs. $(1/P_a)$ provides A_{XM}. For a gas/solid system leading mainly to one adsorbed species, the validity of Equation (3) has been ascertained by different authors [33,35,36,39,50,51] and we have confirmed this point for the L CO species on the reduced metal particles of Cu/Al_2O_3 [62,63] and Ir/Al_2O_3 [64]. If the adsorption leads to different adsorbed species such as L, B and 3FC CO species on Pt/Al_2O_3 (inset Figure 1), the ascertainment of the Beer-Lambert law presents difficulties, due to the fact that at 300 K, volumetric/gravimetric methods provide the total amount of adsorbed species: $Q_T = \Sigma\,Q_X$ whereas A_X depends of a specific X species. For the adsorption of CO on Pt particles, FTIR spectroscopy shows that for 1% CO/He and $T_a > 550$ K, mainly the L CO species is present on the surface due to the difference in the heats of adsorption of the L, B and 3FC CO species. In these conditions, the validity of Equation (3) has been ascertained performing carbon mass balance with a mass spectrometer at the introduction of 1% CO/He taking into account that CO is involved in different processes: adsorption; dissociation and reaction with OH groups of the support forming CO_2 and H_2 [65]. These results and literature data have led us applying the Beer-Lambert law for all adsorbed species X_{ads} providing from Equation (3) the experimental curve $\theta X_{ex} = f(T_a)$ at different adsorption pressures P_a. For instance, symbols ■ in Figure 2 provide the evolution of the coverage of the linear CO species θL_{ex} on the reduced 2.9% Pt/Al_2O_3 catalyst for $P_a = 1$ kPa considering the data in Figure 1. Similarly, symbols ▲ and ● give the experimental coverage of the B and 3FC CO species (inset Figure 1) for $P_a = 1$ kPa from experiments similar to Figure 1 with a higher amount of catalyst to improve the accuracy of the measurements and after the decomposition of their overlapped IR bands [54]. Note that for the B CO species, A_M in Equation (3) is obtained at 300 K for $P_a \geq 10$ kPa [34]. The experimental data in Figure 2 are compared to theoretical curves providing the individual heats of adsorption and the mathematical expressions of interest for the adsorption equilibriums of L, B and 3FC CO species.

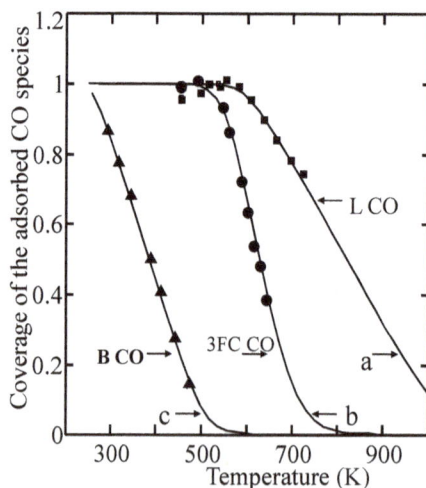

Figure 2. Comparison between experimental and theoretical curves $\theta_X = f(T_a)$ at $P_a = 1$ kPa for the different X CO species on 2% Pt/Al_2O_3: ■, ●, and ▲ experimental data for the L, 3FC and B CO species respectively; (a), (b) and (c) theoretical curve according to Equations (4) and (5) for the L, 3FC and B CO species (see the text for the E_0 and E_1 values of each adsorbed species).

3.4. Exploitation of the Experimental Curves θXₑₓ = f(Tₐ, Pₐ) According to the AEIR Method

Considering our interest for the development of the EMA of catalytic reactions, the aims of the exploitation of the experimental data were the measurement of the individual heats of adsorption of the coadsorbed species via the validation of mathematical expressions for the adsorption coefficients and adsorption models provided by the classical theories of adsorption. It is a fact that the Langmuir model (Equation (2)) is rarely representative of experimental data for strongly adsorbed species on conventional catalysts due to the heterogeneity of the adsorption sites/adsorbed species. However, its extension to heterogeneous surfaces, via the integral approach and the distribution functions [15,16] and referenced therein, provides different equations for others adsorption models. For instance assuming a linear decrease in the heat of adsorption of an adsorbed species with the increase in its coverage, the integral approach leads to the generalized Temkin equation model [66] for non dissociative adsorption:

$$\theta_{th} = \frac{RT_a}{\Delta E} \times \ln\left(\frac{1 + K(E_0)\ P_a}{1 + K(E_1)\ P_a}\right) \qquad (4)$$

where E_0 ($K(E_0)$) and E_1 ($K(E_1)$) are the heats of adsorption (adsorption coefficient) at low and high coverages and $\Delta E = F_0 - F_1$. The statistical thermodynamics and the absolute rate theory provide the adsorption coefficient as a function of the partition functions of the gaseous molecule, the adsorbing site and adsorbed molecule [19,67–69]. In the temperature range of gas/solid catalytic reactions ≈300–900 K, the partition function of a gaseous molecule is dominated by those of translation, rotation and vibration whereas that of the localized adsorbed molecule is dominated by those of rotation and vibration. In many cases, the ratio of the partition functions of rotation and vibration of the gaseous and adsorbed species can be reasonably approximated to ≈1 [30,70] leading to the mathematical expression of the adsorption coefficient:

$$K(T_a) = \frac{h^3}{(2\ \pi\ m\ k)^{3/2}\ k}\ \frac{1}{T_a^{5/2}}\ \exp\left(\frac{E_d - E_a}{R}\ \frac{1}{T_a}\right) \qquad (5)$$

where h is Planck's constant, k is Bolztmann's constant, m is the mass of the molecule, E_d and E_a are the activation energies of desorption and adsorption respectively, while $E = E_d - E_a$ is the heat of adsorption. Note that as commented by Tompkin [19], the attainment of the adsorption equilibrium implicates a surface diffusion of the adsorbed species: this is compatible with a localized adsorbed species considering that localized means that the lifetime of the adsorbed species at the site is longer than its time in flight on the surface.

To obtain the individual heats of adsorption of coadsorbed species, the experimental evolutions of the coverage of each species (Figure 2) are compared to theoretical curves obtained by using Equations (4) and (5) and selecting a couple of E_0 and E_1 values leading to an overlap between experimental and theoretical curves (note that for $E_0 \approx E_1$ the theoretical curve obtained from Equation (4) is overlapped with that from the Langmuir model (Equation (2)). For instance in Figure 2, the curves a, b, c which overlap the experimental data for the L, B and 3FC CO species are obtained considering the following couples of heats of adsorption (E_0, E_1) in kJ/mol at low and high coverages: (206, 115), (94, 45) and (135, 104). This shows that the AEIR method allows one determining the heats of adsorption of an adsorbed species from a single isobar. The practice shows that the choice of the E_1 and E_0 values is limited to short ranges (≈±5 kJ/mol) otherwise the experimental and theoretical curves are clearly distinct. This accuracy is due to the fact that E_1 and E_0 determine the temperature leading to the decrease in the coverage from 1 and the slope of the linear section of the isobar respectively [53].

3.5. Heats of Adsorption from the AEIR Method and Isosteric Heats of Adsorption

Using three isobars similar to Figure 1 for P_{CO} = 1000, 100 and 10 Pa [53], the validity of the AEIR procedure has been ascertained for the L CO species by showing that the E_θ values obtained from Equations (4) and (5) are consistent with the isosteric heats of adsorption (Equation (2)) which is independent on the adsorption model. Similar conclusions have been obtained for different adsorbed CO species on metal particles [54,62,64]. This indicates that Equations (4) and (5) provide a very well representation of the properties of adsorbed CO species whereas as compared to the isosteric heats of adsorption a single isobar is needed using the AEIR method to obtain the individual heats of adsorption of coadsorbed species.

It is a fact that the same adsorption model (localized adsorbed species and Temkin's model) allows fitting numerous experimental data dedicated to the heats of adsorption of L and B CO species formed by the adsorption of CO on supported metal particles on metal oxides as indicated in Table 1.

Table 1. Heats of adsorption at low (E_0) and high (E_1) coverages of the Linear and Bridged CO species adsorbed on different metal supported particles on metal oxides by using the AEIR method.

Metal Particles on Alumina	Heat of Adsorption of Adsorbed CO Species in kJ/mol				Ref.
	Linear CO Species		Bridged CO Species		
	E_1	E_0	E_1	E_0	
Pt$^\circ$	115	206	45	94	[53–55]
Pd$^\circ$	54	92	92	168	[71]
Rh$^\circ$	103	195	75	125	[72]
Ir$^\circ$	115	225			[64]
Ru$^\circ$	115	175			[73]
Cu$^\circ$	57	82	78	125	[62]
Au$^\circ$	47	74			[61]
Ag$^\circ$	58	76	84	88	[58]
Ni$^\circ$	100	153	106	147	[74]
Fe$^\circ$	79	105			[75]
Co$^\circ$-C *	93	165			[57]

* Co$^\circ$ sites modified by C deposition from the CO dissociation.

The versatility of the Temkin model is probably due to the fact that it is the best representation of the heterogeneity of the surface for strongly adsorbed species on conventional catalysts. Classically the heterogeneity for adsorbed species is ascribed to either a difference in the adsorption properties of the sites (biographical or intrinsic heterogeneity) or an interaction between adsorbed species (induced heterogeneity) [66]. The Temkin model is one of the proposals [76,77] to represent by a equation the evolution of the coverage of a gas on a heterogeneous surface as a function of the adsorption temperature and pressure. Different studies have considered the contribution of each type of heterogeneity on the modeling of the coverage in particular the induced heterogeneity due to lateral interaction (Ref. [78] and references therein). However, Temkin [66] noted that the two types of heterogeneities can be simultaneously operant and that a single equation must be representative of this situation to prevent an excessive mathematical complexity. This has been justified by different authors [79–81]. Moreover, considering that the heats of adsorption of an adsorbed species at low (E_0) and high (E_1) coverages have limited values such as 206 kJ/mol and 115 kJ/mol for the L CO species on Pt particles (Table 1), the comparison of a linear (Temkin model) and an exponential (Freundlich model) decrease in the heats of adsorption with the increase in the coverage according to $E_T (\theta) = [E_0 - (E_0 - E_1) \theta]$ and $E_F(\theta) = E_0 \exp [-\theta \ln(E_0/E_1)]$ respectively, shows that the highest difference (6.6 kJ/mol at coverage 0.5) is in the range of the accuracy of the measurements.

3.6. Development of the AEIR Method

The AEIR method has been applied to different gas/catalyst systems such as: NO on 2.7% Pt/Al$_2$O$_3$ [82]; aromatic hydrocarbons on SiO$_2$ [70] and NH$_3$ [18,83,84] and H$_2$O [85] on different TiO$_2$ based solids. For this last application it has been observed that some IR bands provide $\theta_{ex} = f(T_a)$

curves which are not consistent with Equations (4) and (5). This is ascribed to the fact that the IR band selected for the measurement is due to the contributions of two adsorbed species having different heats of adsorption. In this situation $\theta_{ex} = f(T_a)$ gives the evolution of the average coverage of the two adsorbed species. A development of the AEIR method allows one obtaining the individual heats of adsorption of the two species as shown for the adsorption of NH_3 on the Lewis sites (named NH_{3ads-L} species) of TiO_2 P25 from Degussa [18] which is of particular interest because different IR bands can be used for the measurement of the individual heats of adsorption of the coadsorbed NH_{3ads-L} species supporting the development of the method. For instance, Figure 3 gives the evolution of the IR bands in the range 2000–1100 cm^{-1} of the NH_3 species adsorbed on TiO_2 with the increase in T_a for 0.1% NH_3/He.

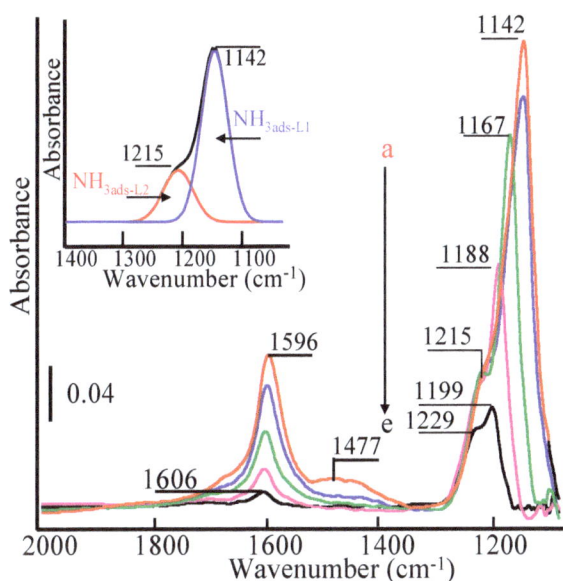

Figure 3. Impact of the adsorption temperature T_a on the IR bands of the adsorbed NH_3 species on TiO_2-P25 using 0.1% NH_3/He: (a–e) T_a = 300, 373, 473, 573 and 673 K. **Inset:** Decomposition of the δ_s IR bands of the $NH_{3ads-L2}$ and $NH_{3ads-L1}$ species at 300 K.

At 300 K, the overlapped IR bands at 1142 and 1215 cm^{-1} in Figure 3 are ascribed to the δ_s deformations of two adsorbed NH_3 species on different Lewis sites L1 and L2 named $NH_{3ads-L1}$ and $NH_{3ads-L2}$ respectively [18] whereas their δ_{as} deformations contribute to the IR band at 1596 cm^{-1}. Moreover, in Figure 3, the broad IR band at 1477 cm^{-1} and the shoulder at 1680 cm^{-1} are ascribed to the antisymmetric and symmetric deformation of NH_4^+ species formed by the adsorption of NH_3 on Brønsted sites [18] and references therein. The increase in T_a to 713 K for 0.1% NH_3/He leads to the decrease in the different IR bands: those of the NH_4^+ species disappear at $\approx$423 K indicating weakly adsorbed species whereas those of the NH_{3ads-L} species are present at 713 K. The individuals heats of adsorption of the $NH_{3ads-L1}$ and $NH_{3ads-L2}$ species have been obtained after decomposition of the two δ_s IR bands as shown in the inset of Figure 3 for T = 300 K. Considering similar IR absorption coefficients for the two NH_{3ads-L} species and taking into account that they are at full coverage at 300 K for 0.1% NH_3/He, the decomposition indicates that the L1 and L2 sites represent 70% and 30% of the Lewis sites of TiO_2 P25 respectively.

After decomposition at each adsorption temperature, the square and triangle symbols in Figure 4 give from Equation (3), the $\theta_{ex} = f(T_a)$ curves in isobaric conditions of $NH_{3ads-L1}$ and $NH_{3ads-L2}$ respectively.

Curves a and b which overlap the experimental data are obtained using Equations (4) and (5) with the following couples of (E_0 and E_1) values in kJ/mol (112, 56) and (160, 104) for the $NH_{3ads-L1}$ and $NH_{3ads-L2}$ species respectively. The circle symbols in Figure 4 give $\theta_{ex} = f(T_a)$ using the δ_{as} IR band at 1596 cm^{-1} at 300 K which is common to the two NH_{3ads-L} species. Equations (4) and (5) do not allow one obtaining a theoretical curve overlapped with the experimental data in the full coverage whatever the set of E_0 and E_1 values. This is consistent with the fact that the two NH_{3ads-L} species have significantly different heats of adsorption. However, the individual heats of adsorption of the $NH_{3ads-L1}$ and $NH_{3ads-L2}$ species can be determined by comparison of the experimental data with the theoretical average coverage provided by [18]:

$$\theta_{th}(T_a, P_a) = x_1\, \theta_{L1}(T_a, P_a) + x_2\, \theta_{L2}(T_a, P_a) \tag{6}$$

where $\theta_{L1}(T_a, P_a)$ and $\theta_{L2}(T_a, P_a)$ are the theoretical coverages of $NH_{3ads-L1}$ and $NH_{3ads-L2}$ respectively provided by Equations (4) and (5) and x_1 and x_2 represent the contribution (in fraction) of each NH_{3ads-L} species to the IR band at saturation of the L1 and L2 sites. For instance, curve c in Figure 4, which overlaps the experimental data is obtained using in Equation (6): $x_1 = 0.73$ and $x_2 = 0.27$, $E_{L1}(1) = 56$ kJ/mol, $E_{L1}(0) = 105$ kJ/mol, $E_{L2}(1) = 105$ kJ/mol, $E_{L2}(0) = 160$ kJ/mol. The heats of adsorption are consistent with those obtained using the δ_s IR bands (curves a and b in Figure 4) whereas x_1 and x_2 are consistent with the values provided by decomposition of the δ_s IR band at 300 K.

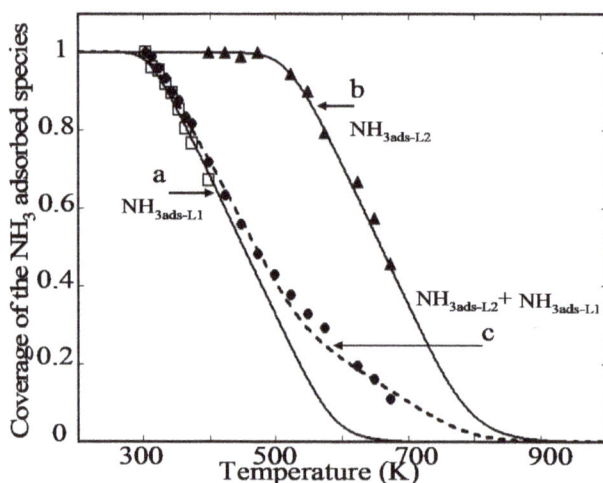

Figure 4. Heats of adsorption of the adsorbed NH_3 species on TiO_2-P25 using the AEIR method. ▲ and □ coverages of the $NH_{3ads-L2}$ and $NH_{3ads-L1}$ adsorbed species respectively using their δ_s IR bands; full lines: (a) and (b) theoretical coverages of the $NH_{3ads-L1}$ and $NH_{3ads-L2}$ species respectively using Equations (4) and (5) with $E_{L1}(1) = 56$ kJ/mol, $E_{L1}(0) = 102$ kJ/mol, $E_{L2}(1) = 102$ kJ/mol and $E_{L2}(0) = 160$ kJ/mol; ● average coverage of the $NH_{3ads-L1}$ and $NH_{3ads-L2}$ species using their common δ_{as} IR band at 1596 cm^{-1}; (c) theoretical evolution of the average coverage of the NH_{3ads-L} species using Equation (6) considering $x_1 = 0.73$ and $x_2 = 0.27$ for the $NH_{3ads-L1}$ and $NH_{3ads-L2}$ species respectively and with heats of adsorption of $E_{L1}(1) = 56$ kJ/mol, $E_{L1}(0) = 105$ kJ/mol, $E_{L2}(1) = 105$ kJ/mol and $E_{L2}(0) = 160$ kJ/mol (see the text for more details).

This procedure of the AEIR method using Equation (6) is of particular interest for the measurement of the individual heats of adsorption of NH_{3ads-L} species on sulfated TiO_2 containing catalysts because the strong $\nu(S=O)$ IR band of the sulfate groups prevents using the δ_s IR band of the NH_{3ads-L} species and only their common δ_{as} IR band can be exploited by the AEIR method [83]. Similarly the deposition of V_2O_5 and/or WO_3 on TiO_2-P25 [84] decreases the IR transmission of the solid below $\approx$1200 cm^{-1}

preventing using the decomposition of the δ_s IR band of the NH_{3ads-L} species as shown in Figure 5 for the adsorption of 0.1% NH_3/He on 6% WO_3/TiO_2. Moreover, this deposition leads to the presence of strong Brønsted and Lewis sites as shown by the evolutions of the IR bands of the NH_{3ads-L} (i.e., IR band at 1602 cm^{-1}) and NH_4^+ (1445 cm^{-1}) species in Figure 5 during the increase in T_a indicating that the two species are present at $T_a > 673$ K. The IR spectra in Figure 5 and the AEIR method give from Equation (6) the individual heats of adsorption of (a) two NH_{3ads-L} species using the IR band at 1602 cm^{-1} and (b) two NH_4^+ species using the IR band at 1445 cm^{-1} [84].

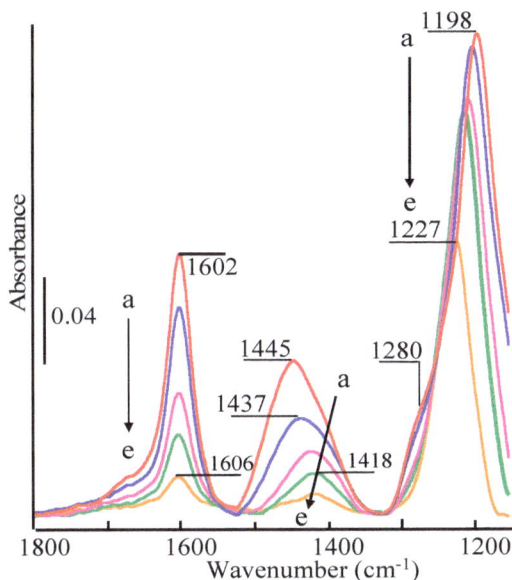

Figure 5. IR bands of adsorbed NH_3 species after adsorption of 0.1% NH_3/He on 6% WO_3/TiO_2-P25 pretreated at 713 K in helium as a function of adsorption temperature T_a: (a–e) $T_a = 300, 373, 498, 573$ and 673 K.

3.7. Application of the AEIR Method to Different Topics Relevant of Heterogeneous Catalysis

The AEIR method can be applied using either IR transmission or reflectance mode. In IR transmission, the microreactor IR cell must associate a low path length and high temperatures [46]. In DRIFT mode, temperature gradients in the solid sample and thermal/chemical stability of the sealing material/IR windows (i.e., presence of H_2O) must be taken into account [56,75]. Indeed, the level of the performances of the IR cells is dependent on the heats of adsorption of the adsorbed species of interest: strongly adsorbed species such as L CO species on Pt particles impose using high temperatures to observe the decrease in their coverage in isobaric conditions. Note that the AEIR method does not impose a variation of the coverage in the range 1–0: experiments in the temperature range corresponding to the beginning of the decrease in the coverage with spectra representative of the linear section of the isobar (Figures 2 and 4) provide the heats of adsorption of the adsorbed species [53].

The design of the experiments associated to the AEIR method is simple as compared to others analytical procedures dedicated to the measurement of the heats of adsorption. This permits the application of the method to study the impacts of different parameters associated to the catalyst preparation on the individual heats of adsorption of coadsorbed species. For instance, for the adsorption of CO on metal particles supported on metal oxides, the AEIR method permits to study the impacts of: the precursors of the metal particles [86,87]; the nature of the support [88,89]; the metal dispersion [55] and the deposition of additives (i.e., K on Pt/Al$_2$O$_3$ [90]). Similarly the AEIR method allows studying the geometric and electronic effects due to the formation of bimetallic particles via the changes in the nature of the adsorbed CO species and in their heats of adsorption respectively. For instance, the AEIR method reveals that the insertion of Sn in Pd particles [91] leads to the total disappearance of the bridged CO species due to a geometric effect. Moreover, the heat of adsorption of the L CO species on Pd$°$ sites, which varies linearly with its coverage from E$_0$ = 92 kJ/mol to E$_1$ = 54 kJ/mol on monometallic particles, is slightly modified on the Pd-Sn bimetallic particles: E$_0$ = 90 kJ/mol and E$_1$ = 50 k/mol, indicating a very small electronic effect. Similarly, Meunier et al. have used the AEIR method in DRIFT mode to study the impacts of the insertion of Zn on Pd$°$ sites of a Pd/CeO$_2$ catalyst [92]. The insertion on Zn suppresses the B CO species due to a geometric effect (as for the Pd-Sn particles [91]) whereas the heat of adsorption of the L CO species for Pd-Zn particles reduced at 773 K varies from E$_0$ = 105 $\pm$ 5 kJ/mol to E$_1$ = 68 $\pm$ 5 kJ/mol revealing a modest electronic effect of Zn. The same group has used the AEIR method to study the electronic effect of the insertion on Sn in Pt$°$ particles supported on Al$_2$O$_3$ revealing a strong electronic effect with a heat of adsorption at low coverage half of that on the monometallic particles [93].

The AEIR method is well adapted to the study of the individual heats of adsorption of adsorbed CO species on metal particles which is of interest either for the characterization of the solid using CO as a probe or for the EMA of catalytic reactions involving CO as reactant. This explains the interest of different groups for its application. For instance, Collins et al. [94] have applied the method to measure the individual heats of adsorption of one L and two B CO species on the Pd sites of a 2% Pd/SiO$_2$ catalyst: the values of the heats of adsorption of the main B CO species E$_0$ = 168 kJ/mol and E$_1$ = 62 kJ/mol are consistent with those measured on a Pd/Al$_2$O$_3$ catalyst [71] (see Table 1). Similarly, Chen et al. [95] have used the AEIR procedure in DRIFT mode to measure the heats of adsorption of two coadsorbed L$_1$ and L$_2$ CO species on Cu particles supported on SiO$_2$: IR bands at 2134 and 2119 cm^{-1} respectively, in relationship with the water-gas shift reaction. The E$_0$ and E$_1$ values are of 51 kJ/mol and 39 kJ/mol for the L1 and 70 kJ/mol and 46 kJ/mol for L$_2$ CO species. These values are consistent with those measured on a Cu/Al$_2$O$_3$ catalyst [62] (see Table 1). Rioux et al. [96] have used the AEIR method in diffuse reflectance mode for the measurement of the heat of adsorption of the L CO species on 2.69% Pt/SiO$_2$ (mesoporous silica: SBA-15, Pt particles 2.9 nm) using 10% CO/He showing a linear decrease with the increase in the coverage from E$_0$ = 167 kJ/mol to E$_1$ = 125 kJ/mol. The E$_1$ value is consistent with that measured on 2.9% Pt/Al$_2$O$_3$ (Table 1) while that a low coverage is slightly lower probably due to the nature of the support and the type of Pt particles due to the preparation method.

Diemant et al. [97] have used Equations (4) and (5) for the exploitation of the coverage of a L CO species on planar Au/TiO$_2$ model catalysts with different particle sizes obtained from Polarization Modulation-IRAS (PM-IRAS) spectra. They show that the evolutions of the experimental coverages in isobaric condition (P$_{CO}$ $\approx$ 10 mbar) are consistent with the theoretical curves. They reveal the significant impact of the particles size Φ on the heats of adsorption of the L CO species: i.e., E$_0$ decreases from 74 kJ/mol to 62 kJ/mol for Φ in the range of 2–4 nm [97]. The values at low and high coverages for Φ = 2 nm: 74 kJ/mol and 40 kJ/mol are consistent with those determined on a conventional Au/TiO$_2$ catalyst [61] (see Table 1). This shows that the AEIR method may allow us studying the impacts of the material gap between conventional catalysts and model surfaces on a key parameter of catalytic reactions.

Thus the AEIR method is often used to reinforce studies using the adsorption of a gas as a probe of the surface properties of a catalyst by providing via the individual heats of adsorption of the adsorbed species a quantification of the strength of the sites. For instance, this was one of the aims studying the modifications of the Lewis and Brønsted acidic sites by the deposition of WO$_3$ and V$_2$O$_5$ groups on sulfated and sulfate free TiO$_2$ supports species [18,83,84]. However, the AEIR method has been clearly developed as contribution to the EMA of catalytic reactions. In this field one of its interest is that it allows one studying the impact of the presence of a second gas (reactant or not) such as O$_2$, H$_2$ and H$_2$O on the heats of adsorption of adsorbed CO species on Pt/Al$_2$O$_3$ [98]. Particularly, the method provides experimental data on the change in the coverages of the different adsorbed species due to the coadsorption. These data can be modeled using theoretical coverages obtained from the Temkin formalism for competitive chemisorption without [15] and with [16] transformation of the reactants which is a key step of the EMA of a catalytic reaction such as for CO/H_2 on Pt/Al_2O_3 [15,16].

4. Conclusions

The AEIR method developed and applied during the last twenty years for the characterization of individual heats of adsorption of coadsorbed species formed by the adsorption of a gas on a solid catalyst, constitutes a tool for the development of the experimental microkinetic approach of gas/solid heterogeneous catalysis using conventional powdered catalysis. For each adsorbed species, the method allows, from the evolution of their characteristic IR bands in isobaric conditions to measure their individual heats of adsorption at different coverages via the validation of mathematical expressions of the adsorption coefficients and adsorption models. These data allow an accurate modeling of the two first surface elementary steps of any gas/solid catalytic reaction taking into account the diversity of the adsorption sites on a conventional catalyst. The design of the experiment for the AEIR method is easy as compared to others classical methods: a single isobar is needed using either a pellet of catalyst in IR transmission mode or sized catalyst particles for IR reflectance mode. This facilitates the use of the method for the study of the impacts of the modifications of the catalyst on the heats of adsorption such as the natures of the precursors, supports and additives and the formation of bimetallic particles. Moreover, the fact that the procedure can be applied on model surfaces [97] permits studying the impact of the material gap on an important thermodynamic parameters controlling the coverage of the surface during a catalytic reaction.

Funding: This research received no external funding.

Acknowledgments: I thank the different co-workers and Ph.D. students who have contributed to the development of the AEIR method and who appear as co-authors in the different references of the present work more particularly O. Dulaurent, A. Bourane, S. Derrouiche, P. Gravejat, J. Couble, and F. Giraud. Thanks are due to the industrial groups which appear in the different references for their financial supports more particularly FAURECIA which has allowed initiating the development of this method.

Conflicts of Interest: The authors declare no conflict of interest.

References

1. Dumesic, J.A.; Rudd, D.F.; Aparicio, L.M.; Rekoske, J.E.; Treviño, A.A. *The Microkinetics of Heterogeneous Catalysis*; ACS Professional Reference Book; An American Chemical Society Publication: Washington, DC, USA, 1993.
2. Stoltze, P. Microkinetic Simulation of Catalytic Reactions. *Prog. Surf. Sci.* **2000**, *65*, 65–150. [CrossRef]
3. Lynggaard, H.; Andreasen, A.; Stegelmann, C.; Stoltze, P. Analysis of Simple Kinetic Models in Heterogeneous Catalysis. *Prog. Surf. Sci.* **2004**, *77*, 71–137. [CrossRef]
4. Fishtik, I.; Callaghan, A.; Datta, R. Reaction Route Graphs. I. Theory and Algorithm. *J. Phys. Chem. B* **2004**, *108*, 5671–5682. [CrossRef]
5. Thybaut, J.W.; Marin, G.B. Single-Event MicroKinetics: Catalyst Design for Complex Reaction Networks. *J. Catal.* **2013**, *308*, 352–362. [CrossRef]
6. Van Helden, P.; van den Berg, J.A.; Coetzer, R.L.J.A. Statistical Approach to Microkinetic Analysis. *Ind. Eng. Chem. Res.* **2012**, *51*, 6631–6640. [CrossRef]
7. Hinrichsen, O.; Rosowski, E.; Muhler, M.; Ertl, G. The Microkinetic of Ammonia Synthesis Catalyzed by Cesium-promoted Supported Ruthenium. *Chem. Eng. Sci.* **1996**, *51*, 1683–1690. [CrossRef]
8. Dahl, S.; Sehested, J.; Jacobsen, C.J.H.; Törnqvist, E.; Chorkendorff, I. Surface Science Based Microkinetic Analysis of Ammonia Synthesis over Ruthenium Catalysts. *J. Catal.* **2000**, *192*, 391–399. [CrossRef]
9. Mhadeshwar, A.B.; Kitchin, J.R.; Barteau, M.A.; Vlachos, D.G. The Role of Adsorbate–adsorbate Interactions in the Rate Controlling Step and the most Abundant Reaction Intermediate of NH_3 Decomposition on Ru. *Catal. Lett.* **2004**, *96*, 13–22. [CrossRef]
10. Xu, J.; Clayton, R.; Balakotaiah, V.; Harold, M.P. Experimental and Microkinetic Modeling of Steady-state NO Reduction by H_2 on $Pt/BaO/Al_2O_3$ Monolith Catalysts. *Appl. Catal. B Environ.* **2008**, *77*, 395–408. [CrossRef]

11. Storsæter, S.; Chen, D.; Holmen, A. Microkinetic Modelling of the Formation of C1 and C2 Products in the Fischer–Tropsch Synthesis over Cobalt Catalysts. *Surf. Sci.* **2006**, *600*, 2051–2063. [CrossRef]
12. Stegelmann, C.; Stoltze, P. Microkinetic Analysis of Transient Ethylene Oxidation Experiments on Silver. *J. Catal.* **2004**, *226*, 129–137. [CrossRef]
13. Bourane, A.; Bianchi, D. Oxidation of CO on a Pt/Al$_2$O$_3$ Catalyst: From the Surface Elementary Steps to Light-Off Tests. 1. Kinetic Study of the Oxidation of the Linear CO Species. *J. Catal.* **2001**, *202*, 34–44. [CrossRef]
14. Bourane, A.; Bianchi, D. Oxidation of CO on a Pt/Al$_2$O$_3$ Catalyst: Form the Elementary Steps to Light-off Tests. 5—Experimental and Kinetic Modeling of Light-off Tests in Excess of O$_2$. *J. Catal.* **2004**, *222*, 499–510. [CrossRef]
15. Couble, J.; Bianchi, D. Experimental Microkinetic Approach of the CO/H$_2$ Reaction on Pt/Al$_2$O$_3$ using the Temkin Formalism. 1. Competitive Chemisorption between Adsorbed CO and Hydrogen Species in the Absence of Reaction. *J. Catal.* **2017**, *352*, 672–685. [CrossRef]
16. Couble, J.; Bianchi, D. Experimental Microkinetic Approach of the CO/H$_2$ Reaction on Pt/Al$_2$O$_3$ using the Temkin Formalism. 2. Coverages of the Adsorbed CO and Hydrogen species during the Reaction and Rate of the CH$_4$ Production. *J. Catal.* **2017**, *352*, 686–698. [CrossRef]
17. Haaland, D. Infrared Sudies of CO adsorbed on Pt/Al$_2$O$_3$: Evidence for CO Bonded in 3-Fold Coordination. *Surf. Sci.* **1987**, *185*, 1–14. [CrossRef]
18. Giraud, F.; Geantet, C.; Guilhaume, N.; Gros, S.; Porcheron, L.; Kanniche, M.; Bianchi, D. Experimental Microkinetic Approach of De-NO$_x$ by NH$_3$ on V$_2$O$_5$/WO$_2$/TiO$_2$ Catalysts: Part 1—Individual Heats of Adsorption of Adsorbed NH$_3$ species on a Sulfate-free TiO$_2$ Support using adsorption isobars. *J. Phys. Chem. C* **2014**, *118*, 15664–15676. [CrossRef]
19. Tompkins, F.C. *Chemisorption of Gases on Metal*; Academic Press: London, UK, 1978.
20. Cardona-Martinez, N.; Dumesic, J.A. Application of Adsorption Microcalorimetry to the Study of heterogeneous catalysis. *Adv. Catal.* **1992**, *38*, 149–237.
21. Spiewak, B.E.; Dumesic, J.A. Microcalorimetric measurements of differential heats of adsorption on reactive catalyst surfaces. *Thermochim. Acta* **1996**, *290*, 43–53. [CrossRef]
22. Podkolzin, S.G.; Shen, J.; de Pablo, J.J.; Dumesic, J.A. Equilibrated Adsorption of CO on Silica-Supported Pt Catalysts. *J. Phys. Chem. B* **2000**, *104*, 4169–4180. [CrossRef]
23. Watwe, R.M.; Spiewak, B.E.; Cortright, R.D.; Dumesic, J.A. Density Functional Theory (DFT) and Microcalorimetric Investigations of CO adsorption on Pt clusters. *Catal. Lett.* **1998**, *51*, 139–147. [CrossRef]
24. Falconer, J.L.; Schwarz, J.A. Temperature-Programmed Desorption and Reaction: Applications to supported Catalysts. *Catal. Rev. Sci. Eng.* **1983**, *25*, 141–227. [CrossRef]
25. Lemaitre, J.C. Temperature Programmed methods. In *Characterization of Heterogeneous Catalysts*; Marcel Dekker: New York, NY, USA, 1984.
26. Gorte, R.J. Design Parameters for Temperature Programmed Desorption from Porous Catalysts. *J. Catal.* **1982**, *75*, 164–174. [CrossRef]
27. Demmin, R.A.; Gorte, R.J. Design Parameters for Temperature-Programmed Desorption from a Packed Bed. *J. Catal.* **1984**, *90*, 32–39. [CrossRef]
28. Rieck, J.S.; Bell, A.T. Influence of Adsorption and Mass Transfer Effects on Temperature-Programmed Desorption from Porous Catalysts. *J. Catal.* **1984**, *85*, 143–153. [CrossRef]
29. Efsthathiou, A.; Bennett, C.O. Enthalpy and Entropy of H$_2$ Adsorption on Rh/Al$_2$O$_3$ Measured by Temperature-Programmed Desorption. *J. Catal.* **1990**, *124*, 116–126. [CrossRef]
30. Derrouiche, S.; Bianchi, D. Heats of Adsorption Using Temperature Programmed Adsorption Equilibrium: Application to the Adsorption of CO on Cu/Al$_2$O$_3$ and H$_2$ on Pt/Al$_2$O$_3$. *Langmuir* **2004**, *20*, 4489–4497. [CrossRef] [PubMed]
31. Xia, X.; Strunk, J.; Litvinov, S.; Muhler, M. Influence of Re-adsorption and Surface Heterogeneity on the Microkinetic Analysis of Temperature-Programmed Desorption Experiments. *J. Phys. Chem. C* **2007**, *111*, 6000–6008. [CrossRef]
32. Kottke, M.L.; Greenler, R.G.; Tompkins, H.G. An Infrared Spectroscopy Study of Carbon Monoxide adsorbed on polycristalline gold using the reflection-Absorption Technique. *Surf. Sci.* **1972**, *32*, 231–243. [CrossRef]
33. Richardson, H.H.; Baumann, C.; Ewing, G.E. Infrared Spectroscopy and Thermodynamic Measurement of CO on NaCl Films. *Surf. Sci.* **1987**, *185*, 15–35. [CrossRef]

34. Truong, C.M.; Rodriguez, J.A.; Goodman, D.W. CO Adsorption Isotherms on Cu(100) at Elevated Pressures and Temperatures using Infrared Reflection Absorption Spectroscopy. *Surf. Sci. Lett.* **1992**, *271*, L385–L391. [CrossRef]

35. Kuhn, W.K.; Szanyi, J.; Goodman, D.W. Adsorption Isobars for CO on Pd/Ta(110) at Elevated Pressures and Temperatures using Infrared Reflection-Absorption Spectroscopy. *Surf. Sci.* **1994**, *303*, 377–385. [CrossRef]

36. Szanyi, J.; Goodman, D.W. CO Oxidation on Palladium. 1. A Combined Kinetic-Infrared Reflection Absorption Spectroscopic Study of Pd(100). *J. Phys. Chem.* **1994**, *98*, 2972–2977. [CrossRef]

37. Szanyi, J.; Kuhn, W.K.; Goodman, D.W. CO Oxidation on Palladium. 2. A Combined Kinetic-Infrared Reflection Absorption Spectroscopic Study of Pd(111). *J. Phys. Chem.* **1994**, *98*, 2978–2981. [CrossRef]

38. Kuhn, W.K.; Szanyi, J.; Goodman, D.W. CO Adsorption on Pd(111): The Effects of Temperature and Pressure. *Surf. Sci. Lett.* **1992**, *274*, L611–L618. [CrossRef]

39. Paukshtis, E.A.; Soltanov, R.I.; Yurchenko, E.N. Determination of the Strength of Aprotic Acidic Centers on Catalyst Surfaces from the IR Spectra of Adsorbed Carbon Monoxide. *React. Kinet. Catal. Lett.* **1981**, *16*, 93–96. [CrossRef]

40. Paukshtis, E.A.; Soltanov, R.I.; Yurchenko, E.N. Determination of the Strength of Lewis Acid Centers via IR Spectroscopic Measurement of Adsorbed Pyridine. *React. Kinet. Catal. Lett.* **1982**, *19*, 105–108. [CrossRef]

41. Beebe, T.P.; Gelin, P.; Yates, J.T., Jr. Infrared Spectroscopic Observation of Surface Bonding in Physical Adsorption: The Physical Adsorption of CO on SiO$_2$ Surfaces. *Surf. Sci.* **1984**, *148*, 526–550. [CrossRef]

42. Ballinger, T.H.; Yates, J.T., Jr. IR Spectroscopic Detection of Lewis Acid Sites on Al$_2$O$_3$ Using Adsorbed CO. Correlation with Al-OH Group Removal. *Langmuir* **1991**, *7*, 3041–3045. [CrossRef]

43. Garrone, E.; Bolis, V.; Fubini, B.; Morterra, C. Thermodynamic and Spectroscopic Characterization of Heterogeneity among Adsorption Sites: CO on Anatase at Ambient Temperature. *Langmuir* **1989**, *5*, 892–899. [CrossRef]

44. Bolis, V.; Morterra, C.; Fubini, B.; Ugliengo, P.; Garrone, E. Temkin-Type Model for the Description of Induced Heterogeneity: CO Adsorption on Group 4 Transition Metal Dioxides. *Langmuir* **1993**, *9*, 1521–1528. [CrossRef]

45. Garrone, E.; Fubini, B.; Bonelli, B.; Onida, B.; Otero Areán, C. Thermodynamics of CO Adsorption on the Zeolite Na-ZSM-5 A Combined Microcalorimetric and FTIR Spectroscopic Study. *Phys. Chem. Chem. Phys.* **1999**, *1*, 513–518. [CrossRef]

46. Chafik, T.; Dulaurent, O.; Gass, J.L.; Bianchi, D. Heat of Adsorption of Carbon Monoxide on a Pt/Rh/CeO$_2$/Al$_2$O$_3$ Three-Way Catalyst Using in-Situ Infrared Spectroscopy at High Temperatures. *J. Catal.* **1998**, *179*, 503–514. [CrossRef]

47. Soma-Noto, Y.; Sachtler, W.M.H. Infrared Spectra of Carbon Monoxide Adsorbed on Supported Palladium and Palladium-Silver Alloys. *J. Catal.* **1974**, *32*, 315–324. [CrossRef]

48. Barth, R.; Pitchai, R.; Anderson, R.L.; Verykios, X.E. Thermal Desorption-Infrared Study of Carbon Monoxide Adsorption by Alumina-Supported Platinum. *J. Catal.* **1989**, *116*, 61–70. [CrossRef]

49. Bourane, A.; Dulaurent, O.; Chandes, K.; Bianchi, D. Heats of Adsorption of the Linear CO Species on a Pt/Al$_2$O$_3$ Catalyst using FTIR Spectroscopy: Comparison between TPD and Adsorption Equilibrium Procedures. *Appl. Catal. A Gen.* **2001**, *214*, 193–202. [CrossRef]

50. Kellner, C.S.; Bell, A.T. Studies of Carbon Monoxide Hydrogenation over Alumina Supported Ruthenium. *J. Catal.* **1981**, *71*, 296–307. [CrossRef]

51. Kohler, M.A.; Cant, N.W.; Wainwright, M.S.; Trim, D.L. Infrared Spectroscopic Studies of Carbon Monoxide Adsorbed on a Series of Silica-Supported Copper Catalysts in Different Oxidation States. *J. Catal.* **1989**, *117*, 188–201. [CrossRef]

52. Clarke, D.B.; Suzuki, I.; Bell, A.T. An Infrared Study of the Interaction of CO and CO$_2$ with Cu/SiO$_2$. *J. Catal.* **1993**, *142*, 27–36. [CrossRef]

53. Dulaurent, O.; Bianchi, D. Adsorption Isobars for CO on a Pt/Al$_2$O$_3$ Catalyst at High Temperatures using FTIR Spectroscopy: Isosteric Heat of Adsorption and Adsorption Model. *Appl. Catal. A Gen.* **2000**, *196*, 271–280. [CrossRef]

54. Bourane, A.; Dulaurent, O.; Bianchi, D. Heats of Adsorption of Linear and Multibound Adsorbed CO Species on a Pt/Al$_2$O$_3$ Catalyst Using in Situ Infrared Spectroscopy under Adsorption Equilibrium. *J. Catal.* **2000**, *196*, 115–125. [CrossRef]

55. Bourane, A.; Bianchi, D. Heats of Adsorption of the Linear CO species on Pt/Al$_2$O$_3$ using Infrared Spectroscopy: Impact of the Pt Dispersion. *J. Catal.* **2003**, *218*, 447–452. [CrossRef]

56. Couble, J.; Gravejat, P.; Gaillard, F.; Bianchi, D. Quantitative Analysis of Infrared Spectra of Adsorbed Species using Transmission and Diffuse Reflectance Modes. Case Study: Heats of Adsorption of CO on TiO$_2$ and CuO/Al$_2$O$_3$. *Appl. Catal. A Gen.* **2009**, *371*, 99–107. [CrossRef]

57. Couble, J.; Bianchi, D. Heats of Adsorption of Linearly Adsorbed CO Species on Co^{2+} and Co$^\circ$ Sites of Reduced Co/Al$_2$O$_3$ Catalysts in Relationship with the CO/H$_2$ Reaction. *Appl. Catal. A* **2012**, *445–446*, 1–13. [CrossRef]

58. Gravejat, P.; Derrouiche, S.; Farrussengn, D.; Lombaert, K.; Mirodatos, C.; Bianchi, D. Heats of Adsorption of Linear and Bridged CO Species Adsorbed on a 3% Ag/Al$_2$O$_3$ Catalyst Using in situ FTIR Spectroscopy under Adsorption Equilibrium. *J. Phys. Chem. C* **2007**, *111*, 9496–9503. [CrossRef]

59. Müslehiddinoglu, J.; Vannice, M.A. CO Adsorption on Supported and Promoted Ag Epoxidation Catalysts. *J. Catal.* **2003**, *213*, 305–320. [CrossRef]

60. Dulaurent, O.; Bianchi, D. Adsorption Model and Heats of Adsorption for Linear CO Species Adsorbed on ZrO$_2$ and Pt/ZrO$_2$ using FTIR Spectroscopy. *Appl. Catal. A Gen.* **2001**, *207*, 211–219. [CrossRef]

61. Derrouiche, S.; Gravejat, P.; Bianchi, D. Heats of Adsorption of Linear CO Species Adsorbed on the Au$^\circ$ and Ti$^+$ Sites of a 1% Au/TiO$_2$ Catalyst Using in Situ FTIR Spectroscopy under Adsorption Equilibrium. *J. Am. Chem. Soc.* **2004**, *126*, 13010–13015. [CrossRef] [PubMed]

62. Derrouiche, S.; Courtois, X.; Perrichon, V.; Bianchi, D. Heats of Adsorption of CO on a Cu/Al$_2$O$_3$ Catalyst Using FTIR Spectroscopy at High Temperatures and under Adsorption Equilibrium Conditions. *J. Phys. Chem. B* **2000**, *104*, 6001–6011.

63. Zeradine, S.; Bourane, A.; Bianchi, D. Comparison of the Coverage of the Linear CO Species on Cu/Al$_2$O$_3$ Measured under Adsorption Equilibrium Conditions by Using FTIR and Mass Spectroscopy. *J Phys. Chem. B* **2001**, *105*, 7254–7267. [CrossRef]

64. Bourane, A.; Nawdali, M.; Bianchi, D. Heats of Adsorption of the Linear CO Species Adsorbed on a Ir/Al$_2$O$_3$ Catalyst Using in Situ FTIR Spectroscopy under Adsorption Equilibrium. *J. Phys. Chem. B* **2002**, *106*, 2665–2671. [CrossRef]

65. Bourane, A.; Dulaurent, O.; Bianchi, D. Comparison of the Coverage of the Linear CO Species on Pt/Al$_2$O$_3$ Measured under Adsorption Equilibrium Conditions by Using FTIR and Mass Spectroscopy. *J. Catal.* **2000**, *195*, 406–411. [CrossRef]

66. Temkin, M.I. The Kinetics of some Industrial Heterogeneous Catalytic Reactions. *Adv. Catal.* **1979**, *28*, 173–291.

67. Glasstone, S.; Laidler, K.J.; Eyring, F. *The Theory of Rate Processes*; McGraw-Hill Inc.: New York, NY, USA; London, UK, 1941.

68. Laidler, K.J. Chap 5: The Absolute Rates of Surface Reactions. In *Catalysis*; Hemmett, P.H., Ed.; Reinhold Publishing Corporation: New York, NY, USA, 1954; Volume 1.

69. Hill, T.L. *An Introduction to Statistical Thermodynamics*; Addison-Wesley Publishing Company, Inc.: Boston, MA, USA, 1962.

70. Hachimi, A.; Chafik, T.; Bianchi, D. Adsorption Models and Heat of Adsorption of Adsorbed Ortho Di-methyl Benzene species on Silica by using Temperature Programmed Adsorption Equilibrium methods. *Appl. Catal. A Gen.* **2008**, *335*, 220–229. [CrossRef]

71. Dulaurent, O.; Chandes, K.; Bouly, C.; Bianchi, D. Heat of Adsorption of Carbon Monoxide on a Pd/Al$_2$O$_3$ Solid Using in Situ Infrared Spectroscopy at High Temperatures. *J. Catal.* **1999**, *188*, 237–251. [CrossRef]

72. Dulaurent, O.; Chandes, K.; Bouly, C.; Bianchi, D. Heat of Adsorption of Carbon Monoxide on a Pd/Rh Three-Way Catalyst and on a Rh/Al$_2$O$_3$ Solid. *J. Catal.* **2000**, *192*, 262–272. [CrossRef]

73. Dulaurent, O.; Nawdali, M.; Bourane, A.; Bianchi, D. Heat of Adsorption of Carbon Monoxide on a Ru/Al$_2$O$_3$ Catalyst using Adsorption Equilibrium Conditions at High Temperatures. *Appl. Catal. A Gen.* **2000**, *201*, 271–279. [CrossRef]

74. Derrouiche, S.; Bianchi, D. Heats of Adsorption of the Linear and Bridged CO species on a Ni/Al$_2$O$_3$ Catalyst by Using the AEIR Method. *Appl. Catal. A Gen.* **2006**, *313*, 208–217. [CrossRef]

75. Couble, J.; Bianchi, D. Heat of adsorption of the linear CO species adsorbed on reduced Fe/Al$_2$O$_3$ catalysts using the AEIR method in diffuse reflectance mode. *Appl. Catal. A Gen.* **2011**, *409*, 28–38. [CrossRef]

76. Lombardo, S.J.; Bell, A.T. A Review of Theoretical Models of Adsorption, Diffusion, Desorption, and Reaction of Gases on Metal Surfaces. *Surf. Sci. Rep.* **1991**, *13*, 1–72. [CrossRef]

77. Foo, K.Y.; Hammed, B.H. Insights into the Modeling of Adsorption Isotherm Systems. *Chem. Eng. J.* **2010**, *156*, 2–10. [CrossRef]

78. Murzin, D.Y. Modeling of Adsorption and Kinetics in Catalysis over Induced Nonuniform Surfaces: Surface Electronic Gas Model. *Ind. Eng. Chem. Res.* **1995**, *34*, 1208–1218. [CrossRef]

79. Yang, C.H. Statistical Mechanical Aspects of Adsorption Systems Obeying the Temkin Isotherm. *J. Phys. Chem.* **1993**, *97*, 7097–7101. [CrossRef]

80. Ritter, J.A.; Kapoor, A.; Yang, R.T. Localized Adsorption with Lateral Interaction on Random and Patchwise Heterogeneous Surfaces. *J. Phys. Chem.* **1990**, *94*, 6785–6791. [CrossRef]

81. Ritter, J.A.; Al-Muhtaseb, S.A. New Model That Describes Adsorption of Laterally Interacting Gas Mixtures on Random Heterogeneous Surfaces. 1. Parametric Study and Correlation with Binary Data. *Langmuir* **1998**, *14*, 6528–6538. [CrossRef]

82. Bourane, A.; Dulaurent, O.; Salasc, S.; Sarda, C.; Bouly, C.; Bianchi, D. Heats of Adsorption of Linear NO Species on a Pt/Al_2O_3 Catalyst Using in Situ Infrared Spectroscopy under Adsorption Equilibrium. *J. Catal.* **2001**, *204*, 77–88. [CrossRef]

83. Giraud, F.; Geantet, C.; Guilhaume, N.; Loridant, S.; Gros, S.; Porcheron, L.; Kanniche, M.; Bianchi, D. Experimental Microkinetic Approach of De-NO$_x$ by NH_3 on $V_2O_5/WO_2/TiO_2$ Catalysts. 2: Impact of Superficial Sulfate and/or V_xO_y groups on the Heats of Adsorption of Adsorbed NH_3 species. *J. Phys. Chem. C* **2014**, *118*, 15677–15692. [CrossRef]

84. Giraud, F.; Geantet, C.; Guilhaume, N.; Loridant, S.; Gros, S.; Porcheron, L.; Kanniche, M.; Bianchi, D. Experimental Microkinetic Approach of De-NO$_x$ by NH_3 on $V_2O_5/WO_2/TiO_2$ Catalysts. 3: Impact of Superficial WO$_z$ and V_xO_y/WO$_z$ Groups on the Heats of Adsorption of Adsorbed NH_3 species. *J. Phys. Chem. C* **2015**, *119*, 15401–15413. [CrossRef]

85. Giraud, F.; Couble, J.; Geantet, C.; Guilhaume, N.; Puzenat, E.; Gros, S.; Porcheron, L.; Kanniche, M.; Bianchi, D. Experimental Microkinetic Approach of De-NO$_x$ by NH_3 on $V_2O_5/WO_2/TiO_2$ Catalysts. 4. Individual Heats of Adsorption of Adsorbed H_2O Species on Sulfate-Free and Sulfated TiO_2 Supports. *J. Phys. Chem. C* **2015**, *119*, 16089–16105. [CrossRef]

86. Nawdali, M.; Bianchi, D. The impact of the Ru precursor on the adsorption of CO on Ru/Al_2O_3: Amount and reactivity of the adsorbed species. *Appl. Catal. A Gen.* **2002**, *231*, 45–54. [CrossRef]

87. Derrouiche, S.; Perrichon, V.; Bianchi, D. Impact of the Residual Chlorine on the Heat of Adsorption of the Linear CO Species on Cu/Al_2O_3 Catalysts. *J. Phys. Chem. B* **2003**, *107*, 8588–8591. [CrossRef]

88. Dulaurent, O.; Chandes, K.; Bouly, C.; Bianchi, D. Heat of Adsorption of Carbon Monoxide on Various Pd-Containing Solids Using in Situ Infrared Spectroscopy at High Temperatures. *J. Catal.* **2000**, *192*, 273–285. [CrossRef]

89. Pillonel, P.; Derrouiche, S.; Bourane, A.; Gaillard, F.; Vernoux, P.; Bianchi, D. Impact of the support on the heat of adsorption of the linear CO species on Pt-containing catalysts. *Appl. Catal. A Gen.* **2005**, *278*, 223–231. [CrossRef]

90. Derrouiche, S.; Gravejat, P.; Bassou, B.; Bianchi, D. Impact of Potassium on the Heats of Adsorption of Adsorbed CO species on Supported Pt Particles by Using the AEIR Method. *Appl. Surf. Sci.* **2007**, *253*, 5894–5898. [CrossRef]

91. Jbir, I.; Couble, J.; Khaddar-Zine, S.; Ksibi, Z.; Meunier, F.; Bianchi, D. Individual Heat of Adsorption of Adsorbed CO Species on Palladium and Pd–Sn Nanoparticles Supported on Al_2O_3 by Using Temperature-Programmed Adsorption Equilibrium Methods. *ACS Catal.* **2016**, *6*, 2545–2558. [CrossRef]

92. Meunier, F.; Maffre, M.; Schuurmann, Y.; Colussi, S.; Trovarelli, A. Acetylene semi-hydrogenation over Pd-Zn/CeO$_2$: Relevance of CO adsorption and methanation as descriptors of selectivity. *Catal. Commun.* **2018**, *105*, 52–55. [CrossRef]

93. Moscu, A.; Schuurman, Y.; Veyre, L.; Thieuleux, C.; Meunier, F. Direct evidence by in situ IR CO monitoring of the formation and the surface segregation of a Pt–Sn alloy. *Chem. Commun.* **2014**, *50*, 8590–8592. [CrossRef] [PubMed]

94. Collins, S.E.; Baltanas, M.A.; Bonivardi, A.L. Heats of adsorption and activation energies of surface processes measured by infrared spectroscopy. *J. Mol. Catal. A* **2008**, *281*, 73–78. [CrossRef]

95. Chen, C.-S.; Lai, T.-W.; Chen, C.-C. Effect of Active Sites for a Water–Gas Shift Reaction on Cu Nanoparticles. *J. Catal.* **2010**, *273*, 18–28. [CrossRef]

96. Rioux, R.M.; Hoefelmeyer, J.D.; Grass, M.; Song, H.; Niesz, K.; Yang, P.; Somorjai, G.A. Adsorption and Co-adsorption of Ethylene and Carbon Monoxide on Silica-Supported Monodisperse Pt Nanoparticles: Volumetric Adsorption and Infrared Spectroscopy Studies. *Langmuir* **2008**, *24*, 198–207. [CrossRef] [PubMed]

97. Diemant, T.; Hartmann, H.; Bansmann, J.; Behm, R.J. CO adsorption energy on planar Au/TiO$_2$ model catalysts under catalytically relevant conditions. *J. Catal.* **2007**, *252*, 171–177. [CrossRef]

98. Bourane, A.; Dulaurent, O.; Bianchi, D. Heats of Adsorption of the Linear CO Species Adsorbed on a Pt/Al$_2$O$_3$ Catalyst in the Presence of Coadsorbed Species Using FTIR Spectroscopy. *Langmuir* **2001**, *17*, 5496–5502. [CrossRef]

catalysts

MDPI

Review

Kinetic Modeling of Catalytic Olefin Cracking and Methanol-to-Olefins (MTO) over Zeolites: A Review

Sebastian Standl [1,2,*] and Olaf Hinrichsen [1,2]

[1] Department of Chemistry, Technical University of Munich, Lichtenbergstraße 4,
    85748 Garching near Munich, Germany; olaf.hinrichsen@ch.tum.de
[2] Catalysis Research Center, Technical University of Munich, Ernst-Otto-Fischer-Straße 1,
    85748 Garching near Munich, Germany
*   Correspondence: sebastian.standl@ch.tum.de

Received: 2 November 2018; Accepted: 27 November 2018; Published: 5 December 2018

Abstract: The increasing demand for lower olefins requires new production routes besides steam cracking and fluid catalytic cracking (FCC). Furthermore, less energy consumption, more flexibility in feed and a higher influence on the product distribution are necessary. In this context, catalytic olefin cracking and methanol-to-olefins (MTO) gain in importance. Here, the undesired higher olefins can be catalytically converted and, for methanol, the possibility of a green synthesis route exists. Kinetic modeling of these processes is a helpful tool in understanding the reactivity and finding optimum operating points; however, it is also challenging because reaction networks for hydrocarbon interconversion are rather complex. This review analyzes different deterministic kinetic models published in the literature since 2000. After a presentation of the underlying chemistry and thermodynamics, the models are compared in terms of catalysts, reaction setups and operating conditions. Furthermore, the modeling methodology is shown; both lumped and microkinetic approaches can be found. Despite ZSM-5 being the most widely used catalyst for these processes, other catalysts such as SAPO-34, SAPO-18 and ZSM-23 are also discussed here. Finally, some general as well as reaction-specific recommendations for future work on modeling of complex reaction networks are given.

Keywords: kinetics; kinetic model; microkinetics; cracking; methanol-to-olefins (MTO); zeolite; ZSM-5; ZSM-23; SAPO-18; SAPO-34

1. Introduction

Propene is one of the crucial building blocks originating from the petrochemical industry [1]. After ethene, it is the second most-produced crude oil derivative [2]. In 2014, its global demand was quantified as 89×10^6 t [2]. Around 90% of the worldwide supply is produced via fluid catalytic cracking (FCC) or steam cracking [3], the latter being the process with the highest energy demand in the chemical industry [4]. Besides the economic disadvantages, the enormous CO_2 emissions represent another problem [5,6]. Moreover, the high-temperature process allows almost no product adjustment and the shift from higher feedstocks to ethane as feed further reduces C_3 yields [7]. In FCC, propene is a byproduct because this process is aimed at gasoline production [8].

An increase in propene demand is predicted [9,10]; see, for example, a recent review from Blay et al. [3]. Thus, alternative catalytic processes are necessary. Cracking of higher olefins [3], methanol-to-hydrocarbons (MTH) [11], olefin metathesis [12,13], propane dehydrogenation [14,15], oxidative dehydrogenation of propane [16] or ethene-to-propene [3,17] are amongst the most prominent alternative processes.

Kinetic modeling is an indispensable tool for assessing reaction kinetics, heat management, product distribution and reactor performance [18,19]. The application range of kinetic models depends on their complexity: many different strategies exist between the simplest approach, a power-law model and the highest level of detail, a microkinetic model. Models with less complexity are created relatively quickly and do not require much computational power, but they are restricted in terms of their possible applications. On the other hand, the preparation of a microkinetic model is time-consuming and complicated, but it can be used to gain insight into intermediates and preferred reaction pathways, for extrapolation, transfer to other systems and optimization of both catalysts and the process [18,20,21].

When dealing with hydrocarbon conversion over zeolites as catalytic materials, reaction networks are extremely large because of the many different isomers. This is why kinetic modeling of these processes is challenging; without suitable assumptions, derivations and simplifications, no reasonable solutions can be achieved. Nevertheless, the importance of such models is especially high because propene, which is the desired compound in many processes, is an intermediate and not a final product.

This review focuses on the kinetic modeling of two important alternative pathways for propene production: cracking of higher olefins and methanol-to-olefins (MTO) as a special case of MTH. Most studies were performed on either ZSM-5 or SAPO-34, but other zeolite types are also discussed. All examples presented here are deterministic kinetic models and involve three essential features: gathering of experimental data, creation of a reaction network that leads to the model equations and fitting of the kinetic parameters by comparing the modeled results with the obtained data. Although both catalysts and experimental details are mentioned, the emphasis of this review is on the modeling methodology: How is the reaction network created? Which assumptions are made? How many and what types of compounds are included? Is there any mechanistic background considered in deriving the rate equations? How is the adsorption process treated? Which software is used for parameter estimation? Are any details of the numeric routine given? How many fitting responses and parameters are necessary?

To the best of our knowledge, such an overview does not exist for the two processes mentioned above. Indeed, two reviews of MTO kinetic models do exist, namely those of by Khadzhiev et al. [22] and by Keil [23] are available. The latter, however, was published in 1999; since then, both mechanistic understanding of the reaction and computational power have developed rapidly leading to the proposal of a variety of new models. On the other hand, the work by Khadzhiev et al. [22] from 2015 is a useful overview of various kinetic MTO studies, but only a few models are selected. Furthermore, the focus is not on the underlying reaction networks and modeling methodologies. Especially for MTO, there is a wide range of options for representing the reactivity using a model. This review should elucidate that almost every literature study is unique because of different assumptions and methodologies. For this reason, we attempt to establish some general advantages and disadvantages of the approaches in the concluding remarks, ending with a suggestion on the choice of methodology and the suitability of assumptions.

The criteria mentioned above mean that numerous studies are excluded from this review. Firstly, all kinetic approaches published before 2000 are ignored. Apart from the fact that they have already been discussed in the helpful review of Keil [23], most of these examples focus not on MTO, but on methanol-to-gasoline (MTG) where temperatures are lower to increase the yield of the gasoline fraction. In addition to the first kinetic description by Chen and Reagan [24], this includes the models of Chang [25], Ono and Mori [26], Mihail et al. [27,28], Schipper and Krambeck [29], Sedrán et al. [30,31], Schönfelder et al. [32] and Bos et al. [33]. Noteworthy are the comparably large reaction network in [27,28] and the elevated temperatures in [32,33] which are within the MTO range. In addition to the mentioned review of Keil [23], some of the models are compared in [30,34].

Secondly, first principle and ab initio studies are not covered because no actual fitting to experimental data is performed. Nevertheless, this theory gives important insight into mechanistic details which is why some examples should be mentioned here. Where zeolite chemistry is concerned, there are many publications by the van Speybroeck group. In addition to reviews about the theory [35] and MTO [36,37], several aspects of the MTO reactivity are investigated in detail: for example, the influence of adsorption effects [38] and especially of water [38,39], the methylation of aromatics [38,40], the methylation of olefins [41,42] and the formation as well as the reactivity of surface methyl groups [43] are analyzed. Furthermore, general mechanistic details [39,44] and the relationship between catalyst properties, the morphology of the catalyst and product compositions can be elucidated [45]. Similar investigations exist for the cracking of paraffins [46,47] and olefins [48–50] using different zeolites.

Thirdly, publications with kinetic parameters resulting from simple Arrhenius plots without any underlying reaction network are not discussed here.

Fourthly, no hydrocracking is reviewed here as some steps of the underlying chemistry are different. For example, initial physisorption on the catalytic surface takes place with a paraffin and not with an olefin. Next, the catalyst is bifunctional in hydrocracking, meaning that the first reaction step leads to a dehydrogenation of the paraffin. From now on, the surface reactions of the resulting olefin are comparable to the mechanisms in olefin cracking. Finally, the product olefin is hydrogenated yielding the corresponding paraffin. In ideal hydrocracking, all hydrogen assisted steps at the metal phase are assumed to be *quasi*-equilibrated, so the kinetically relevant reactions are comparable to the ones in olefin cracking. However, there are also conditions where this ideal scenario is not realized. In the literature, several microkinetic studies for hydrocracking using the single-event methodology are available [51–68]. Other approaches are possible and useful especially for complex feeds such as a Fischer–Tropsch product mixture or vacuum gas oil [69–74].

Fifthly, alternative approaches such as the stochastic method by Shahrouzi et al. [75] are ignored because they are too different to be compared with deterministic models.

In summary, this review presents and compares kinetic models for olefin cracking and MTO with the emphasis on reaction network complexity and methodology. This overview should help in finding suitable approaches for the particular requirements of future studies.

2. Theoretical Background

As mentioned in Section 1, the focus of this review is the comparison of kinetic modeling methodologies in order to find suitable solutions for future studies of complex hydrocarbon conversion. For this reason, the theoretical part is restricted to the most important facts without going into details. The cited literature should be referred to for more detailed information about kinetic modeling fundamentals, zeolites and underlying reaction mechanisms because these topics are discussed only in brief.

2.1. Thermodynamics

In contrast to the other topics of this section, thermodynamics are broadly analyzed here for several reasons. Many kinetic models require thermodynamic data, e.g., for the calculation of equilibrated or backward reactions. A correct implementation of equilibrium constants is crucial for the model performance; thus, the underlying theory and calculation procedures should be shown in the following. The results are compared with literature correlations. Thermodynamic equilibrium distributions are evaluated for olefin cracking as well as MTO. This is helpful as first step in order to find intermediate and stable products. Finally, insight into the influence of typical reaction conditions on equilibrium distributions might help in understanding overall reactivity. Thermodynamic equilibria are obtained by minimization of the total Gibb's free energy $G_t (T)$ (see Equation (1)) [76–78]:

$$G_t (T) = \sum_j \mu_j (T) \, n_j,$$
(1)

$$\text{with } \mu_j (T) = \mu_j^\circ (T) + R T \ln \left(\frac{f_j}{f_j^\circ} \right).$$
(2)

Equation (1) yields an absolute value in joules, equal to the sum of all considered species j with their chemical potential $\mu_j (T)$ given as a molar value multiplied by the number of moles n_j of compound j when equilibrium is reached. In this state, the total number of moles n_t may differ from the initial value, thus n_t is not constant. For an ideal gas, the fugacity f_j equals the partial pressure p_j, whereas f_j° is equivalent to a well-defined standard pressure p°. According to IUPAC [79], p° is set equal to 10^5 Pa. Although a standard temperature T° is defined as 273.15 K, the superscript $^\circ$ for thermo-physical properties only relates to the standard pressure [79]. The standard chemical potential $\mu_j^\circ (T)$ in Equation (2) is equal to the standard Gibb's energy of formation $\Delta_f G^\circ (T)$. Thus, the relation in Equation (3) is obtained,

$$\mu_j (T) = \Delta_f G_j^\circ (T) + R T \ln \left(\frac{p_t}{p^\circ} \right) + R T \ln \left(\frac{n_j}{n_t} \right).$$
(3)

When the total pressure p_t equals the standard pressure p°, the term in the middle of Equation (3) can be omitted. Values of $\Delta_f G^\circ (T)$ are tabulated in standard references [80], in several collections published by Alberty [81–95] or they can be calculated using group additivity methods [96–103]. According to the Gibbs–Helmholtz equation [76], $\Delta_f G^\circ (T)$ remains a function of temperature. When no suitable values are found in literature, $\Delta_f G^\circ (T)$ can be calculated via Equation (4). Since no standard entropy of formation exists, the sum over all elements el must be subtracted from $S_j^\circ (T)$; the former value is obtained by multiplying the standard entropy of the respective element $S_{el}^\circ (T)$ by the number of atoms $N_{el,j}$ which are part of compound j.

$$\Delta_f G_j^\circ (T) = \Delta_f H_j^\circ (T) - T \left(S_j^\circ (T) - \sum_{el} N_{el,j} \, S_{el}^\circ (T) \right),$$
(4)

$$\text{with } \Delta_f H_j^\circ (T) = \Delta_f H_j^\circ (298.15\,\text{K}) + \int_{298.15\,\text{K}}^{T} c_{p,j}(T)\, dT, \tag{5}$$

$$\text{and } S_j^\circ (T) = S_j^\circ (298.15\,\text{K}) + \int_{298.15\,\text{K}}^{T} \frac{c_{p,j}(T)}{T}\, dT. \tag{6}$$

The temperature dependence of the heat capacity can be described via polynomial approximations [104,105]. For this review, $\Delta_f G^\circ (T)$ values as a function of temperature are extracted from literature for ethene ($C_2^=$) to octenes ($C_8^=$) [88], for methanol [91] and for water [80]. These are fitted to a second degree polynomial using *polyfit* within MATLAB. With the resulting coefficients, $\Delta_f G^\circ (T)$ can be evaluated for each desired temperature. For dimethyl ether (DME), heat capacity values from [106] are fitted with the same routine. In combination with $\Delta_f H^\circ (298.15\,\text{K})$ from [107] and $S^\circ (298.15\,\text{K})$ from [108] as well as heat capacity and $S^\circ (298.15\,\text{K})$ values for carbon, hydrogen and oxygen from [80], $\Delta_f G^\circ (T)$ is calculated with the help of Equations (5) and (6). Two cases are analyzed here: a mixture of ethene to octenes and the system methanol/DME/water. These should represent the olefin cracking case and the MTO feed, respectively. The resulting equilibria as a function of temperature can be seen in Figure 1. They are obtained by minimizing Equation (1) using *fmincon* in MATLAB. Here, the *sqp* algorithm is applied which yields stable solutions independent of the starting values for the molar composition.

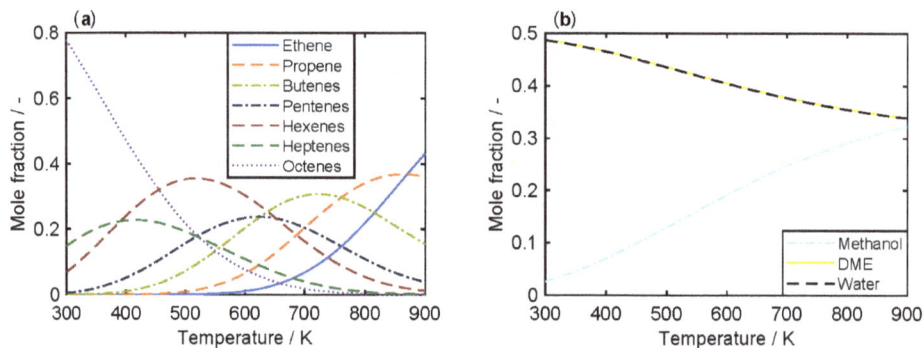

Figure 1. Composition of an equilibrated mixture as a function of temperature at standard pressure $p_t = p^\circ$: (**a**) for $C_2^=$ to $C_8^=$; and (**b**) for the system methanol/DME/water.

Figure 1a shows a clear trend towards lower olefins at high temperatures. For an MTO feed, the equimolar fraction of DME and water decreases when the temperature is raised. During the conversion of methanol to DME and water, the number of moles remains constant, which is why a change in pressure does not effect the equilibrium. On the other hand, the influence of pressure on the olefin distribution is depicted in Figure 2a for a characteristic cracking temperature of 650 K.

It is obvious that thermodynamics favor the generation of higher olefins when the total pressure is increased. Figure 2b summarizes the results for the desired product propene: for maximum yields, the pressure should be as low and the temperature as high as possible. However, the optimum conditions taken from Figure 2 deviate from an applicable industrial case. Usually, the equilibrated olefin distribution does not depict the process, because propene is an intermediate product here. This makes a proper description of reaction kinetics inevitable.

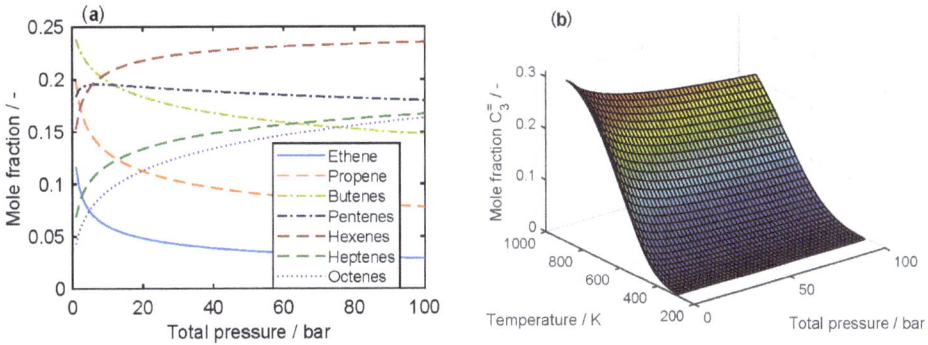

Figure 2. Composition of an equilibrated mixture for $C_2^=$ to $C_8^=$: (**a**) as a function of total pressure at 650 K; and (**b**) as mole fraction of propene at equilibrium conditions as a function of both temperature and total pressure.

In this context, the thermodynamic equilibrium constant K^{TD} of the system methanol/DME/water is especially important because it can be incorporated into a model, e.g., to describe the equilibrated feed. In general, this value is accessible via the Gibb's free energy of reaction $\Delta_r G^\circ (T)$ [76]. This relation is shown in Equation (7) using the exothermic reaction $2\text{MeOH} \rightleftharpoons \text{DME} + \text{H}_2\text{O}$ as an example,

$$K^{TD} = \exp\left(-\frac{\Delta_r G^\circ (T)}{RT}\right) = \frac{p\,(\text{DME})\,p\,(\text{H}_2\text{O})}{p\,(\text{MeOH})^2}. \tag{7}$$

In the following, some literature correlations for this constant are shown. Figure 3 compares these approaches with our own solution from Figure 1.

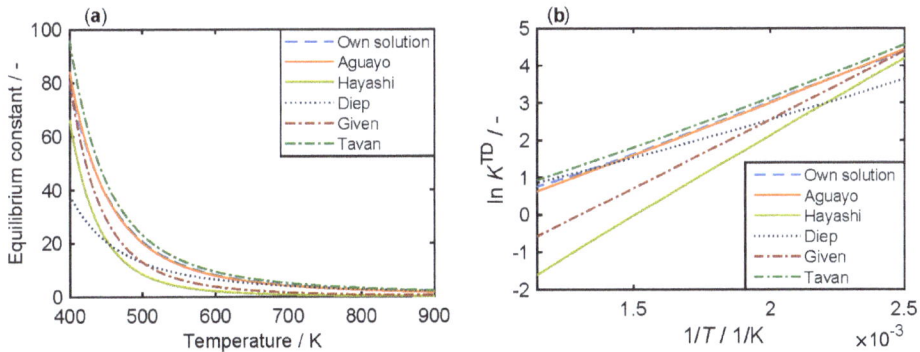

Figure 3. Equilibrium constants for the system methanol/DME/water, taken from different references [109–113] and compared with our own solution according to Figure 1, as a function of temperature: (**a**) with a regular scale; and (**b**) with a logarithmic scale.

Figure 3 shows that only the correlation published by Aguayo et al. [109] closely matches the solution derived from thermodynamics. This correlation is represented by Equation (8):

$$K^{TD} = \exp\left(-9.76 + \frac{3200\,\text{K}}{T} + 1.07\ln\left(\frac{T}{\text{K}}\right) - 6.57 \times 10^{-4}\frac{T}{\text{K}} + 4.90 \times 10^{-8}\frac{T^2}{\text{K}^2} + \frac{6050\,\text{K}^2}{T^2}\right). \tag{8}$$

In the high temperature range, i.e., above 600 K, the correlations of Tavan and Hasanvandian [113] and Diep and Wainwright [111] also yield satisfying results (see Equations (9) and (10), respectively):

$$K^{TD} = \exp\left(\frac{4019\,K}{T} + 3.707\ln\left(\frac{T}{K}\right) - 2.783 \times 10^{-3}\frac{T}{K} + 3.8 \times 10^{-7}\frac{T^2}{K^2} - \frac{65\,610\,K^3}{T^3} - 26.64\right), \quad (9)$$

$$K^{TD} = \exp\left(\frac{2835.2\,K}{T} + 1.675\ln\left(\frac{T}{K}\right) - 2.39 \times 10^{-4}\frac{T}{K} - 0.21 \times 10^{-6}\frac{T^2}{K^2} - 13.360\right). \quad (10)$$

By contrast, use of the correlations of Given [112] and Hayashi and Moffat [110] shown in Equations (11) and (12), respectively, is recommended only for temperatures not significantly greater than 400 K,

$$K^{TD} = \exp\left(\frac{30\,564\,J\,mol^{-1}}{R\,T} - 4.8\right), \quad (11)$$

$$K^{TD} = \exp\left(\left(\frac{-6836\,K}{T} + 3.32\ln\left(\frac{T}{K}\right) - 4.75 \times 10^{-4}\frac{T}{K} - 1.1 \times 10^{-7}\frac{T^2}{K^2} - 10.92\right)\right. \quad (12)$$
$$\left.\frac{4.1868\,J\,mol^{-1}\,K^{-1}}{-R}\right).$$

The correlations of Gayubo et al. [114], Schiffino and Merrill [115] and Khademi et al. [116] are not shown here because their application leads to high deviation from the results in Figure 3. The equations of Gayubo et al. [114] and Hayashi and Moffat [110] are of the same form, but different values are used by the former group [114]. The authors refer to the review by Spivey [117] who used the equation by Hayashi and Moffat [110] with the original values.

2.2. Kinetic Modeling

A kinetic model describes the relation between rate r_l of a certain reaction l and the concentration of one or several reactants i [18,118–121]. The latter can be expressed as partial pressure $p\,(i)$, as mole concentration per volume $C\,(i)$, as mole fraction $y\,(i)$, or as mass fraction $w\,(i)$. In the following, a subscript C in $p_C\,(i)$, $y_C\,(i)$ and $w_C\,(i)$ means that only carbon containing species are considered. The value $y_C\,(i)$ of a certain compound is determined by multiplying its number of carbon atoms by the number of molecules of this type and comparing this value with the total number of carbon atoms.

In this review, only those models are investigated where the influence of transport phenomena can be neglected. According to the seven steps of heterogeneous catalysis [122], the description is then simplified to adsorption, surface reaction and desorption. Adsorption is an exothermic step, in which the reactant interacts with the catalyst. It is divided into physisorption and chemisorption [123]. The former describes an undirected, unselective and comparably weak interaction, often with the catalyst surface, which is mainly caused by van der Waals forces. The chemisorption is highly selective and is formed for example through a chemical bond between reactant and active center. Here, the adsorption enthalpy is significantly higher compared to physisorption [123]. The reverse process to adsorption is desorption. From thermodynamics, it follows that high pressures and low temperatures favor adsorption. There are different strategies for describing these effects mathematically. A common approach is the Langmuir (L) isotherm in Equation (13), which depends on the temperature T [118,124],

$$\theta_i\,(T) = \frac{K_i^{ads}\,(T)\,p\,(i)}{1 + K_i^{ads}\,(T)\,p\,(i)}, \quad (13)$$

with the relative coverage θ_i of species i on the catalyst surface and a specific adsorption equilibrium constant K_i^{ads}. In the form of Equation (13), an underlying assumption is that adsorption and desorption are *quasi*-equilibrated. Furthermore, a uniform surface, no interaction between adsorbed species, monolayer adsorption and non-dissociative adsorption are assumed. In addition to the Langmuir isotherm, other approaches also exist [125].

In the following, typical kinetic expressions are introduced: power law, Langmuir, Langmuir–Hinshelwood (LH), Eley–Rideal (ER) and Hougen–Watson (HW). It should be underlined that for these examples, the surface reaction is assumed to be the slowest step, whereas all sorption processes are treated as *quasi*-equilibrated. Although this is a common scenario, conditions where adsorption or desorption becomes kinetically relevant are also possible. In the following, non-dissociative and competing adsorption of all species is assumed, thereby deviating from the classical formulations of the kinetic expressions found in the literature. At this point, it is important to mention that there is no unique mechanism for any of the preceding kinetic expressions because the resulting equation always depends on the assumptions. This is why all kinetic equations in this review are denoted as type of a certain mechanism.

The simplest way to construct a kinetic model is using power law expressions [124,126]. Equation (14) is typical of a monomolecular reaction:

$$r_l = k_l \, p \, (i)^\kappa \, . \tag{14}$$

Here, the rate constant k_l as well as the reaction order κ are unknown. They can be obtained by fitting the model to experimental data [63]. The reaction order does not need to correspond to the stoichiometric coefficient of species i in step l. Especially in power law models, the former value is often determined as a purely empirical value without any physical meaning.

The level of detail is increased by choosing one of the following basic mechanistic approaches. When such a scheme is applied, the reactions are assumed to be elementary in most cases, meaning that the reaction order equals the stoichiometric coefficient.

For monomolecular reactions, the adsorption of the reactant can be described via an L type of isotherm which leads to the kinetic description in Equation (15) [119,124,127]:

$$r_l = \frac{k_l \, K_i^{ads} \, p \, (i)}{1 + \sum_j K_j^{ads} \, p \, (j)} \, . \tag{15}$$

A similar description is obtained for bimolecular reactions where both reactants i and v must be adsorbed before the reaction takes place. The approach in Equation (16) is often referred to as an LH type of mechanism [120,124]:

$$r_l = \frac{k_l \, K_i^{ads} \, p \, (i) \, K_v^{ads} \, p \, (v)}{\left(1 + \sum_j K_j^{ads} \, p \, (j)\right)^2} \, . \tag{16}$$

In the classical LH expression, which is frequently shown, only the two reactants are included for the inhibiting adsorption term in the denominator. In contrast, Equation (16) considers all adsorbing species in the system which is closer to the HW type of mechanism [120,121,127,128]. The latter usually consists of three parts, describing the reaction kinetics (rate constant), the potential (concentrations as well as difference from the thermodynamic equilibrium, if applicable) and inhibition through competing adsorption. Equation (17) describes an example of a monomolecular reversible reaction of reactant i which leads to the two products v and w. Because both reactants of the backward step adsorb before reaction, it is a combination of LH and HW types of mechanism. The equilibrium constant can either be calculated from thermodynamics (K^{TD}) or estimated as an unknown parameter (K_l):

$$r_l = \frac{k_l \, K_i^{ads} \, p \, (i) - \frac{k_l}{K} \, K_v^{ads} \, p \, (v) \, K_w^{ads} \, p \, (w)}{1 + \sum_j K_j^{ads} \, p \, (j)} \, . \tag{17}$$

A bimolecular reaction where only one of the reactants i has to be adsorbed while the second compound v reacts directly from the gas phase is known as an ER type of mechanism [121] (see Equation (18)), again with a combination of an HW type of mechanism:

$$r_l = \frac{k_l \, K_i^{ads} \, p\,(i) \, p\,(v)}{1 + \sum_j K_j^{ads} \, p\,(j)}.$$ (18)

Besides the description via relative, i.e., dimensionless, quantities for the coverage, absolute concentration values of adsorbed surface species can be applied by multiplying θ_i by the total concentration of acid sites. For the well-defined zeolites, this value is usually known. Consequently, the rate and equilibrium constants remain as unknown parameters.

The temperature dependence of the rate constants is expressed via the Arrhenius approach in Equation (19) [121] which introduces the activation energy E_a:

$$k = A \, \exp\left(-\frac{E_a}{R\,T}\right).$$ (19)

The coherence given by Eyring [129] is shown in modified form [130] in Equation (20). The preexponential factor A contains the Boltzmann constant k_B, the Planck constant h and the entropy change from reactant to transition state $\Delta_\ddagger S^\circ$. Furthermore, the value $\Delta_\ddagger \nu$ resembles the difference in number of moles between activated complex and reactant state; it is required to correctly relate activation enthalpy and energy,

$$k = \frac{k_B \, T}{h} \, \exp\left(\frac{\Delta_\ddagger S^\circ}{R}\right) \, \exp\left(1 - \Delta_\ddagger \nu\right) \, \exp\left(-\frac{E_a}{R\,T}\right).$$ (20)

Usually, both preexponential factor and activation energy must be estimated. Reparameterization according to Equation (21) is often performed to reduce the correlation between these two values [63,131]:

$$k = k^{ref} \, \exp\left(-\frac{E_a}{R}\left(\frac{1}{T} - \frac{1}{T^{ref}}\right)\right) = A^{ref} \, \exp\left(-\frac{E_a}{R\,T^{ref}}\right) \, \exp\left(-\frac{E_a}{R}\left(\frac{1}{T} - \frac{1}{T^{ref}}\right)\right).$$ (21)

Alternatively, the approach in Equation (22) can be used [132]:

$$k = \exp\left(\left(\ln A^{ref} - \frac{E_a}{R\,T^{ref}}\right) - \frac{E_a}{R}\left(\frac{1}{T} - \frac{1}{T^{ref}}\right)\right).$$ (22)

The reference temperature T^{ref} should be within the investigated range and is often chosen as the average, although detailed guidelines for its proper estimation exist [131,133].

Another option is to additionally consider the temperature dependence of the preexponential factor (see Equation (23)):

$$k = A^{ref} \, \frac{T}{T^{ref}} \, \exp\left(-\frac{E_a}{R\,T^{ref}}\right) \, \exp\left(-\frac{E_a}{R}\left(\frac{1}{T} - \frac{1}{T^{ref}}\right)\right).$$ (23)

The preexponential factor of a reaction can also be calculated prior to the fitting process to reduce the number of unknown parameters [19,51,134]. For this purpose, reliable assumptions for the entropy change $\Delta_\ddagger S^\circ$ are required [18].

During estimation of adsorption or reaction equilibrium constants, reparameterization is applicable in a manner analogous to that shown in Equation (24) [63,128]:

$$K = K^{ref} \, \exp\left(-\frac{\Delta H^\circ}{R}\left(\frac{1}{T} - \frac{1}{T^{ref}}\right)\right).$$ (24)

Again, the reference value can be written within the exponential function, as it is done in Equation (25) [132]:

$$K = \exp\left(\left(\frac{\Delta S^{\circ}}{R} - \frac{\Delta H^{\circ}}{R\, T^{\text{ref}}}\right) - \frac{\Delta H^{\circ}}{R}\left(\frac{1}{T} - \frac{1}{T^{\text{ref}}}\right)\right) \tag{25}$$

For kinetic models, it is crucial to differentiate the rate r_l of a reaction step l from the net rate of production $R(i)$ of a certain species i [120]. The latter is obtained by summing up all reaction rates where the compound i is consumed or produced. Each rate must be multiplied by the stoichiometric coefficient $\nu_l(i)$ of i in step l, as shown in Equation (26):

$$R(i) = \sum_l \nu_l(i)\, r_l. \tag{26}$$

From these remarks, it follows that stoichiometry should be considered for three points: for the formulation of reaction rates ($2C_4^=$ to $C_8^=$ instead of $C_4^=$ to $C_8^=$), for the reaction order as long as elementary reactions are assumed ($p\,(C_4^=)^2$ instead of $p\,(C_4^=)$) and for the net rate of production ($-2\,k_l\, p\,(C_4^=)^2$ instead of $-k_l\, p\,(C_4^=)^2$). However, in this review, it is shown that approaches deviating from this suggestion exist, which nevertheless can still yield a model with high agreement, although it is purely empirical.

The net rate of production is required to obtain the molar flow rate $F(i)$ of a certain species i along the reactor. For this, integration over the catalyst mass W is performed. If not further specified, Equation (27) for a one-dimensional, pseudo-homogeneous, isothermal plug flow reactor applies for all examples in this review [121]:

$$\frac{\mathrm{d}F(i)}{\mathrm{d}W} = R(i). \tag{27}$$

An objective function compares the difference between modeled and measured output [54,121]. Several values are suitable, for example, molar flow rates, mass flow rates or mole fractions. The latter option is chosen for the example in Equation (28) where $y_{j,k}$ characterizes the experimental and $\hat{y}_{j,k}$ the modeled mole fraction, respectively. In this common approach, the objective function equals the sum of squared residuals SSQ which should be minimized during parameter estimation [54]; a more generalized least-squares criterion can be found in [121],

$$SSQ = \sum_k^{N_{\text{Exp}}} \sum_j^{N_{\text{Res}}} \omega_j \left(y_{j,k} - \hat{y}_{j,k}\right)^2. \tag{28}$$

Evaluation is performed with all experimental data points N_{Exp} and all fitting responses N_{Res}. The latter value comprises all species j which should be used for parameter estimation; however, this need not match the number of lumps in the event that one or several lumps are to be explicitly excluded during fitting. In combination with the number of estimated parameters N_{Par} of the model, its degree of freedom dof can be calculated according to Equation (29):

$$dof = N_{\text{Exp}}\, N_{\text{Res}} - N_{\text{Par}}. \tag{29}$$

Equation (28) contains a weighting factor ω_j which is accessible through replicate experiments: these lead to the experimental errors whose covariance matrix can be inverted, thereby leading to ω_j which equals the diagonal elements [121]. Without replicate experiments, the necessary values can be obtained via Equation (30) [54] using the molar flow rate F_j,

$$\omega_j = \frac{\left(\sum_k^{N_{\text{Exp}}} F_{j,k}\right)^{-1}}{\sum_j^{N_{\text{Res}}} \left(\sum_k^{N_{\text{Exp}}} F_{j,k}\right)^{-1}}. \tag{30}$$

It is explicitly highlighted below if other approaches are used to calculate the weighting factor.

2.3. Zeolites

Originally, the term zeolite referred solely to aluminosilicates consisting of SiO_4 and AlO_4^- units. In the meantime, other materials with similar structural features have been included in the definition. All have a common crystalline and tetrahedral structure [135]. Two of their properties are especially important in the context of catalysis. Firstly, they have well-defined channels and intersections whose cross sections are often within the range of molecular size. Through this, a shape selectivity during reactions is achieved: the small openings can prevent certain molecules from entering or leaving the channels while the pore structure also influences the reaction transition states [136]. Secondly, they contain acid sites which is why they are also called solid acids. Brønsted acidity arises when aliovalent cations such as Al^{3+} and Si^{4+} are connected via oxygen [135]. The resulting negativity of the framework is balanced by additional cations. The incorporation of H^+ creates a Brønsted acid hydroxyl group situated between aluminum and silicon. The oxygen itself acts as a Lewis base by providing electrons for the non-fully coordinated metal cations. The latter are typical Lewis acid sites, either within the framework or as extra-framework cations [137]. For industrial use, zeolites are often mixed with binders which provide a mesoporous surrounding of the crystals. This can also affect the catalytic performance [138,139]. The resulting extrudates are then formed to the desired pellet shape.

Within a zeolite crystal, the tetrahedral units represent the primary building units. Their systematic arrangement leads to a block consisting of several tetrahedra which is referred to as a secondary building unit (SBU) [135]. The SBU is characteristic of a certain zeolite because it is found periodically within the framework. A three-letter code is used to differentiate the various frameworks [140]. Another important property is the channel opening which is defined by the number of cations: a ten-membered ring means that the opening is formed by ten cations connected via ten oxygen atoms. Eight-, ten- and twelve-membered rings are classified into small-, medium- and large-pore zeolites, respectively [135]. The Structure Commission of the International Zeolite Association presents an overview of the different zeolites online [141]. Details about morphology, synthesis and characterization can be found elsewhere [135–137,140,142]. Here, only the four examples discussed within the review are briefly described.

ZSM-5 [143] is the second most applied zeolite in industry [142]. The framework code is MFI and the SBU consists of a pentasil unit, which itself is composed of eight rings with five cations each. It is a medium-pore zeolite where two types of pores can be found: straight channels along the (010) direction and zigzag channels along the (100) direction [142,144]. These cross each other at intersections; a three-dimensional pore network is obtained with the dimensions 5.1×5.5 Å (straight channel) and 5.3×5.6 Å (zigzag channel) [135]. In general, ZSM-5 shows strong acidity, high activity and stability and a pronounced shape selectivity during hydrocarbons conversion [135].

ZSM-23 [145] with framework code MTT is another example of a medium-pore zeolite [135]. Its channels are one-dimensional with an opening of 5.2×4.5 Å [141]. This accelerates deactivation, but also yields more higher and branched olefins, which are suitable for gasoline production [146,147].

The aluminophosphates consist of tetrahedral AlO_4^- and PO_4^+ units. Consequently, this framework is neutral [137]. When P is replaced with Si, a negative charge is introduced which creates acid sites. This leads to the silicoaluminophosphates with SAPO-34 and SAPO-18 as examples. Their strict framework ordering allows only even-numbered rings as pore openings [142]. While the Si/Al ratio must be greater than one for the aluminosilicates [137], it is usually less than one for SAPO. The structure of SAPO-34 [148] is similar to chabazite (framework code CHA) meaning it is a small-pore zeolite with an opening of 3.8 Å [149] and a three-dimensional cage structure. On the one hand, its moderate acidity in combination with the shape selectivity leads to high yields of lower olefins. On the other hand, SAPO catalysts are prone to rapid deactivation effects [150].

SAPO-18 [151] is isomorphic to SAPO-34, while the framework is of AEI type [149]. It belongs to the small-pore zeolites with openings similar to SAPO-34 [141]. Although these two SAPO examples also have the same Si/(Si+Al+P) ratio, the amount of Brønsted acid sites is significantly lower for SAPO-18. Hence, it has a longer lifetime [151,152]. Moreover, synthesis of SAPO-18 is simpler and cheaper than for SAPO-34 [151].

2.4. Reaction Mechanisms

Many references about olefin cracking and MTO exist in the literature. Here, only the most important facts are summarized. In addition, the discussion of each reaction network in the individual sections also gives hints about the underlying mechanism.

2.4.1. Olefin Cracking

Several studies analyze the mechanism and product distribution of olefin cracking over zeolites [153–160]. A helpful overview can be found in [3]. The conversion of higher olefins to mainly propene using a recycle reactor concept comparable to MTP is viable [161,162]; however, thus far, no commercial process has been achieved.

Two main pathways must be differentiated: monomolecular cracking and dimerization with subsequent cracking [163]. The former is only possible for olefins having carbon numbers greater than or equal to [164]. Ethene, propene and butenes must undergo a dimerization step first. The resulting higher olefin can either crack in a second step or it can react in another dimerization. This leads to a complex interconversion scheme [158]. Figure 3 shows that thermodynamics favor lower olefins at high temperatures. One reason is that dimerization is exothermic, while the cracking reaction is endothermic. In addition, adsorption effects might be important for this observation because adsorption is less favored at higher temperatures [165].

The general description of cracking within a model and especially the difference between a simpler approach and microkinetics is well illustrated by the monomolecular cracking of 2-methyl-hept-2-ene.

In a simple model considering stoichiometry, the reaction would be formulated as $C_7^= \rightleftharpoons C_3^= + C_4^=$ with the corresponding rate equation $r = k\,p\,(C_7^=)$. However, this ignores both the backward reaction and the adsorption of $C_7^=$ prior to the reaction. Due to of the latter fact, the estimated rate constant is an apparent value that includes adsorption effects. This could lead to negative activation energies, especially when more than one reactant is required in the adsorbed state [166].

The microkinetic approach for this reaction is shown in Scheme 1. Here, the adsorption step must be considered first: it is described as a two-step chemisorption at a Brønsted acid site, which can be divided into π complex formation and protonation [167]. The resulting intermediate is depicted as a carbenium ion here, however, it should be noted that alkoxides are also proposed as stable intermediates [167–169].

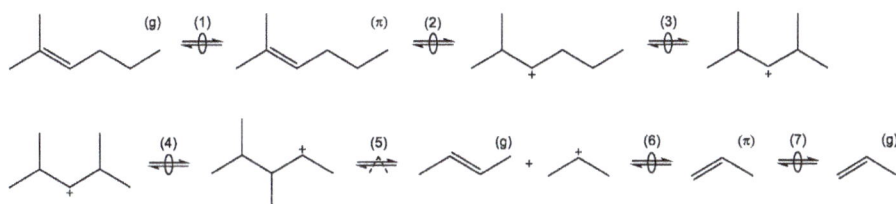

Scheme 1. Elementary reactions occurring during cracking of olefins connected to a complete catalytic cycle: formation of π complex (1), protonation (2), isomerization through PCP branching (3), isomerization through methyl shift (4), cracking (5), deprotonation (6) and desorption (7). The superscripts (g) and (π) represent olefins in the gas phase and bound in a π complex, respectively.

The intermediate obtained after protonation in Scheme 1 illustrates why microkinetic models differentiate each isomer: in the initial form, monomolecular cracking would be energetically less desired because of the formation of a primary intermediate [164]. By contrast, the molecule undergoes two isomerization reactions: an additional side group is formed by branching via a protonated cyclopropane (PCP) transition state, whereas a subsequent methyl shift changes the position of this side group. Cracking to a secondary propyl intermediate is now possible. The other product, but-2-ene, is released directly to the gas phase. Finally, the desorption to propene takes place. As illustrated in Scheme 1, all steps are reversible and only the cracking or dimerization as a backward reaction are of kinetic relevance [164]. By contrast, the adsorption as well as isomerization reactions are often assumed to be *quasi*-equilibrated [54,164]. In addition to PCP branching and methyl shift, hydride shifts also exist as isomerization steps.

Apart from these olefin interconverting steps, side reactions also exist [170,171]. The most important pathway produces both paraffins and aromatics. It starts with a hydride transfer from an olefin to a protonated intermediate. The latter is converted to a paraffin, whereas the former converts into a protonated olefin. Provided the chain is long enough, a cyclization reaction takes place, yielding a cyclic olefin subsequent to a deprotonation [170]. Through two additional hydride transfers and deprotonations, an aromatic structure is obtained. This mechanism results in a ratio of 3:1 of paraffins to aromatics. However, the latter can form polymerized species, leading to coke, which also allows different ratios. The formation of methane is attributed to thermal cracking effects. Further elementary steps occurring during olefin interconversion, especially when many cyclic compounds are involved, are beyond the focus of this review and can be found for example in [20,52,53,58,172].

2.4.2. Methanol-to-Olefins

Since the conversion of methanol to hydrocarbons was discovered accidentally by two independent research teams at Mobil [173], many scientists have tried to determine the exact mechanism. At first, the focus of this process was on the production of high-octane compounds (MTG), but the product spectrum always contained high quantities of olefins, which is why MTO was introduced [174]. A commercial solution for increasing propene yields by recycling the higher olefins is called methanol-to-propylene (MTP) [175]. The product composition depends heavily on the conditions, setup and catalyst [11]. Nevertheless, some general features of methanol conversion over zeolites are shown here.

When pure methanol is led over an acid zeolite, the reaction to form DME and water proceeds comparably quickly. The thermodynamic equilibrium in Figure 1b is rapidly achieved. Several studies exist that consider the exact mechanism of this reaction [176–182]. The dissociative mechanism suggests that after methanol is chemisorbed, its dehydration leads to a surface methyl group. In a subsequent step, the latter reacts with a second methanol molecule to a protonated DME which finally desorbs. In the associative route, DME is produced directly without forming the surface methyl group as an intermediate.

Mechanistically, the formation of the first C-C bond, i.e., the conversion of C_1 (methanol) and C_2 (DME) oxygenates to higher hydrocarbons, has been under debate for decades. It is still not fully understood, although some recent contributions underline the importance of formaldehyde in this context [183]. Previously, other mechanisms were proposed: the oxonium ylide mechanism, the carbene mechanism, the carbocationic mechanism, the free radical mechanism and the consecutive type mechanism. These are summarized and discussed in the review by Stöcker [174] and also in [10]. Despite the unresolved mechanism, the autocatalytic nature of MTO with a pure methanol feed is well-known [24,26]. During the initiation phase, the conversion of oxygenates is almost zero because the formation of the first C-C bond is slow. After a certain contact time, the conversion rate increases: the first hydrocarbons are formed, this accelerates the conversion of oxygenates which again produces more hydrocarbons. Figure 4 clearly shows the resulting S-shape of the curve which only slows down when the concentration of oxygenates becomes too low.

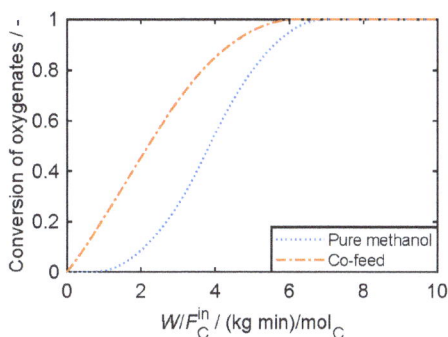

Figure 4. Typical case of oxygenates conversion as a function of contact time for a pure methanol feed and for a feed comprising olefins co-fed with methanol.

An important concept for MTO chemistry is the hydrocarbon pool proposed by Dahl and Kolboe [184–186]. According to this theory, adsorbed or trapped hydrocarbon species which are not further defined act as a co-catalyst through ongoing methylation and dealkylation reactions, the latter releasing mainly ethene and propene. It was suggested that the pool species are somehow similar to coke, i.e., polymethylated aromatic compounds formed during the early reaction stage [184]. Much research has gone into determining the exact structure of the aromatic compounds [11,187–191]. Furthermore, two mechanistic pathways for methanol consumption and subsequent olefin dealkylation were suggested [189]. In the side-chain mechanism [192–196], one of the side chains of the aromatic compound is continuously growing until it is dealkylated as olefin. By contrast, in the paring mechanism [197], the growth of the aromatic compound causes complex structural rearrangements which also lead to olefin release.

The trapped aromatics are especially important for small pore catalysts like SAPO-18 or SAPO-34. However, different characteristics were determined over ZSM-5. Here, olefins are the main source of methanol consumption by continuous methylation and cracking [153,198]. This observation led to the proposal of the dual-cycle theory [199,200]: it was found that yields of ethene correlate with those of aromatics in contrast to propene. Consequently, their formation routes must be mechanistically separated. Whereas the aromatic hydrocarbon pool is similar to the one described above, an olefin hydrocarbon pool also plays an important role. Here, olefins grow through methylation reactions and crack down to lower olefins again. Over ZSM-5, this is the main route towards propene, especially at high temperatures. Again, whether the olefins are methylated in a stepwise mechanism via a surface methyl group [26,176] or in a concerted step [201] remains the subject of much debate. Furthermore, it has been observed that DME can also perform methylation reactions, a step which exhibits lower barriers than methanol [202,203].

The formation of side products is not restricted to the evolution of polymethylated aromatics. The mechanism described in Section 2.4.1 for olefin cracking is also valid for MTO. However, it was observed that, when methanol is present, the side product formation is significantly higher than for the pure cracking case [171]; in addition, methane formation is pronounced [204]. This led to the proposal of a methanol-induced hydrogen transfer [205,206] where again formaldehyde plays an important role. Figure 5 shows the MTO reaction network on a ZSM-5 catalyst in a simplified way. This illustration emphasizes the dual-cyle mechanism, with the olefin based cycle on the left side and the aromatic based cycle on the right side. The latter is not further specified, i.e., it characterizes both the less methylated aromatics which are found in the product spectrum as well as the heavier compounds trapped in the pores.

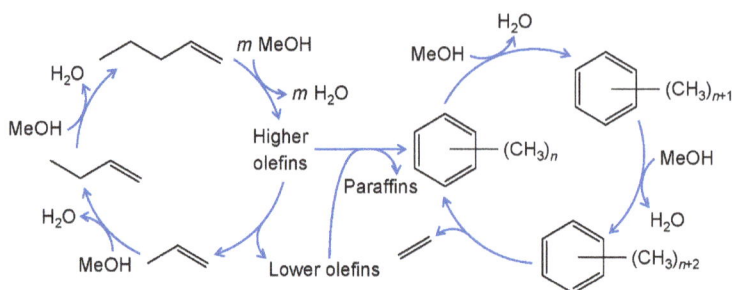

Figure 5. Simplified reaction network for MTO over ZSM-5 with the olefin based cycle on the left and the aromatics based cycle on the right side; the latter produces mainly ethene, whereas aromatics and paraffins are formed both through olefin interconversion reactions and a methanol induced pathway, adapted from [11].

As mentioned above, the undesired higher olefins are recycled and co-fed in the commercial MTP process [175,207]. This changes the underlying chemistry drastically [208–213]: the slow formation of the first C-C bond is obsolete because higher hydrocarbons are available straightaway. Consequently, no initiation phase is observed; the oxygenates conversion increases immediately from the beginning as depicted in Figure 4.

Several reviews [11,23,36,37,144,174,190,214–216] and overviews [217] provide more details. Current research is focused on a wide range of issues, i.e., the exact mechanism of methylation [38,40–42,201,202,218–224], catalyst properties [45,222,225–228], reaction conditions [210,211,229] and deactivation [203,230–234]. Finally, an overview of the current state of MTO commercialization is given in [191].

3. Kinetic models for Olefin Cracking

The different models are grouped according to crucial characteristics. The ones by Epelde et al. [235], Ying et al. [236] and Huang et al. [166] depict manifold olefin interconversion reactions over ZSM-5. All three have comparable numeric approaches and especially the studies by Ying et al. [236] and Huang et al. [166] are very similar, although the latter includes a mechanistic approach. The next section groups the models by Borges et al. [165] and Oliveira et al. [237] over ZSM-5. Here, the focus is not on a complete olefin interconversion picture, but on describing the feed consumption rate [165] and on considering the different acid strengths of the sites [237]. The next two sections both contain only one model: there is no other microkinetic study of olefin cracking except for von Aretin et al. [164] and the model by Zhou et al. [77] is the only example over SAPO-34. An overview of the analyzed models can be found in Tables 1 and 2; the horizontal lines divide the different sections. At the end of each section, a short summary paragraph compares the models and shows advantages and disadvantages of the approach. Table 1 contains information about the catalysts used in the studies, whereas Table 2 lists the experimental conditions and details about the modeling approach. In the following text, only special features of both catalyst and setup are mentioned. The description focuses on the underlying reaction network and the derivation of the model. In Table 2, the maximum experimental contact time is given in the same unit as in the original publication. This value is always based on the inlet molar flow rate which is either expressed as molar flow rate of carbon (subscript C) or of all species (subscript t). For the kinetic parameters, only the subscript C is used when the values are explicitly related to carbon units; otherwise, no subscript is shown.

Table 1. Properties of the different catalysts which were used for the kinetic models of olefin cracking; besides the zeolite type, its silicon-to-aluminum ratio (Si/Al), its total number of acid sites plus determination method, its ratio of Brønsted to Lewis acid sites (BAS/LAS) and its surface area according to the method by Brunauer–Emmett–Teller (BET) are shown. Furthermore, the time-on-stream (TOS) after which the kinetic data were taken, the particle size (d_P) and information about whether an extrudate or pure powder was used are presented. The line separates the different subsections. A hyphen represents missing information.

Model	Zeolite Type	Si/Al	Total Acidity	BAS/LAS	BET	TOS	d_P	Extrudate
Epelde [235]	ZSM-5 (1%$_{wt}$ K)	280 [1]	0.033 mmol g_{cat}^{-1} (t-BA) [2]	-	194 m^2 g_{cat}^{-1}	5 h [3]	150–300 μm	25/30/45%$_{wt}$ (Zeolite/Bentonite/Alumina)
Ying [236]	ZSM-5	103 [1]	0.21 mmol g_{cat}^{-1} (NH$_3$)	-	340 m^2 g_{cat}^{-1}	0.03 h	420–841 μm	Yes [4]
Huang [166]	ZSM-5	200 [1]	0.012 mmol g_{cat}^{-1} (NH$_3$)	1.35 at 423 K	301.1 m^2 g_{cat}^{-1}	0–10 h [5]	125–149 μm	70/30%$_{wt}$ (Zeolite/Alumina)
Borges [165]	ZSM-5	30	-	-	416 m^2 g_{cat}^{-1} [6]	0–1.4 h	-	No
Oliveira [237]	ZSM-5 (0, 0.52, 0.65 and 0.80%$_{wt}$ Na) [7]	30	-	-	416 m^2 g_{cat}^{-1} [8]	0–1.4 h	-	No
von Aretin [164]	ZSM-5	90	0.174 mmol g_{cat}^{-1} (C$_5$H$_5$N)	4.27	454 m^2 g_{cat}^{-1}	> 6 h	400–500 μm	No
Zhou [77]	SAPO-34 [9]	0.25 [10]	1.25 mmol g_{cat}^{-1} (NH$_3$)	-	550 m^2 g_{cat}^{-1}	0.02 h	3.2 μm [11]	No

(1) Value of the zeolite, i.e., without binder. (2) Calculated under the assumption that the total acidity of the extrudate is equal to one fourth of the pure zeolite's acidity according to the annotations in [235]. (3) Results extrapolated to 0 h TOS. (4) No further information available (commercial catalyst from Süd-Chemie), but the patent cited in [236] reveals the use of alumina as binder. (5) Regeneration after 10 h TOS. (6) Extracted from a subsequent publication [237]. (7) Corresponds to catalysts where 0%, 2.4%, 3.0% and 3.2%, respectively, of the protons were exchanged with Na ions. (8) Surface areas of 396, 394 and 386 m^2 g_{cat}^{-1}, respectively, with higher Na amounts. (9) SAPO-34 and SAPO-18 disordered intergrowth structure revealed by XRD. (10) Value of the gel. (11) Mean size.

Table 2. Experimental conditions and modeling details for the kinetic models of olefin cracking; the feed components, the temperature range (T), the total pressure (p_t), the partial pressure range of the feed olefin (p_{Ol}) and the maximum contact time ((W/F^{in})$_{max}$) with resulting conversion (X_{max}) are listed. Concerning the model, the number of fitted responses (N_{Res}), the number of estimated parameters (N_{Par}), the number of experiments (N_{Exp}) and the degree of freedom (dof) are shown. Finally, it is noted whether the model follows a type of a mechanistical scheme (Mech.), whether adsorption is considered (Ads.) and which side products are included (Side prod.). The line separates the different subsections. A hyphen represents missing information.

Model	Feed	T	p_t	p_{Ol}	(W/F^{in})$_{max}$	X_{max}	N_{Res}	N_{Par}	N_{Exp}	dof	Mech.	Ads.	Side prod.
Epelde [235]	$C_4^=$, He	673–873 K	1.5 bar	0.375–1.35 bar	1.60 g$_{cat}$ h mol$_C^{-1}$	0.75	8 5	23 16	51 51	385 239	No	No	$C_{1-4}^=$, C_{6-8}^{ar}
Ying [236]	$C_3^=$, N$_2$ $C_4^=$, N$_2$ $C_5^=$, N$_2$ $C_6^=$, N$_2$ $C_7^=$, N$_2$	673–763 K	1.013 bar	0.131 bar	0.23 h (1) 0.16 h (1) 0.07 h (1) 0.10 h (1) 0.04 h (1)	0.67 0.65 0.57 0.94 0.98	7	28	115	777	No	No	C_{3-7}, C_{6-7}^{ar}
Huang [166]	3 NW(2) 4 NW(2) 5 NW(2) 6 NW(2) 7 NW(2)	673–763 K	1.013 bar	0.0832 bar 0.0706 bar 0.0601 bar 0.0532 bar 0.0476 bar	1.12 kg$_{cat}$ s mol$_t^{-1}$ 0.73 kg$_{cat}$ s mol$_t^{-1}$ 0.39 kg$_{cat}$ s mol$_t^{-1}$ 0.20 kg$_{cat}$ s mol$_t^{-1}$ 0.03 kg$_{cat}$ s mol$_t^{-1}$	0.33 0.40 0.23 0.46 0.46	6	44	104	580	LH, HW	$C_{2-7}^=$, H$_2$O	No
Borges [165]	$C_2^=$, N$_2$ $C_3^=$, N$_2$ $C_4^=$, N$_2$	473–723 K	1.013 bar	0.05–0.30 bar	9.33 g$_{cat}$ h mol$_C^{-1}$ 1.55 g$_{cat}$ h mol$_C^{-1}$ 0.27 g$_{cat}$ h mol$_C^{-1}$	– – –	3	8	36	100	ER, HW	$C_{2-4}^=$	No
Oliveira [237]	$C_2^=$, N$_2$ $C_3^=$, N$_2$ $C_4^=$, N$_2$	473–723 K	1.013 bar	0.05–0.30 bar	9.33 g$_{cat}$ h mol$_C^{-1}$ 1.55 g$_{cat}$ h mol$_C^{-1}$ 0.27 g$_{cat}$ h mol$_C^{-1}$	– – –	17	20	61	1017	L, ER, HW	$C_{2-8}^=$	C_{2-8}, C_{6-8}^{ar}
von Aretin [164]	$C_5^=$, N$_2$	633–733 K	1.2 bar	0.043–0.070 bar	1.6 kg$_{cat}$ s mol$_t^{-1}$	0.55	5	5	141	700	L, ER, HW	$C_{2-12}^=$	No
Zhou [77]	$C_2^=$ (3) $C_3^=$ (3) $C_4^=$ (3)	723 K	1.013 bar	1.013 bar (3)	0.22 h (4) 0.12 h (4) 0.12 h (4)	0.78 0.65 0.82	8	14	16	114	No	No	C_{1-4}

(1) Contact time defined with mass flow rate of the reactant. (2) Respective n-alcohol was fed instead of the olefin; N = N$_2$, W = H$_2$O, 3 = C$_3^=$, 4 = C$_4^=$, 5 = C$_5^=$, 6 = C$_6^=$, 7 = C$_7^=$ (3) Three measurements diluted with N$_2$ (excluded for parameter estimation). (4) Inverse value of minimum WHSV containing only the mass flow rate of the reactant.

3.1. Studies Focusing on Olefin Interconversion over ZSM-5

3.1.1. Epelde et al.: Eight- and Five-Lump Approach for $C_4^=$ Feeds at Elevated Partial Pressures

Catalyst

The self-synthesized ZSM-5 zeolite has a comparatively high Si/Al ratio (280). According to the authors [235], this was done to attenuate hydrogen transfer so that side product formation is hindered and propene yields are increased. In addition to this, 1%$_{wt}$ K was added to the zeolite which lowers overall acidity and leads to a homogeneous distribution of acid strength. This should reduce side reactions and especially the evolution of coke precursors [238]. The measurements were performed at a time-on-stream (TOS) of 5 h; however, the authors extrapolated the results to 0 h TOS to characterize the reactivity of a fresh catalyst. In a preliminary study [239], the influencing factors of coke evolution were evaluated in detail.

Setup and Conditions

The experimental setup consisted of an automated reaction equipment where the feed components were provided as gases. The continuous fixed bed reactor was located within a furnace chamber whose temperature could be controlled via three test points, one of them being inside the catalyst bed and the other two in the chamber and in the transfer line to the GC, respectively. The stainless steel reactor had an inner diameter of 9 mm. Product analysis was performed using a micro gas chromatograph (GC) equipped with a thermal conductivity detector (TCD) and four columns. Both the feed and the catalyst bed were diluted using helium and SiC, respectively. More details about the setup can be found in the original publications [235,238,239]. In this study, 1-butene was the only reactive feed component analyzed; its partial pressures at the reactor inlet were relatively high.

Reaction Network

The proposed reaction network results from an analysis of kinetic experiments shown elsewhere [239]. The different species are grouped by means of reactivity which yields eight lumps: $C_2^=$, $C_3^=$, $C_4^=$, C_{5+}^{al}, C_1, C_{2-3}, C_4 and C_{6-8}^{ar}. The reaction rates are formulated based on experimental observations of primary and secondary products and evolution of the lump yields with changing conditions; the network with the best fit is chosen. Here, the formation of ethene (k_3 and $k_{8''}$) as well as of the side products (k_4–k_7 and k_{10}) is assumed to be irreversible whereby a minor part of $C_2^=$, C_{2-3}, C_4 and C_{6-8}^{ar} can still react to methane. The remaining steps comprise the interconversion of $C_3^=$ to C_{5+}^{al} hydrocarbons (k_1, k_2, k_8, $k_{8'}$ and k_9) where the only irreversible step is the production of propene out of C_{5+}^{al}. Besides methane formation, ethene does not act as reactant. As it can be seen in Scheme 2, the steps are considered as elementary reactions. Moreover, the stoichiometry is neglected both in the derivation of the rates and in the formulation of the net rates of production. Adsorption effects are not included.

The net rates of production can be obtained by adding all reaction rates where the respective lump is involved (see Scheme 3).

The authors observed only a minor side product formation [238], which is why they reduce the original eight-lump model. All paraffins are grouped together now (C_{1-4}), whereas the aromatics are summarized with the higher aliphatics to the new lump C_{5+}^{HC}. The resulting reaction network can be found in Scheme 4.

$$C_4^= \underset{k_9}{\overset{k_1}{\rightleftharpoons}} C_3^= \qquad r_1 = k_1\, p\,(C_4^=)$$

$k_1^{\mathrm{ref}} = (1112 \pm 9) \times 10^{-3}\,\mathrm{mol_C\,g_{cat}^{-1}\,h^{-1}\,atm^{-1}}$
$E_{a,1} = 3.3 \pm 0.5\,\mathrm{kJ\,mol^{-1}}$

$$C_4^= \underset{k_8'}{\overset{k_2}{\rightleftharpoons}} C_{5+}^{\mathrm{al}} \qquad r_2 = k_2\, p\,(C_4^=)$$

$k_2^{\mathrm{ref}} = (1391 \pm 6) \times 10^{-3}\,\mathrm{mol_C\,g_{cat}^{-1}\,h^{-1}\,atm^{-1}}$
$E_{a,2} = 3.7 \pm 0.5\,\mathrm{kJ\,mol^{-1}}$

$$C_4^= \overset{k_3}{\longrightarrow} C_2^= \qquad r_3 = k_3\, p\,(C_4^=)$$

$k_3^{\mathrm{ref}} = (3.9 \pm 1.0) \times 10^{-2}\,\mathrm{mol_C\,g_{cat}^{-1}\,h^{-1}\,atm^{-1}}$
$E_{a,3} = 31.0 \pm 2.1\,\mathrm{kJ\,mol^{-1}}$

$$C_4^= \overset{k_4}{\longrightarrow} C_{2-3} \qquad r_4 = k_4\, p\,(C_4^=)$$

$k_4^{\mathrm{ref}} = (1.8 \pm 1.2) \times 10^{-2}\,\mathrm{mol_C\,g_{cat}^{-1}\,h^{-1}\,atm^{-1}}$
$E_{a,4} = (3 \pm 1) \times 10^{-2}\,\mathrm{kJ\,mol^{-1}}$

$$C_4^= \overset{k_{4'}}{\longrightarrow} C_4 \qquad r_{4'} = k_{4'}\, p\,(C_4^=)$$

$k_{4'}^{\mathrm{ref}} = (88 \pm 8) \times 10^{-3}\,\mathrm{mol_C\,g_{cat}^{-1}\,h^{-1}\,atm^{-1}}$
$E_{a,4'} = (3 \pm 1) \times 10^{-2}\,\mathrm{kJ\,mol^{-1}}$

$$C_4^= \overset{k_5}{\longrightarrow} C_{6-8}^{\mathrm{ar}} \qquad r_5 = k_5\, p\,(C_4^=)$$

$k_5^{\mathrm{ref}} = (82.5 \pm 9.5) \times 10^{-4}\,\mathrm{mol_C\,g_{cat}^{-1}\,h^{-1}\,atm^{-1}}$
$E_{a,5} = (3 \pm 1) \times 10^{-2}\,\mathrm{kJ\,mol^{-1}}$

$$C_3^= \overset{k_6}{\longrightarrow} C_{2-3} \qquad r_6 = k_6\, p\,(C_3^=)$$

$k_6^{\mathrm{ref}} = (73.2 \pm 8.5) \times 10^{-4}\,\mathrm{mol_C\,g_{cat}^{-1}\,h^{-1}\,atm^{-1}}$
$E_{a,6} = 16.6 \pm 1.2\,\mathrm{kJ\,mol^{-1}}$

$$C_3^= \overset{k_7}{\longrightarrow} C_{6-8}^{\mathrm{ar}} \qquad r_7 = k_7\, p\,(C_3^=)$$

$k_7^{\mathrm{ref}} = (2.4 \pm 1.1) \times 10^{-2}\,\mathrm{mol_C\,g_{cat}^{-1}\,h^{-1}\,atm^{-1}}$
$E_{a,7} = 6.1 \pm 0.7\,\mathrm{kJ\,mol^{-1}}$

$$C_{5+}^{\mathrm{al}} \overset{k_8}{\longrightarrow} C_3^= \qquad r_8 = k_8\, p\,\left(C_{5+}^{\mathrm{al}}\right)$$

$k_8^{\mathrm{ref}} = (2.67 \pm 1.10) \times 10^{-1}\,\mathrm{mol_C\,g_{cat}^{-1}\,h^{-1}\,atm^{-1}}$
$E_{a,8} = 55.9 \pm 1.0\,\mathrm{kJ\,mol^{-1}}$

$$C_{5+}^{\mathrm{al}} \underset{k_2}{\overset{k_{8'}}{\rightleftharpoons}} C_4^= \qquad r_{8'} = k_{8'}\, p\,\left(C_{5+}^{\mathrm{al}}\right)$$

$k_{8'}^{\mathrm{ref}} = (269.9 \pm 5.6) \times 10^{-2}\,\mathrm{mol_C\,g_{cat}^{-1}\,h^{-1}\,atm^{-1}}$
$E_{a,8'} = 55.9 \pm 1.0\,\mathrm{kJ\,mol^{-1}}$

$$C_{5+}^{\mathrm{al}} \overset{k_{8''}}{\longrightarrow} C_2^= \qquad r_{8''} = k_{8''}\, p\,\left(C_{5+}^{\mathrm{al}}\right)$$

$k_{8''}^{\mathrm{ref}} = (2.4 \pm 1.7) \times 10^{-2}\,\mathrm{mol_C\,g_{cat}^{-1}\,h^{-1}\,atm^{-1}}$
$E_{a,8''} = 55.9 \pm 1.0\,\mathrm{kJ\,mol^{-1}}$

$$C_3^= \underset{k_1}{\overset{k_9}{\rightleftharpoons}} C_4^= \qquad r_9 = k_9\, p\,(C_3^=)$$

$k_9^{\mathrm{ref}} = (104.1 \pm 4.2) \times 10^{-2}\,\mathrm{mol_C\,g_{cat}^{-1}\,h^{-1}\,atm^{-1}}$
$E_{a,9} = 0.5 \pm 1.0\,\mathrm{kJ\,mol^{-1}}$

$$C_i^{\mathrm{HC}} \overset{k_{10}}{\longrightarrow} C_1 \qquad r_{10} = k_{10}\, p\,(C_i^{\mathrm{HC}})$$

$k_{10}^{\mathrm{ref}} = (71.1 \pm 3.0) \times 10^{-5}\,\mathrm{mol_C\,g_{cat}^{-1}\,h^{-1}\,atm^{-1}}$
$E_{a,10} = 55.7 \pm 1.0\,\mathrm{kJ\,mol^{-1}}$

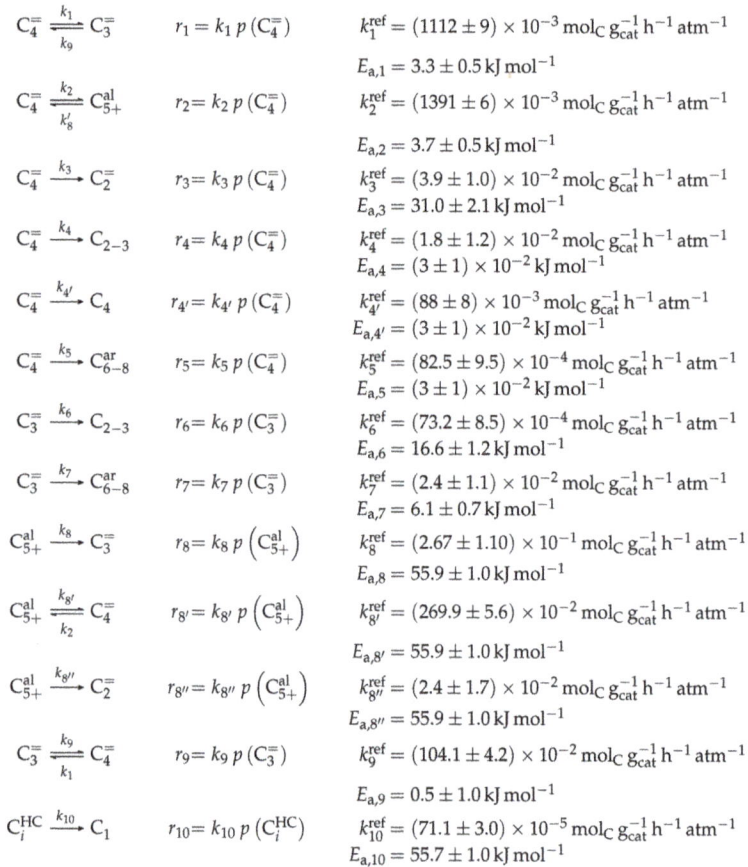

Scheme 2. Reaction network, rate equations and estimated parameters for the model by Epelde et al. [235] (eight lumps) with *i* ranging from 2 to 4 (olefins) or being 2–3, 4 (paraffins), 6–8 (aromatics) or 5+ (aliphatics).

$$R\,(C_2^=) = k_3\, p\,(C_4^=) + k_{8''}\, p\,\left(C_{5+}^{\mathrm{al}}\right) - k_{10}\, p\,(C_2^=)$$

$$R\,(C_3^=) = k_1\, p\,(C_4^=) + k_8\, p\,\left(C_{5+}^{\mathrm{al}}\right) - k_6\, p\,(C_3^=) - k_7\, p\,(C_3^=) - k_9\, p\,(C_3^=) - k_{10}\, p\,(C_3^=)$$

$$R\,(C_4^=) = k_{8'}\, p\,\left(C_{5+}^{\mathrm{al}}\right) + k_9\, p\,(C_3^=) - k_1\, p\,(C_4^=) - k_2\, p\,(C_4^=) - k_3\, p\,(C_4^=) - k_4\, p\,(C_4^=)$$
$$\qquad - k_{4'}\, p\,(C_4^=) - k_5\, p\,(C_4^=) - k_{10}\, p\,(C_4^=)$$

$$R\,\left(C_{5+}^{\mathrm{al}}\right) = k_2\, p\,(C_4^=) - k_8\, p\,\left(C_{5+}^{\mathrm{al}}\right) - k_{8'}\, p\,\left(C_{5+}^{\mathrm{al}}\right) - k_{8''}\, p\,\left(C_{5+}^{\mathrm{al}}\right) - k_{10}\, p\,\left(C_{5+}^{\mathrm{al}}\right)$$

$$R\,(C_1) = k_{10}\, p\,(C_2^=) + k_{10}\, p\,(C_3^=) + k_{10}\, p\,(C_4^=) + k_{10}\, p\,\left(C_{5+}^{\mathrm{al}}\right) + k_{10}\, p\,(C_{2-3}) + k_{10}\, p\,(C_4)$$
$$\qquad + k_{10}\, p\,(C_{5+}^{\mathrm{ar}})$$

$$R\,(C_{2-3}) = k_4\, p\,(C_4^=) + k_6\, p\,(C_3^=) - k_{10}\, p\,(C_{2-3})$$
$$R\,(C_4) = k_{4'}\, p\,(C_4^=) - k_{10}\, p\,(C_4)$$
$$R\,(C_{5+}^{\mathrm{ar}}) = k_5\, p\,(C_4^=) + k_7\, p\,(C_3^=) - k_{10}\, p\,(C_{5+}^{\mathrm{ar}})$$

Scheme 3. Net rates of production of the different lumps for the model by Epelde et al. [235] (eight lumps).

$$C_4^= \underset{k_9}{\overset{k_1}{\rightleftharpoons}} C_3^= \qquad r_1 = k_1\, p\,(C_4^=) \qquad k_1^{\text{ref}} = (1073 \pm 10) \times 10^{-3}\,\text{mol}_C\,g_{\text{cat}}^{-1}\,h^{-1}\,\text{atm}^{-1}$$

$$E_{a,1} = 3.1 \pm 0.6\,\text{kJ}\,\text{mol}^{-1}$$

$$C_4^= \underset{k_8'}{\overset{k_2}{\rightleftharpoons}} C_{5+}^{\text{HC}} \qquad r_2 = k_2\, p\,(C_4^=) \qquad k_2^{\text{ref}} = (163.1 \pm 1.5) \times 10^{-2}\,\text{mol}_C\,g_{\text{cat}}^{-1}\,h^{-1}\,\text{atm}^{-1}$$

$$E_{a,2} = 3.3 \pm 0.6\,\text{kJ}\,\text{mol}^{-1}$$

$$C_4^= \xrightarrow{k_3} C_2^= \qquad r_3 = k_3\, p\,(C_4^=) \qquad k_3^{\text{ref}} = (2.2 \pm 1.0) \times 10^{-2}\,\text{mol}_C\,g_{\text{cat}}^{-1}\,h^{-1}\,\text{atm}^{-1}$$
$$E_{a,3} = 35.6 \pm 2.7\,\text{kJ}\,\text{mol}^{-1}$$

$$C_4^= \xrightarrow{k_4} C_{1-4} \qquad r_4 = k_4\, p\,(C_4^=) \qquad k_4^{\text{ref}} = (95 \pm 8) \times 10^{-3}\,\text{mol}_C\,g_{\text{cat}}^{-1}\,h^{-1}\,\text{atm}^{-1}$$
$$E_{a,4} = (1 \pm 1) \times 10^{-1}\,\text{kJ}\,\text{mol}^{-1}$$

$$C_3^= \xrightarrow{k_6} C_{1-4} \qquad r_6 = k_6\, p\,(C_3^=) \qquad k_6^{\text{ref}} = (2.6 \pm 1.3) \times 10^{-2}\,\text{mol}_C\,g_{\text{cat}}^{-1}\,h^{-1}\,\text{atm}^{-1}$$
$$E_{a,6} = 1.3 \pm 7.5\,\text{kJ}\,\text{mol}^{-1}$$

$$C_{5+}^{\text{HC}} \xrightarrow{k_8} C_3^= \qquad r_8 = k_8\, p\,\left(C_{5+}^{\text{HC}}\right) \qquad k_8^{\text{ref}} = (24.5 \pm 9.8) \times 10^{-2}\,\text{mol}_C\,g_{\text{cat}}^{-1}\,h^{-1}\,\text{atm}^{-1}$$
$$E_{a,8} = 55.5 \pm 1.0\,\text{kJ}\,\text{mol}^{-1}$$

$$C_{5+}^{\text{HC}} \underset{k_2}{\overset{k_{8'}}{\rightleftharpoons}} C_4^= \qquad r_{8'} = k_{8'}\, p\,\left(C_{5+}^{\text{HC}}\right) \qquad k_{8'}^{\text{ref}} = 2.759 \pm 2.796\,\text{mol}_C\,g_{\text{cat}}^{-1}\,h^{-1}\,\text{atm}^{-1}$$
$$E_{a,8'} = 55.5 \pm 1.0\,\text{kJ}\,\text{mol}^{-1}$$

$$C_{5+}^{\text{HC}} \xrightarrow{k_{8''}} C_2^= \qquad r_{8''} = k_{8''}\, p\,\left(C_{5+}^{\text{HC}}\right) \qquad k_{8''}^{\text{ref}} = 0.75 \pm 1.29\,\text{mol}_C\,g_{\text{cat}}^{-1}\,h^{-1}\,\text{atm}^{-1}$$
$$E_{a,8''} = 55.5 \pm 1.0\,\text{kJ}\,\text{mol}^{-1}$$

$$C_3^= \underset{k_1}{\overset{k_9}{\rightleftharpoons}} C_4^= \qquad r_9 = k_9\, p\,(C_3^=) \qquad k_9^{\text{ref}} = (102.0 \pm 5.3) \times 10^{-2}\,\text{mol}_C\,g_{\text{cat}}^{-1}\,h^{-1}\,\text{atm}^{-1}$$
$$E_{a,9} = 2.9 \pm 1.2\,\text{kJ}\,\text{mol}^{-1}$$

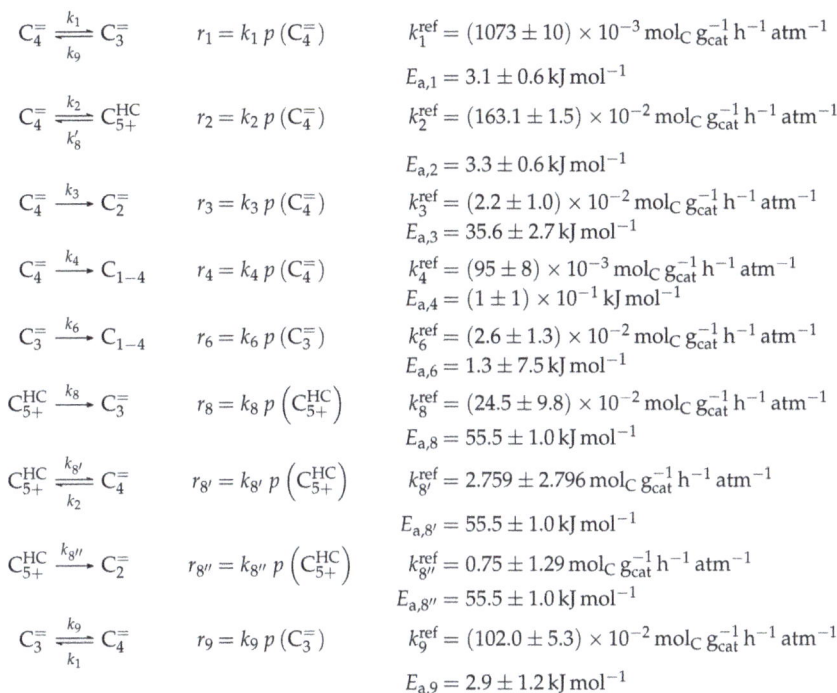

Scheme 4. Reaction network, rate equations and estimated parameters for the model by Epelde et al. [235] (five lumps).

From this, the net rates of production are defined according to Scheme 5.

$$R\,(C_2^=) = k_3\, p\,(C_4^=) + k_{8''}\, p\,\left(C_{5+}^{\text{HC}}\right)$$
$$R\,(C_3^=) = k_1\, p\,(C_4^=) + k_8\, p\,\left(C_{5+}^{\text{HC}}\right) - k_6\, p\,(C_3^=) - k_9\, p\,(C_3^=)$$
$$R\,(C_4^=) = k_{8'}\, p\,\left(C_{5+}^{\text{HC}}\right) + k_9\, p\,(C_3^=) - k_1\, p\,(C_4^=) - k_2\, p\,(C_4^=) - k_3\, p\,(C_4^=) - k_4\, p\,(C_4^=)$$
$$R\,\left(C_{5+}^{\text{HC}}\right) = k_2\, p\,(C_4^=) - k_8\, p\,\left(C_{5+}^{\text{HC}}\right) - k_{8'}\, p\,\left(C_{5+}^{\text{HC}}\right) - k_{8''}\, p\,\left(C_{5+}^{\text{HC}}\right)$$
$$R\,(C_{1-4}) = k_4\, p\,(C_4^=) + k_6\, p\,(C_3^=)$$

Scheme 5. Net rates of production of the different lumps for the model by Epelde et al. [235] (five lumps).

Parameter Estimation

The mole fractions and molar flow rates in this study are expressed in carbon units, whereas, for the reaction rates in Scheme 2, partial pressures are used. Parameter estimation is performed with a multivariable nonlinear regression in MATLAB. The molar flow rates along the reactor are obtained with a fourth-order finite differences approximation, whereas the actual regression is two-part: a self-written routine using the Levenberg–Marquardt algorithm delivers initial values for the final step, the minimization of the objective function via *fminsearch*. The objective function returns the weighted sum of squared residuals between the experimental and theoretical mole fractions. For replicate measurements, an average value is used for the experimental value. The calculation of the weighting factor is different to Equation (30): due to the lacking division by the sum of the weighting factors for all fitting responses, the individual values might exceed one for Epelde et al. [238]. With this methodology, 13 reference rate constants and ten activation energies are estimated. This means the reparameterized Arrhenius approach (see Equation (21)) is used with the reference temperature being the average value of the investigated range (773 K). Steps 4 and 4′ are assumed to have similar activation energies, as well as Steps 8, 8′ and 8″, to reduce the number of estimated parameters. For the five lump version, 16 unknown values exist: nine reference rate constants and seven activation energies. The same simplification for the activation energy of Steps 8, 8′ and 8″ is introduced.

3.1.2. Ying et al.: Eight-Lump Model for Arbitrary Olefin Feeds Including Side Product Formation

Catalyst

As shown in Table 1, not many details about the catalyst are accessible because Ying et al. [236] used a commercial ZSM-5 extrudate sample from Süd-Chemie. The only noteworthy fact is the relatively large particle size (420–841 µm). The measurements were performed with a fresh catalyst.

Setup and Conditions

In the kinetic measurements, different olefins from propene to heptene were analyzed as feed. Ethene was also fed at the beginning of the study. It showed almost no reactivity and was therefore ignored. Whereas propene and butenes could be fed directly as gases, the higher olefins were provided as liquids and had to be evaporated. The temperature was measured within the catalyst bed diluted with silica. For feed dilution, nitrogen was chosen. The continuous fixed bed reactor had an inner diameter of 10 mm, but high volumetric flow rates were applied to prevent film diffusion. For each feed, different maximum contact times and conversions had to be analyzed. However, the latter value was comparable for propene, butenes and pentenes. Both hexene and heptene are very reactive and, therefore, conversion was almost one despite having short contact times. Samples were evaluated with a GC equipped with one column and a flame ionization detector (FID).

Reaction Network

The authors conducted a profound analysis of the selectivity results of each olefinic feed. This insight is used to create the reaction network which consists of seven lumps: $C_2^=$, $C_3^=$, $C_4^=$, $C_5^=$, $C_6^=$, $C_7^=$ and C_i^{SP}. The whole network describes olefin interconversion (k_1–k_8) except for one side product formation step (k_9). It is mentioned that the theoretical $C_7^=$ lump is compared with an experimental result of $C_{7-8}^=$ olefins. The side product lump contains all paraffins and aromatics with arbitrary carbon numbers. As mentioned above, ethene showed negligible reactivity, so the authors assume its formation reactions to be irreversible. The same is applied to the step leading to C_i^{SP} and to the formation of $C_4^=$ and $C_6^=$ out of two pentenes (see Scheme 6). The latter assumption is justified with the missing improvement when the backward reaction is implemented. Stoichiometry is considered and various olefin interconversion reactions are included: there is a clear separation between monomolecular cracking ($C_5^=$, $C_6^=$ and $C_7^=$) and dimerization-cracking reactions ($C_3^=$–$C_7^=$, but especially important for lower olefins). For the dimerization, the highest intermediate included is $C_{10}^=$. The steps are treated as elementary reactions without any adsorption effects. Scheme 6 shows an overview of all reactions covered by Ying et al. [236].

$$2\,C_3^= \underset{k_{-1}}{\overset{k_1}{\rightleftharpoons}} C_6^=$$

$$k_1^{\text{ref}} = 14.42 \pm 0.98 \, \text{m}^6 \, \text{mol}^{-1} \, \text{kg}_{\text{cat}}^{-1} \, \text{h}^{-1}$$
$$k_{-1}^{\text{ref}} = 89.10 \pm 3.49 \, \text{m}^3 \, \text{kg}_{\text{cat}}^{-1} \, \text{h}^{-1}$$

$$r_1 = k_1 \, C \, (C_3^=)^2 - k_{-1} \, C \, (C_6^=)$$
$$E_{a,1} = -15.65 \pm 5.08 \, \text{kJ} \, \text{mol}^{-1}$$
$$E_{a,-1} = 38.68 \pm 2.68 \, \text{kJ} \, \text{mol}^{-1}$$

$$C_3^= + C_6^= \underset{k_{-2}}{\overset{k_2}{\rightleftharpoons}} C_4^= + C_5^=$$

$$k_2^{\text{ref}} = 188.73 \pm 13.90 \, \text{m}^6 \, \text{mol}^{-1} \, \text{kg}_{\text{cat}}^{-1} \, \text{h}^{-1}$$
$$k_{-2}^{\text{ref}} = 73.49 \pm 8.94 \, \text{m}^6 \, \text{mol}^{-1} \, \text{kg}_{\text{cat}}^{-1} \, \text{h}^{-1}$$

$$r_2 = k_2 \, C \, (C_3^=) \, C \, (C_6^=) - k_{-2} \, C \, (C_4^=) \, C \, (C_5^=)$$
$$E_{a,2} = -53.04 \pm 5.36 \, \text{kJ} \, \text{mol}^{-1}$$
$$E_{a,-2} = -26.62 \pm 9.40 \, \text{kJ} \, \text{mol}^{-1}$$

$$C_3^= + C_5^= \underset{k_{-3}}{\overset{k_3}{\rightleftharpoons}} 2\,C_4^=$$

$$k_3^{\text{ref}} = 82.50 \pm 8.11 \, \text{m}^6 \, \text{mol}^{-1} \, \text{kg}_{\text{cat}}^{-1} \, \text{h}^{-1}$$
$$k_{-3}^{\text{ref}} = 29.70 \pm 1.77 \, \text{m}^6 \, \text{mol}^{-1} \, \text{kg}_{\text{cat}}^{-1} \, \text{h}^{-1}$$

$$r_3 = k_3 \, C \, (C_3^=) \, C \, (C_5^=) - k_{-3} \, C \, (C_4^=)^2$$
$$E_{a,3} = -28.12 \pm 7.88 \, \text{kJ} \, \text{mol}^{-1}$$
$$E_{a,-3} = -13.59 \pm 4.01 \, \text{kJ} \, \text{mol}^{-1}$$

$$2\,C_5^= \overset{k_4}{\longrightarrow} C_4^= + C_6^=$$

$$k_4^{\text{ref}} = 7.97 \pm 2.34 \, \text{m}^6 \, \text{mol}^{-1} \, \text{kg}_{\text{cat}}^{-1} \, \text{h}^{-1}$$

$$r_4 = k_4 \, C \, (C_5^=)^2$$
$$E_{a,4} = -16.26 \pm 25.49 \, \text{kJ} \, \text{mol}^{-1}$$

$$C_5^= \overset{k_5}{\longrightarrow} C_2^= + C_3^=$$

$$k_5^{\text{ref}} = 2.52 \pm 1.27 \, \text{m}^3 \, \text{kg}_{\text{cat}}^{-1} \, \text{h}^{-1}$$

$$r_5 = k_5 \, C \, (C_5^=)$$
$$E_{a,5} = 96.73 \pm 26.19 \, \text{kJ} \, \text{mol}^{-1}$$

$$C_6^= \overset{k_6}{\longrightarrow} C_2^= + C_4^=$$

$$k_6^{\text{ref}} = 11.32 \pm 2.35 \, \text{m}^3 \, \text{kg}_{\text{cat}}^{-1} \, \text{h}^{-1}$$

$$r_6 = k_6 \, C \, (C_6^=)$$
$$E_{a,6} = 63.45 \pm 12.05 \, \text{kJ} \, \text{mol}^{-1}$$

$$C_7^= \underset{k_{-7}}{\overset{k_7}{\rightleftharpoons}} C_3^= + C_4^=$$

$$k_7^{\text{ref}} = 474.49 \pm 23.25 \, \text{m}^6 \, \text{mol}^{-1} \, \text{kg}_{\text{cat}}^{-1} \, \text{h}^{-1}$$
$$k_{-7}^{\text{ref}} = 61.46 \pm 9.79 \, \text{m}^3 \, \text{kg}_{\text{cat}}^{-1} \, \text{h}^{-1}$$

$$r_7 = k_7 \, C \, (C_7^=) - k_{-7} \, C \, (C_3^=) \, C \, (C_4^=)$$
$$E_{a,7} = 23.85 \pm 3.44 \, \text{kJ} \, \text{mol}^{-1}$$
$$E_{a,-7} = -43.76 \pm 15.60 \, \text{kJ} \, \text{mol}^{-1}$$

$$C_3^= + C_7^= \underset{k_{-8}}{\overset{k_8}{\rightleftharpoons}} 2\,C_5^=$$

$$k_8^{\text{ref}} = 83.26 \pm 11.76 \, \text{m}^6 \, \text{mol}^{-1} \, \text{kg}_{\text{cat}}^{-1} \, \text{h}^{-1}$$
$$k_{-8}^{\text{ref}} = 10.38 \pm 1.74 \, \text{m}^6 \, \text{mol}^{-1} \, \text{kg}_{\text{cat}}^{-1} \, \text{h}^{-1}$$

$$r_8 = k_8 \, C \, (C_3^=) \, C \, (C_7^=) - k_{-8} \, C \, (C_5^=)^2$$
$$E_{a,8} = -61.87 \pm 16.36 \, \text{kJ} \, \text{mol}^{-1}$$
$$E_{a,-8} = -20.34 \pm 15.55 \, \text{kJ} \, \text{mol}^{-1}$$

$$C_i^= \overset{k_9}{\longrightarrow} C_i^{\text{SP}}$$

$$k_9^{\text{ref}} = 2.14 \pm 0.47 \, \text{m}^3 \, \text{kg}_{\text{cat}}^{-1} \, \text{h}^{-1}$$

$$r_9 = k_9 \, C \, (C_i^=)$$
$$E_{a,9} = -25.87 \pm 19.57 \, \text{kJ} \, \text{mol}^{-1}$$

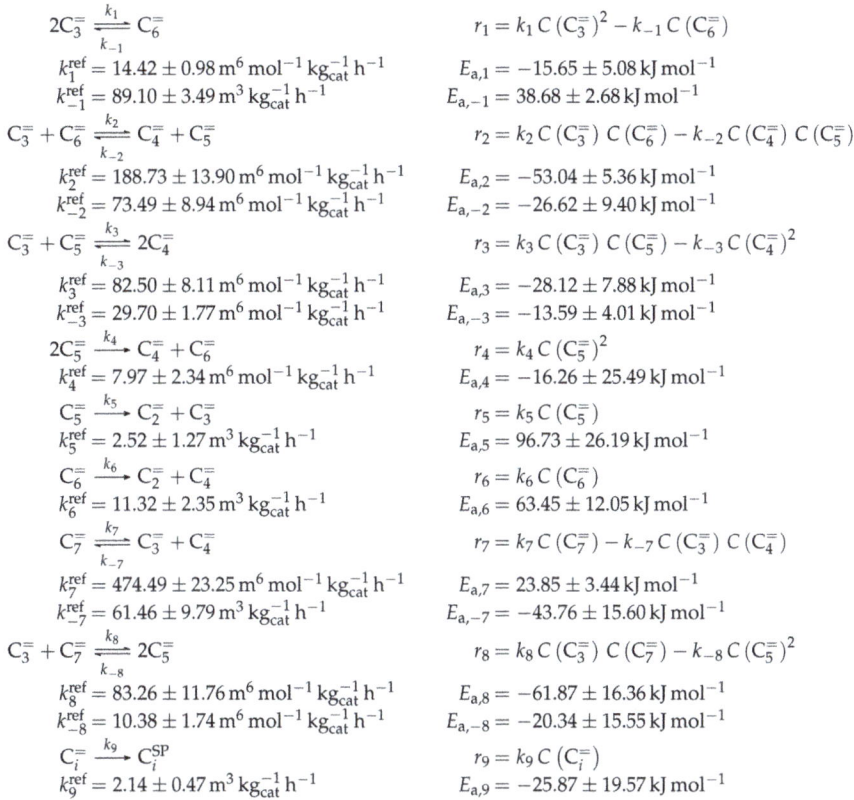

Scheme 6. Reaction network, rate equations and estimated parameters for the model by Ying et al. [236] with i ranging from 3 to 7.

This network leads to the net rates of production listed in Scheme 7.

$$R \, (C_2^=) = k_5 \, C \, (C_5^=) + k_6 \, C \, (C_6^=)$$

$$R \, (C_3^=) = 2\,k_{-1} \, C \, (C_6^=) + k_{-2} \, C \, (C_4^=) \, C \, (C_5^=) + k_{-3} \, C \, (C_4^=)^2 + k_5 \, C \, (C_5^=) + k_7 \, C \, (C_7^=)$$
$$+ k_{-8} \, C \, (C_5^=)^2 - 2\,k_1 \, C \, (C_3^=)^2 - k_2 \, C \, (C_3^=) \, C \, (C_6^=) - k_3 \, C \, (C_3^=) \, C \, (C_5^=)$$
$$- k_{-7} \, C \, (C_3^=) \, C \, (C_4^=) - k_8 \, C \, (C_3^=) \, C \, (C_7^=) - k_9 \, C \, (C_3^=)$$

$$R \, (C_4^=) = k_2 \, C \, (C_3^=) \, C \, (C_6^=) + 2\,k_3 \, C \, (C_3^=) \, C \, (C_5^=) + k_4 \, C \, (C_5^=)^2 + k_6 \, C \, (C_6^=) + k_7 \, C \, (C_7^=)$$
$$- k_{-2} \, C \, (C_4^=) \, C \, (C_5^=) - 2\,k_{-3} \, C \, (C_4^=)^2 - k_{-7} \, C \, (C_3^=) \, C \, (C_4^=) - k_9 \, C \, (C_4^=)$$

$$R \, (C_5^=) = k_2 \, C \, (C_3^=) \, C \, (C_6^=) + k_{-3} \, C \, (C_4^=)^2 + 2\,k_8 \, C \, (C_3^=) \, C \, (C_7^=) - k_{-2} \, C \, (C_4^=) \, C \, (C_5^=)$$
$$- k_3 \, C \, (C_3^=) \, C \, (C_5^=) - 2\,k_4 \, C \, (C_5^=)^2 - k_5 \, C \, (C_5^=) - 2\,k_{-8} \, C \, (C_5^=)^2 - k_9 \, C \, (C_5^=)$$

$$R \, (C_6^=) = k_1 \, C \, (C_3^=)^2 + k_{-2} \, C \, (C_4^=) \, C \, (C_5^=) + k_4 \, C \, (C_5^=)^2 - k_{-1} \, C \, (C_6^=) - k_2 \, C \, (C_3^=) \, C \, (C_6^=)$$
$$- k_6 \, C \, (C_6^=) - k_9 \, C \, (C_6^=)$$

$$R \, (C_7^=) = k_{-7} \, C \, (C_3^=) \, C \, (C_4^=) + k_{-8} \, C \, (C_5^=)^2 - k_7 \, C \, (C_7^=) - k_8 \, C \, (C_3^=) \, C \, (C_7^=) - k_9 \, C \, (C_7^=)$$

$$R \, \left(C_{3-7}^{\text{SP}} \right) = k_9 \, C \, (C_3^=) + k_9 \, C \, (C_4^=) + k_9 \, C \, (C_5^=) + k_9 \, C \, (C_6^=) + k_9 \, C \, (C_7^=)$$

Scheme 7. Net rates of production of the different lumps for the model by Ying et al. [236].

Parameter Estimation

Both the contact time and the reactor model are calculated with mass flow rates, which means that the net rate of production of each lump (Scheme 7) has to be multiplied by its molar mass. The reaction rates (Scheme 6) are expressed with molar concentrations per volume. For parameter fitting, the Levenberg–Marquardt algorithm is used to minimize the objective function. The latter is defined as the unweighted sum of squared residuals between the theoretical and experimental mass fractions. The reparameterized Arrhenius approach according to Equation (21) is used with a reference temperature of 673 K, which is the lowest examined value. As unknown parameters, 14 reference rate constants and 14 activation energies follow from this model.

3.1.3. Huang et al.: Six-Lump Approach for Arbitrary Olefin Feeds Including LH and HW Types of Mechanism

Catalyst

The authors [166] chose a commercial ZSM-5 catalyst by Shanghai Fuyu Company due to its coking resistance and high propene to ethene ratio. As shown in Table 1, the increased Si/Al ratio (200) caused a low number of acid sites ($0.012 \, \mathrm{mmol \, g_{cat}^{-1}}$). A preliminary test revealed that catalyst deactivation was negligible, which is why a broad spectrum of TOS was chosen with a regeneration after each 10 h. With 17 h TOS, the coke selectivity was still below 0.01%.

Setup and Conditions

Huang et al. [166] used a continuous U-shaped fixed bed reactor made of titanium with an inner diameter of 6 mm. Different olefins from propene to heptene were applied as feed, but, in contrast to the study of Ying et al. [236], the corresponding linear 1-alcohols were fed as liquids and evaporated in a pre-heater. The authors stated that the dehydration to the corresponding 1-olefin occurred very quickly when the feed mixture reached the catalyst bed [166]. However, this inevitably caused water release, which can be seen as further diluent, but also interacted with the acid sites of the catalyst. Further feed dilution could be achieved by using nitrogen, whereas the catalyst was diluted 1:5 with an inert not further specified. The reactor was surrounded by a molten salt bath which allowed controlling the temperature, although no thermocouple was available within the catalyst bed. A GC equipped with an FID and one column was used for product analysis. Each data point resulted from a twofold GC sampling. The authors performed two additional experimental series at 713 K and 753 K with a mixture of different olefins as feed. These were not included into parameter fitting, but used to prove the validity of the model not only for single olefins as feed, but also for mixtures. Therefore, the detailed molar composition without inerts was 0.07, 0.235, 0.22, 0.235, 0.12 and 0.12 for $C_2^=$, $C_3^=$, $C_4^=$, $C_5^=$, $C_6^=$ and $C_7^=$, respectively.

Reaction Network

Similar to Ying et al. [236], a detailed study of each olefinic feed was performed. This could be used to derive the reaction network which consists of the following six lumps: $C_2^=$, $C_3^=$, $C_4^=$, $C_5^=$, $C_6^=$ and $C_{7+}^=$; the latter also contains species higher than heptenes. All steps in the network are related to olefin interconversion. Huang et al. [166] allow the highest intermediate to have a carbon number of twelve, so hexene dimerization can occur. Furthermore, they include not only monomolecular cracking and dimerization, but also four trimolecular alkylation reactions, for example, the trimerization of propene to butene and pentene. The network shows no irreversible steps: no evolution of side products is included and, although ethene dimerization is neglected, the ethene formation out of higher olefins is assumed to be reversible. The resulting network contains a huge variety of olefin interconversion reactions and can be found in Scheme 8. For the derivation of the reaction rates, Huang et al. [166] follow a combination of LH and HW types of mechanism. This means the backward reactions are determined with equilibrium constants and the denominator contains the inhibition through competing adsorption. For the latter, all olefins and water are considered. The different reactions are assumed to be elementary and stoichiometry is retained.

For this model, the expressions for the reaction rates are quite complex, which is why Scheme 9 only shows r_1.

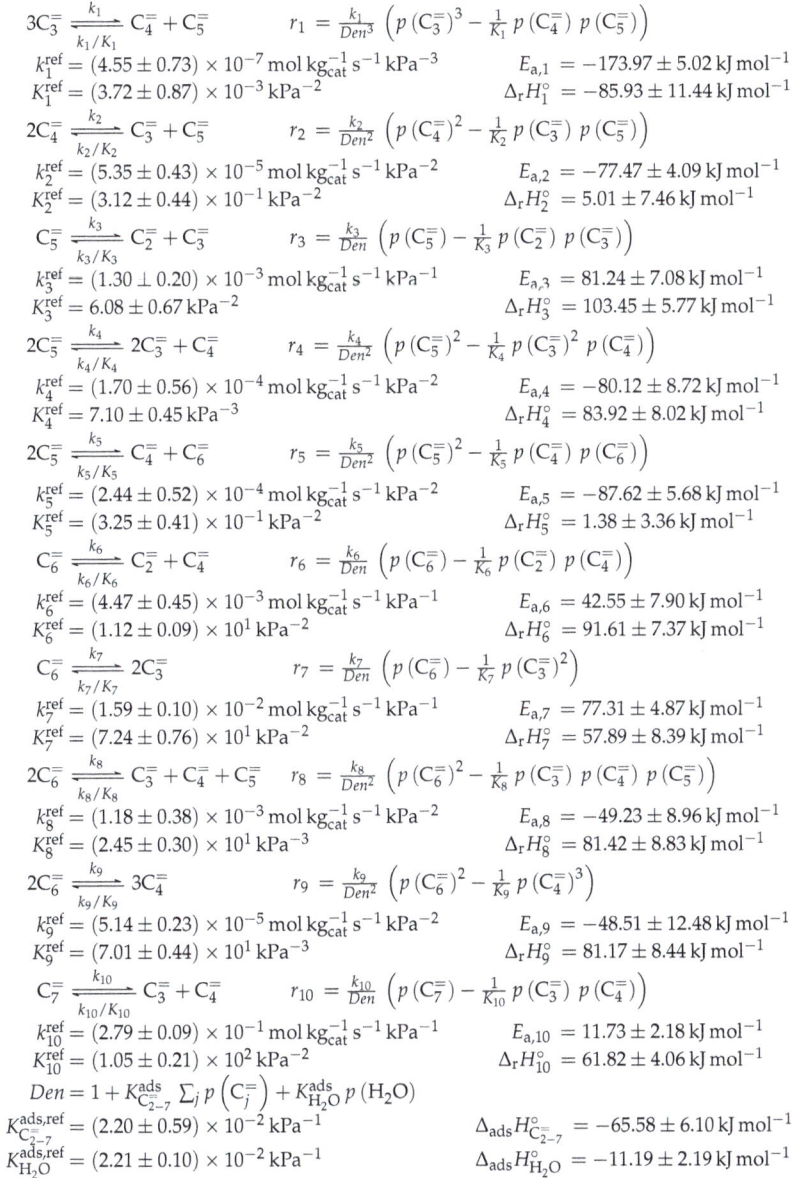

$$3C_3^= \underset{k_1/K_1}{\overset{k_1}{\rightleftharpoons}} C_4^= + C_5^= \qquad r_1 = \frac{k_1}{Den^3}\left(p\,(C_3^=)^3 - \frac{1}{K_1}p\,(C_4^=)\,p\,(C_5^=)\right)$$

$k_1^{ref} = (4.55 \pm 0.73) \times 10^{-7}\,\mathrm{mol\,kg_{cat}^{-1}\,s^{-1}\,kPa^{-3}}$ $\qquad E_{a,1} = -173.97 \pm 5.02\,\mathrm{kJ\,mol^{-1}}$
$K_1^{ref} = (3.72 \pm 0.87) \times 10^{-3}\,\mathrm{kPa^{-2}}$ $\qquad \Delta_r H_1^\circ = -85.93 \pm 11.44\,\mathrm{kJ\,mol^{-1}}$

$$2C_4^= \underset{k_2/K_2}{\overset{k_2}{\rightleftharpoons}} C_3^= + C_5^= \qquad r_2 = \frac{k_2}{Den^2}\left(p\,(C_4^=)^2 - \frac{1}{K_2}p\,(C_3^=)\,p\,(C_5^=)\right)$$

$k_2^{ref} = (5.35 \pm 0.43) \times 10^{-5}\,\mathrm{mol\,kg_{cat}^{-1}\,s^{-1}\,kPa^{-2}}$ $\qquad E_{a,2} = -77.47 \pm 4.09\,\mathrm{kJ\,mol^{-1}}$
$K_2^{ref} = (3.12 \pm 0.44) \times 10^{-1}\,\mathrm{kPa^{-2}}$ $\qquad \Delta_r H_2^\circ = 5.01 \pm 7.46\,\mathrm{kJ\,mol^{-1}}$

$$C_5^= \underset{k_3/K_3}{\overset{k_3}{\rightleftharpoons}} C_2^= + C_3^= \qquad r_3 = \frac{k_3}{Den}\left(p\,(C_5^=) - \frac{1}{K_3}p\,(C_2^=)\,p\,(C_3^=)\right)$$

$k_3^{ref} = (1.30 \pm 0.20) \times 10^{-3}\,\mathrm{mol\,kg_{cat}^{-1}\,s^{-1}\,kPa^{-1}}$ $\qquad E_{a,3} = 81.24 \pm 7.08\,\mathrm{kJ\,mol^{-1}}$
$K_3^{ref} = 6.08 \pm 0.67\,\mathrm{kPa^{-2}}$ $\qquad \Delta_r H_3^\circ = 103.45 \pm 5.77\,\mathrm{kJ\,mol^{-1}}$

$$2C_5^= \underset{k_4/K_4}{\overset{k_4}{\rightleftharpoons}} 2C_3^= + C_4^= \qquad r_4 = \frac{k_4}{Den^2}\left(p\,(C_5^=)^2 - \frac{1}{K_4}p\,(C_3^=)^2\,p\,(C_4^=)\right)$$

$k_4^{ref} = (1.70 \pm 0.56) \times 10^{-4}\,\mathrm{mol\,kg_{cat}^{-1}\,s^{-1}\,kPa^{-2}}$ $\qquad E_{a,4} = -80.12 \pm 8.72\,\mathrm{kJ\,mol^{-1}}$
$K_4^{ref} = 7.10 \pm 0.45\,\mathrm{kPa^{-3}}$ $\qquad \Delta_r H_4^\circ = 83.92 \pm 8.02\,\mathrm{kJ\,mol^{-1}}$

$$2C_5^= \underset{k_5/K_5}{\overset{k_5}{\rightleftharpoons}} C_4^= + C_6^= \qquad r_5 = \frac{k_5}{Den^2}\left(p\,(C_5^=)^2 - \frac{1}{K_5}p\,(C_4^=)\,p\,(C_6^=)\right)$$

$k_5^{ref} = (2.44 \pm 0.52) \times 10^{-4}\,\mathrm{mol\,kg_{cat}^{-1}\,s^{-1}\,kPa^{-2}}$ $\qquad E_{a,5} = -87.62 \pm 5.68\,\mathrm{kJ\,mol^{-1}}$
$K_5^{ref} = (3.25 \pm 0.41) \times 10^{-1}\,\mathrm{kPa^{-2}}$ $\qquad \Delta_r H_5^\circ = 1.38 \pm 3.36\,\mathrm{kJ\,mol^{-1}}$

$$C_6^= \underset{k_6/K_6}{\overset{k_6}{\rightleftharpoons}} C_2^= + C_4^= \qquad r_6 = \frac{k_6}{Den}\left(p\,(C_6^=) - \frac{1}{K_6}p\,(C_2^=)\,p\,(C_4^=)\right)$$

$k_6^{ref} = (4.47 \pm 0.45) \times 10^{-3}\,\mathrm{mol\,kg_{cat}^{-1}\,s^{-1}\,kPa^{-1}}$ $\qquad E_{a,6} = 42.55 \pm 7.90\,\mathrm{kJ\,mol^{-1}}$
$K_6^{ref} = (1.12 \pm 0.09) \times 10^1\,\mathrm{kPa^{-2}}$ $\qquad \Delta_r H_6^\circ = 91.61 \pm 7.37\,\mathrm{kJ\,mol^{-1}}$

$$C_6^= \underset{k_7/K_7}{\overset{k_7}{\rightleftharpoons}} 2C_3^= \qquad r_7 = \frac{k_7}{Den}\left(p\,(C_6^=) - \frac{1}{K_7}p\,(C_3^=)^2\right)$$

$k_7^{ref} = (1.59 \pm 0.10) \times 10^{-2}\,\mathrm{mol\,kg_{cat}^{-1}\,s^{-1}\,kPa^{-1}}$ $\qquad E_{a,7} = 77.31 \pm 4.87\,\mathrm{kJ\,mol^{-1}}$
$K_7^{ref} = (7.24 \pm 0.76) \times 10^1\,\mathrm{kPa^{-2}}$ $\qquad \Delta_r H_7^\circ = 57.89 \pm 8.39\,\mathrm{kJ\,mol^{-1}}$

$$2C_6^= \underset{k_8/K_8}{\overset{k_8}{\rightleftharpoons}} C_3^= + C_4^= + C_5^= \qquad r_8 = \frac{k_8}{Den^2}\left(p\,(C_6^=)^2 - \frac{1}{K_8}p\,(C_3^=)\,p\,(C_4^=)\,p\,(C_5^=)\right)$$

$k_8^{ref} = (1.18 \pm 0.38) \times 10^{-3}\,\mathrm{mol\,kg_{cat}^{-1}\,s^{-1}\,kPa^{-2}}$ $\qquad E_{a,8} = -49.23 \pm 8.96\,\mathrm{kJ\,mol^{-1}}$
$K_8^{ref} = (2.45 \pm 0.30) \times 10^1\,\mathrm{kPa^{-3}}$ $\qquad \Delta_r H_8^\circ = 81.42 \pm 8.83\,\mathrm{kJ\,mol^{-1}}$

$$2C_6^= \underset{k_9/K_9}{\overset{k_9}{\rightleftharpoons}} 3C_4^= \qquad r_9 = \frac{k_9}{Den^2}\left(p\,(C_6^=)^2 - \frac{1}{K_9}p\,(C_4^=)^3\right)$$

$k_9^{ref} = (5.14 \pm 0.23) \times 10^{-5}\,\mathrm{mol\,kg_{cat}^{-1}\,s^{-1}\,kPa^{-2}}$ $\qquad E_{a,9} = -48.51 \pm 12.48\,\mathrm{kJ\,mol^{-1}}$
$K_9^{ref} = (7.01 \pm 0.44) \times 10^1\,\mathrm{kPa^{-3}}$ $\qquad \Delta_r H_9^\circ = 81.17 \pm 8.44\,\mathrm{kJ\,mol^{-1}}$

$$C_7^= \underset{k_{10}/K_{10}}{\overset{k_{10}}{\rightleftharpoons}} C_3^= + C_4^= \qquad r_{10} = \frac{k_{10}}{Den}\left(p\,(C_7^=) - \frac{1}{K_{10}}p\,(C_3^=)\,p\,(C_4^=)\right)$$

$k_{10}^{ref} = (2.79 \pm 0.09) \times 10^{-1}\,\mathrm{mol\,kg_{cat}^{-1}\,s^{-1}\,kPa^{-1}}$ $\qquad E_{a,10} = 11.73 \pm 2.18\,\mathrm{kJ\,mol^{-1}}$
$K_{10}^{ref} = (1.05 \pm 0.21) \times 10^2\,\mathrm{kPa^{-2}}$ $\qquad \Delta_r H_{10}^\circ = 61.82 \pm 4.06\,\mathrm{kJ\,mol^{-1}}$

$$Den = 1 + K_{C_{2-7}^=}^{ads}\,\Sigma_j\,p\,(C_j^=) + K_{H_2O}^{ads}\,p\,(H_2O)$$

$K_{C_{2-7}^=}^{ads,ref} = (2.20 \pm 0.59) \times 10^{-2}\,\mathrm{kPa^{-1}}$ $\qquad \Delta_{ads}H_{C_{2-7}^=}^\circ = -65.58 \pm 6.10\,\mathrm{kJ\,mol^{-1}}$
$K_{H_2O}^{ads,ref} = (2.21 \pm 0.10) \times 10^{-2}\,\mathrm{kPa^{-1}}$ $\qquad \Delta_{ads}H_{H_2O}^\circ = -11.19 \pm 2.19\,\mathrm{kJ\,mol^{-1}}$

Scheme 8. Reaction network, rate equations and estimated parameters for the model by Huang et al. [166] with j ranging from 2 to 7.

$$R\left(C_2^=\right) = r_3 + r_6$$
$$R\left(C_3^=\right) = r_2 + r_3 + 2\,r_4 + 2\,r_7 + r_8 + r_{10} - 3\,r_1$$
$$R\left(C_4^=\right) = r_1 + r_4 + r_5 + r_6 + r_8 + 3\,r_9 + r_{10} - 2\,r_2$$
$$R\left(C_5^=\right) = r_1 + r_2 + r_8 - r_3 - 2\,r_4 - 2\,r_5$$
$$R\left(C_6^=\right) = r_5 - r_6 - r_7 - 2\,r_8 - 2\,r_9$$
$$R\left(C_7^=\right) = -\,r_{10}$$

Scheme 9. Net rates of production of the different lumps for the model by Huang et al. [166].

Parameter Estimation

The mole fractions shown in the figures [166] are based only on hydrocarbons, whereas the rate expressions in Scheme 8 are defined with partial pressures. The estimated parameters are obtained via nonlinear regression which is used to minimize the objective function. The latter returns the weighted sum of squared residuals between measured and predicted mole fractions. The weighting is performed in a relatively simple manner: the respective feed component is multiplied by 0.25 and the remaining components by 1. In a subsequent study [240], the authors gave some explanations on numerics: the integration is performed with a fourth–fifth-order Runge–Kutta method provided by *ode45* in MATLAB, whereas the Levenberg–Marquardt algorithm is used for minimizing the objective function. The olefin adsorption constant is assumed to be independent of chain length, so only one reference constant and one adsorption enthalpy are fitted. The interaction between water and the catalyst is reduced to a competitive adsorption, which also requires the estimation of these two values. Finally, the equilibrium constants of the backward reactions are fitted and not calculated from thermodynamics, because the lumps resemble isomer distributions which are difficult to characterize with single values. This causes 44 estimated parameters: ten reference rate constants, ten activation energies, twelve reference equilibrium constants, ten reaction enthalpies and two adsorption enthalpies. The reparameterized approach according to Equations (21) and (24) is used both for rate and for equilibrium constants with a reference temperature of 733 K, which is in the upper third of the investigated range.

3.1.4. Summary

All three examples comprise several steps of olefin interconversion reactions. Whereas Huang et al.'s model [166] is experimentally covered only for lower conversions where side product formation can be neglected, this aspect is included for Epelde et al.'s [235] model as well as Ying et al.'s [236] model. The former model differentiates paraffins and aromatics in four lumps (eight-lump version), whereas the latter only has one general side product lump. On the other hand, the HW type of mechanism used by Huang et al. [166] yields a comparably robust model, although performance could be further improved by using different adsorption constants for all carbon numbers. Moreover, the high number of estimated parameters can cause numerical difficulties during estimation. For the two other models, both adsorption effects and a mechanistic approach are missing. In addition, feed partial pressures are relatively high for Epelde et al. [235]; consequently, extrapolation to lower values might be difficult. This is the reason why the authors could not notice any improvement when using an HW type of mechanism [235]. Furthermore, use of this model is restricted to butenes as feed, whereas the other two examples can be applied to different olefins and also to mixtures as feed. This feature is derived from their reaction networks, which contain a high number of pure olefin interconversion steps. Conclusions concerning the mechanism is difficult for Epelde et al. [235] because their network neglects stoichiometry and, in the five lump version, combines final and intermediate products in one lump. However, it is suitable to describe conversion of butenes over ZSM-5 modified with potassium.

3.2. Studies Focusing on Feed Olefin Consumption over ZSM-5

3.2.1. Borges et al.: Three-Lump Approach for Oligomerization of $C_2^=$ to $C_4^=$ Feed Olefins

Catalyst

A commercial ZSM-5 powder by Zeolyst International with a rather low Si/Al ratio of 30 was used here [241]. As shown in Table 1, no further details are available. Measurements were performed with a TOS between 0 and 1.4 h; no deactivation was observed during this period. Furthermore, no coke could be detected during heating up the catalyst to 973 K under air and analyzing the effluent with a thermogravimetry (TG)/differential scanning calorimetry (DSC) combination. This was attributed to the mild conditions and steric hindrances of coke evolution [237].

Setup and Conditions

In the study of Borges et al. [165], a continuous fixed bed reactor was used; no additional information about the setup is given. Ethene, propene and 1-butene were provided as gases and fed separately, each of them diluted with nitrogen. The products were analyzed via a GC containing a single column and an FID.

Reaction Network

This work focuses on the consumption of a certain feed olefin through oligomerization. Thus, no interconversion reactions are implemented, the model consists of only one rate equation, which is equal to the net rate of production of either $C_2^=$, $C_3^=$ or $C_4^=$ (see Scheme 10). For the values of R, stoichiometry is not retained. Although the actual rate is written as dimerization, the authors account for the oligomerization through allowing also higher intermediates to participate in this reaction step: one reactant is always the feed component (e.g., $C_2^=$), whereas the other reactant is either also the feed molecule or a multiple of it (e.g., $C_2^=$, $C_4^=$, $C_6^=$...). In the derivation of Scheme 10, it is assumed that the sum of partial pressures equals the inlet partial pressure of the feed component p^{in} $(C_i^=)$ throughout the whole reactor. This allows expressing the partial pressures of all reactants via the conversion X and p^{in} $(C_i^=)$. Furthermore, irreversible elementary reactions are underlaid. This work is an example where ethene dimerization is included. Scheme 10 is a combination of ER and HW types of mechanism, so adsorption effects are included for one of the reacting olefins (superscript "ads"), whereas the other olefin reacts directly from the gas phase. In the numerator, adsorption equilibrium and rate constant as well as the total number of acid sites C_t are summarized to a composite value k_{i-1}^{co}. The scope of describing the feed olefin consumption via oligomerization means that no cracking and no side reactions are considered, although the corresponding interconversion and side products are observed.

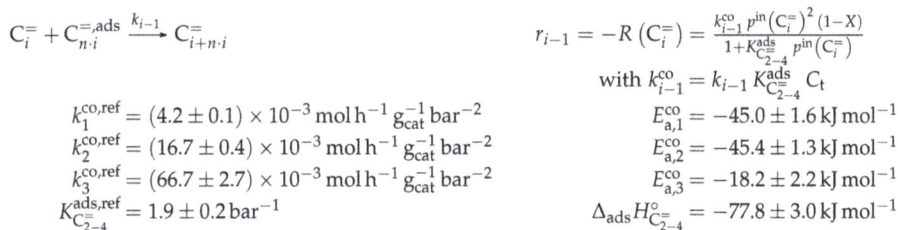

$$C_i^= + C_{n\cdot i}^{=,ads} \xrightarrow{k_{i-1}} C_{i+n\cdot i}^=$$

$$r_{i-1} = -R\left(C_i^=\right) = \frac{k_{i-1}^{co}\, p^{in}\left(C_i^=\right)^2 (1-X)}{1 + K_{C_{2-4}^=}^{ads}\, p^{in}\left(C_i^=\right)}$$

$$\text{with } k_{i-1}^{co} = k_{i-1}\, K_{C_{2-4}^=}^{ads}\, C_t$$

$$k_1^{co,ref} = (4.2 \pm 0.1) \times 10^{-3}\,\text{mol}\,\text{h}^{-1}\,\text{g}_{cat}^{-1}\,\text{bar}^{-2}$$
$$k_2^{co,ref} = (16.7 \pm 0.4) \times 10^{-3}\,\text{mol}\,\text{h}^{-1}\,\text{g}_{cat}^{-1}\,\text{bar}^{-2}$$
$$k_3^{co,ref} = (66.7 \pm 2.7) \times 10^{-3}\,\text{mol}\,\text{h}^{-1}\,\text{g}_{cat}^{-1}\,\text{bar}^{-2}$$
$$K_{C_{2-4}^=}^{ads,ref} = 1.9 \pm 0.2\,\text{bar}^{-1}$$

$$E_{a,1}^{co} = -45.0 \pm 1.6\,\text{kJ}\,\text{mol}^{-1}$$
$$E_{a,2}^{co} = -45.4 \pm 1.3\,\text{kJ}\,\text{mol}^{-1}$$
$$E_{a,3}^{co} = -18.2 \pm 2.2\,\text{kJ}\,\text{mol}^{-1}$$
$$\Delta_{ads}H_{C_{2-4}^=}^{\circ} = -77.8 \pm 3.0\,\text{kJ}\,\text{mol}^{-1}$$

Scheme 10. Reaction network, rate equations, net rate of production of the different lumps and estimated parameters for the model by Borges et al. [165] with i ranging from 2 to 4 and n being a positive integer such that $n \cdot i$ is a multiple of i.

Parameter Estimation

For the three reaction rates in Scheme 10, partial pressures are used. The rates are fitted to measured data points using Microsoft Excel. Here, a non-linear least-squares regression is performed to minimize the sum of squared residuals between experiments and model. In contrast to most other studies, the objective function evaluates catalytic activity and not mole fractions or comparable values. The catalytic activity relates the final conversion with the initial molar flow rate of the olefin and the catalyst mass. For the parameter estimation, no weighting factors are included. The adsorption of the different olefins is realized via the same constant. This is justified by a reference to literature studies and by own hybrid Hartree–Fock (HF) and DFT calculations which show a significant difference only for ethene. Although the model describes not only the dimerization of two feed molecules, but also of multiples of it and a feed molecule, all rate constants for a certain feed are assumed to be the same. These as well as the adsorption equilibrium constant are expressed via values at the reference temperature of 648 K, which is the mean of the experimentally covered range. The kinetic rate constants are composite values (superscript "co") which include the rate constant itself, the adsorption constant and the molar concentration of total acid sites per catalyst mass C_t. Finally, eight parameters are estimated with the experimental data: three reference rate constants, three activation energies, one reference equilibrium constant and one adsorption enthalpy.

3.2.2. Oliveira et al.: 17-Lump Model for $C_2^=$ to $C_4^=$ Feeds Considering Heterogeneity in Acid Sites

Catalyst

This model [237] is a subsequent work to Borges et al. [165]. Thus, the same ZSM-5 zeolite powder was used (see Table 1). However, focus of this study was creating a kinetic model which has the heterogeneity of the acid sites implemented. The authors investigated the coherence between acidity and activity earlier [241] and found a linear relationship between the activation energy of ammonia desorption resembling acid strength and of several surface reactions as well as of the adsorption enthalpy. These results were further confirmed by ab initio calculations [237]. This contradicts the approach of Thybaut et al. [54], where a difference in acid strength is fully attributed to the adsorption properties, whereas the kinetic descriptors, i.e., preexponential factor and activation energy, are independent of the catalyst properties. However, in Thybaut et al.'s study [54], an average acidity was assumed for each catalyst, whereas, for Oliveira et al. [237], several sites with different strength were defined. Further variety within the catalyst samples was achieved by exchanging 0%, 2.4%, 3.0% and 3.2% of the protons with Na. The number of acid sites as well as their strength decreased with higher Na contents. In contrast, for Thybaut et al. [54], the strength of the acid sites increases when their amount is lowered. An explanation could be that the different catalysts were synthesized already with the reduced number of acid sites, whereas, for Oliveira et al. [237], some of the sites' protons were exchanged with Na after synthesis which might especially affect the ones with highest strength.

Setup and Conditions

The same apparatus and similar conditions as for Borges et al. [241] were used (see Table 2).

Reaction Network

As stated above, this kinetic model aims at simulating the olefin interconversion on four different ZSM-5 samples where each had a uniform distribution of acid strength additionally. For this, 17 lumps are introduced: $C_2^=$, $C_3^=$, $C_4^=$, $C_5^=$, $C_6^=$, $C_7^=$, $C_8^=$, C_2, C_3, C_4, C_5, C_6, C_7, C_8, C_6^{ar}, C_7^{ar} and C_8^{ar}. In contrast to the previous study, cracking is also considered as a backward reaction to the dimerization. Furthermore, the irreversible evolution of side products is included. The corresponding rate equations can be found in Scheme 11 where θ_h represents the fraction of acid sites having the activation energy of ammonia desorption of $E_{a,h}^{NH_3}$. The total number of acid sites is included in the preexponential fitting parameter α of each reaction type. The linear relationship between $E_{a,h}^{NH_3}$ and activation energy or adsorption enthalpy is expressed via the parameters β and δ. The carbon number dependence of the non-equilibrated steps is implemented with a hyperbolic tangent function and the additional parameters γ, ϕ and φ. As in the previous study [165], the sum of partial pressures should always be equal to the inlet partial pressure of the feed component C_w. In Scheme 11, one olefin is always in adsorbed state (superscript "ads"), whereas, when applicable, the other one is in the gas phase. Consequently, the dimerization and aromatization steps are combined ER and HW types of mechanism. For the monomolecular cracking reactions, the L and HW types of mechanism are coupled. All steps should occur as elementary reactions. Ethene dimerization reactions are covered by this model. Although Scheme 11 proposes that three olefins are converted to three paraffins per evolution of one aromatic molecule, stoichiometry is retained neither in the reaction rate nor in the net rate of production.

$$C_i^{=,ads} + C_v^{=} \underset{k_{i+v}^{cr}}{\overset{k_{i,v}^{dim}}{\rightleftharpoons}} C_{i+v}^{=,ads}$$

$$C_i^{=,ads} + 3C_v^{=} \xrightarrow{k_{i,v}^{ar}} C_i^{ar,ads} + 3C_v$$

$$r_l^{dim/cr} = \frac{k_l^{dim,co}\, p(C_i^{=})\, p(C_v^{=}) - k_l^{cr,co}\, p(C_{i+v}^{=})}{1 + \sum_h K_{h,C_{2-8}^{=}}^{ads}\, p^{in}(C_w^{=})}$$

$$r_l^{ar} = \frac{k_l^{ar,co}\, p(C_i^{=})\, p(C_v^{=})}{1 + \sum_h K_{h,C_{2-8}^{=}}^{ads}\, p^{in}(C_w^{=})}$$

with $k^{dim,co} = \sum_h \theta_h\, \alpha^{dim} \exp\left(\beta^{dim}\left(E_a^{0,dim} - \delta^{dim} E_{a,h}^{NH_3}\right)\right) K_{h,C_{2-8}^{=}}^{ads} \exp\left(-\frac{E_a^{0,dim} - \delta^{dim} E_{a,h}^{NH_3}}{RT}\right)$

$k^{cr,co} = \sum_h \theta_h\, \alpha^{cr} \exp\left(\beta^{cr}\left(E_a^{0,cr} - \delta^{cr} E_{a,h}^{NH_3}\right)\right) K_{h,C_{2-8}^{=}}^{ads} \exp\left(-\frac{E_a^{0,cr} - \delta^{cr} E_{a,h}^{NH_3}}{RT}\right)$

$k^{ar,co} = \sum_h \theta_h\, \alpha^{ar} \exp\left(\beta^{ar}\left(E_a^{0,ar} - \delta^{ar} E_{a,h}^{NH_3}\right)\right) v^{\varphi_{ar}} K_{h,C_{2-8}^{=}}^{ads} \exp\left(-\frac{E_a^{0,ar} - \delta^{ar} E_{a,h}^{NH_3}}{RT}\right)$

$E_a^{0,cr} = \frac{E_a^{cr}}{\tanh(\gamma^{cr}(i+v))}\left(1 + \phi^{cr}|i-v|\right)$ $\quad E_a^{0,dim} = \frac{E_a^{dim}}{\tanh(\gamma^{dim} v)}$ $\quad E_a^{0,ar} = \frac{E_a^{ar}}{\tanh(\gamma^{ar} v)}$

$K_{h,C_{2-8}^{=}}^{ads} = K_{C_{2-8}^{=}}^{ads,ref} \exp\left(-\frac{\Delta_{ads} H_{C_{2-8}^{=}}^{\circ} - \delta^{ads} E_{a,h}^{NH_3}}{RT}\right)$

$\alpha^{dim} = 1.05 \times 10^5\ \mathrm{mol\,min^{-1}\,atm^{-1}\,g_{cat}^{-1}}$	$\beta^{dim} = 1.27 \times 10^{-5}$	$\delta^{dim} = 0.005$
$E_a^{dim} = 103.5\ \mathrm{kJ\,mol^{-1}}$	$\gamma^{dim} = 0.523$	
$\alpha^{cr} = 8.16 \times 10^2\ \mathrm{mol\,min^{-1}\,g_{cat}^{-1}}$	$\beta^{cr} = 7.73 \times 10^{-5}$	$\delta^{cr} = 0.334$
$E_a^{cr} = 126.1\ \mathrm{kJ\,mol^{-1}}$	$\gamma^{cr} = 0.168$	$\phi^{cr} = 0.029$
$\alpha^{ar} = 1.81 \times 10^5\ \mathrm{mol\,min^{-1}\,atm^{-1}\,g_{cat}^{-1}}$	$\beta^{ar} = 3.72 \times 10^{-7}$	$\delta^{ar} = 0.374$
$E_a^{ar} = 130.9\ \mathrm{kJ\,mol^{-1}}$	$\gamma^{ar} = 0.164$	$\varphi^{ar} = 1.909$
$K_{C_{2-8}^{=}}^{ads,ref} = 4.86 \times 10^{-17}\ \mathrm{atm^{-1}}$	$\Delta_{ads} H_{C_{2-8}^{=}}^{\circ} = 132.9\ \mathrm{kJ\,mol^{-1}}$	$\delta^{ads} = 0.634$

Scheme 11. Reaction network, rate equations and estimated parameters for the model by Oliveira et al. [237] with i and v ranging from 2 to 6 and $i + v$ being less than or equal to 8 for the dimerization/cracking reactions; for the aromatization, i is between 6 and 8 and v between 2 and 8; the carbon number of the feed olefin is characterized by w and can be between 2 and 8.

For each of the lumps, the net rate of production is defined. Because of the many combination possibilities, Scheme 12 is written in generalized form.

$$R\left(C_i^{=}\right) = -\sum_l r_l^{dim/cr}\left(C_i^{=,ads}, C_v^{=}; C_{i+v}^{=,ads}\right) - \sum_l r_l^{dim/cr}\left(C_v^{=,ads}, C_i^{=}; C_{v+i}^{=,ads}\right)$$
$$- \sum_l r_l^{ar}\left(C_i^{=,ads}, 3C_v^{=}; C_i^{ar,ads}, 3C_v\right) - \sum_l r_l^{ar}\left(C_v^{=,ads}, 3C_i^{=}; C_v^{ar,ads}, 3C_i\right)$$

$$R\left(C_i\right) = \sum_l r_l^{ar}\left(C_v^{=,ads}, 3C_i^{=}; C_v^{ar,ads}, 3C_i\right)$$

$$R\left(C_i^{ar}\right) = \sum_l r_l^{ar}\left(C_i^{=,ads}, 3C_v^{=}; C_i^{ar,ads}, 3C_v\right)$$

Scheme 12. Net rates of production of the different lumps for the model by Oliveira et al. [237] with i ranging from 2 to 8 for $R\left(C_i^{=}\right)$, from 6 to 8 for $R\left(C_i^{ar}\right)$ and from 2 to 8 for $R\left(C_i\right)$, respectively; the same rules as in Scheme 11 apply for the different indices of the reaction rates.

Parameter Estimation

The reaction rates in Scheme 11 are defined with partial pressures. As in the previous study, the objective function compares catalytic activities between experiment and model. The discrepancy minimized via a nonlinear least squares regression using MATLAB and without any weighting. The solver *ode15s* is applied to integrate the differential equations. With ab initio HF calculations, the authors could show that only the size of the gas phase olefin is crucial for the activation energy, an effect which is included via γ, ϕ and φ. Again, it is assumed that all olefins have the same value for the adsorption constant. Both the latter and the rate constants are expressed via values at the reference temperature, a value which is not mentioned. Finally, 20 parameters are estimated: three activation energies, three preexponential factors, six values to correlate acid strength and activation energies, five factors for the carbon number dependence, one reference equilibrium constant, one adsorption enthalpy and one factor to correlate acid strength and adsorption enthalpy.

3.2.3. Summary

The model by Borges et al. [165] is an effective way to describe the consumption of $C_2^=$ to $C_4^=$ feeds. However, due to the negligence of interconversion and side reactions, its application is restricted to low conversion in contrast to the examples in Section 3.1.3. On the other hand, computational effort is less for Borges et al. [165]. The limitation of low conversion is improved by the subsequent model by Oliveira et al. [237] where more variability in reactivity is given, but where also more parameters are required. Here, the description of side product formation is also possible. For both models, agreement could be increased by considering the carbon number dependence of adsorption constants. Furthermore, the assumption that the sum of all partial pressures is equal to the inlet partial pressure of the feed component might not always be fulfilled. Nevertheless, the approach by Oliveira et al. [237] is the only one in this review which allows considering the fact that not all sites of a zeolite have the same acid strength.

3.3. Microkinetic Study over ZSM-5

3.3.1. von Aretin et al.: Model for Arbitrary Olefin Feeds Considering all Interconversion Steps with Maximum Carbon Number of Twelve

Catalyst

A commercial ZSM-5 powder provided by Clariant AG was used in this study [164]. As shown in Table 1, the applied particle size is in the upper range. Before the first data point was recorded, the catalyst had to be deactivated for six hours. In this period, a significant loss of activity was observed, whereas it reached an almost constant value for the next ten hours [242].

Setup and Conditions

The experiments were performed with 1-pentene as feed using a continuous fixed bed quartz glass reactor with an inner diameter of 6 mm. During the measurements, two different volumetric flow rates were applied (300 and 400 mL min^{-1}), which caused two total pressures (1.16 and 1.23 bar). Isothermality was guaranteed by using SiC and nitrogen to dilute the catalyst and the feed, respectively. This was monitored by measuring the temperature at the tube wall. The 1-pentene stream had to be evaporated before it was fed into the reactor. The products were analyzed with a GC which had three columns and both an FID and a TCD.

Reaction Network

In contrast to all other examples shown in Table 2, a microkinetic model is formulated here. Lumping is still performed during evaluation of the experiments: The model differentiates each isomer, but summarizes them for comparison with the measurements. Here, the responses $C_2^=$, $C_3^=$, $C_4^=$, $C_5^=$ and $C_{6-12}^=$ are compared during parameter estimation. The model considers all olefins from $C_2^=$ to $C_{12}^=$ which have no or only methyl side groups, two side groups as maximum, no quaternary carbon atoms and which are not derivatives of 2,3-dimethylbutene. As reaction families, protonation, deprotonation, cracking, dimerization and two isomerization reactions (methyl shift and PCP branching) are implemented. This causes 4395 different elementary reactions. However, all steps where the carbon number is kept constant are assumed as *quasi*-equilibrated, meaning that 611 cracking and 293 dimerization reactions remain. Combined with the different protonation possibilities, 1585 pathways of kinetic relevance are considered. A representative reaction is shown in Scheme 13. Ethene is considered as product of irreversible cracking reactions, so no dimerization with ethene is implemented. Chemisorption of the reactants which is a combination of π complex and protonation is explicitly included, meaning that the cracking reactions are formulated with a combination of L and HW types of mechanism. For the dimerization, the second olefin is assumed to react directly out of the gas phase; this corresponds to coupled ER and HW types of mechanism. The equilibrium constants describing the π complex as well as the protonation values are extracted from a theoretical QM-PoT and statistical thermodynamics study by Nguyen et al. [167,243]. In Table 1, the total number of acid sites is shown. However, calculations are performed with the molar concentration of strong Brønsted acid sites per catalyst mass $C_{t,SBAS}$ which equals 0.135 mmol g_{cat}^{-1}. Although the measurements revealed small amounts of pentane, cyclopentane, cyclopentene, methylcyclopentene and aromatics as side products, their formation is not included in the model because the mole fraction of all side products never exceeded 3% [163].

$$C_i^{=,chem}(n) + C_v^= \underset{\tilde{k}_r^{cr}(m;n)}{\overset{\tilde{k}_l^{RDS}\,\tilde{k}_r^{cr}(m;n)}{\rightleftharpoons}} C_{i+v}^{=,chem}(m)$$

$$r_i^{dim} = \frac{n_{e,i}^{dim}\,\tilde{k}_i^{RDS}\,\tilde{k}_r^{cr}(m;n)\,C_{t,SBAS}\,\tilde{K}^{chem}\left(C_i^=;n\right)p\left(C_i^=\right)p\left(C_v^=\right)}{1+\sum_j K^\pi\left(C_j^=\right)p\left(C_j^=\right)}$$

$$r_i^{cr} = \frac{n_{e,i}^{cr}\,\tilde{k}_r^{cr}(m;n)\,C_{t,SBAS}\,\tilde{K}^{chem}\left(C_{i+v}^=;m\right)p\left(C_{i+v}^=\right)}{1+\sum_j K^\pi\left(C_j^=\right)p\left(C_j^=\right)}$$

with $K^\pi\left(C_i^=\right)= \exp\left[\left(-\frac{\alpha j+\beta}{RT}\right)-\left(\frac{\gamma j+\delta}{T}\right)\right]$

$E_a^{cr}(s;p)= 229.9 \pm 0.9\,\mathrm{kJ\,mol^{-1}}$

$E_a^{cr}(t;s)= 171.5 \pm 0.8\,\mathrm{kJ\,mol^{-1}}$

$\tilde{A}^{cr,ref}= (2.7 \pm 0.4)\times 10^{16}\,\mathrm{s^{-1}}$

$$\tilde{K}^{RDS} = \frac{\tilde{k}_l^{TD}\,\tilde{K}^{chem}\left(C_{i-v}^=;m\right)}{\tilde{K}^{chem}\left(C_i^=;n\right)}$$

$E_a^{cr}(s;s)= 200.2 \pm 0.9\,\mathrm{kJ\,mol^{-1}}$

$E_a^{cr}(t;p)= 211.9 \pm 1.5\,\mathrm{kJ\,mol^{-1}}$

Scheme 13. Reaction network, rate equations and estimated parameters for the model by von Aretin et al. [164] with j ranging from 2 to 12 as well as i and v ranging from 2 to 10 and from 3 to 10, respectively and $i + v$ being less than or equal to 12; when i is equal to 2, no dimerization takes place, cracking is thus irreversible in these cases; m and n resemble the types of protonated intermediates: primary, secondary or tertiary; and $\tilde{K}^{chem}$ is composed of K^π, symmetry contributions, an equilibrium constant of isomerization and one of protonation (see [164]). Instead of the original parameters from [164], the slightly changed values from [244] are shown here, where $\tilde{K}^{TD}$ is calculated exclusively with thermodynamic data stemming from Benson's group additivity method [96,97].

From the reaction rates, the net rates of production according to Scheme 14 are obtained. For better clarity, they are divided into olefins in the gas phase and the corresponding chemisorbed intermediates.

$$R\left(C_i^=\right) = \sum_l r_l^{cr}\left(C_{i+v}^{=,chem};C_v^{=,chem},C_i^=\right) - \sum_l r_l^{dim}\left(C_v^{=,chem},C_i^=;C_{i+v}^{=,chem}\right) + R\left(C_i^{=,chem}\right)$$

$$R\left(C_i^{=,chem}\right) = \sum_l r_l^{cr}\left(C_{i+v}^{=,chem};C_i^{=,chem},C_v^=\right) - \sum_l r_l^{cr}\left(C_i^{=,chem};C_v^{=,chem},C_{i-v}^=\right)$$
$$+ \sum_l r_l^{dim}\left(C_v^{=,chem},C_{i-v}^=;C_i^{=,chem}\right) - \sum_l r_l^{dim}\left(C_i^{=,chem},C_v^=;C_{i+v}^{=,chem}\right)$$

Scheme 14. Net rates of production of the different species for the model by von Aretin et al. [164] with i ranging from 2 to 12 for $R\left(C_i^=\right)$ and for $R\left(C_i^{=,chem}\right)$; the same rules as in Scheme 13 apply for the different indices of the reaction rates.

Parameter Estimation

The reaction rates in Scheme 13 require partial pressures. The objective function in this study evaluates the squared residuals between the molar flow rates of model and experiment. This value is minimized with a nonlinear and unweighted regression using the Levenberg–Marquardt algorithm of the routine *lsqnonlin* in MATLAB. The molar flow rates are obtained by applying the solver *ode15s* to the differential equations. As explained above, all adsorption and equilibrium constants are calculated before parameter estimation. Although a broad picture of the olefin interconversion is depicted, only five parameters are necessary. This is possible through the application of the single-event methodology which is extensively described in the literature [20,21,54,245]. Here, the estimated parameters are not related to single reactions, but to reaction families and to types of reactant and product intermediates. It follows that, for a certain reaction family such as cracking, only a handful of different combination possibilities and thus single-event rate constants $\tilde{k}$ exist (see Scheme 13). All symmetry related information is considered via the number of single events n_e, a value which can be calculated for each reaction. The principle of thermodynamic reversibility [246] is applied in the model by von Aretin et al. [164] to express all dimerization reactions with the cracking parameters and an equilibrium constant $\tilde{K}^{RDS}$. For this, the thermodynamic equilibrium constant $\tilde{K}^{TD}$ is required. The five estimated values consist of four activation energies and one preexponential factor; the latter is assumed to be the same for all cracking reactions. During parameter estimation, the reparameterized Arrhenius approach is used, but with the additional temperature dependence of the preexponential factor (see Equation (23)). The reference temperature of 683 K is the mean value of the experimentally investigated range.

3.3.2. Summary

Similar to Huang et al. [166] und Ying et al. [236], the model by von Aretin et al. [164] can be applied to different olefins as feed. Through considering all possible reactions as well as the carbon number dependence of adsorption effects, this approach yields a realistic picture of overall reactivity. It can be further transferred to other catalysts because of a separation of kinetic and catalyst descriptors [54,244]. On the other hand, it has to be underlined that creation of such a model is very complex and time-consuming. Moreover, the computational power which is required is another disadvantage. Finally, the implementation of side product formation missing for this model is significantly more difficult compared to the approaches shown above.

3.4. Study Elucidating the Peculiarities over SAPO-34

3.4.1. Zhou et al.: Eight-Lump Model for $C_2^=$ to $C_4^=$ Feeds Considering Side Product Formation

Catalyst

In this work [77], a self-synthesized zeolite was used. Although the X-ray diffraction (XRD) pattern revealed a 50/50 structure of SAPO-18 and SAPO-34 fragments, it is referred to as SAPO-34 because both zeolite types show the same MTO performance according to the authors. The zeolite powder was sieved to a very fine fraction with a mean size of 3.2 μm. Because of the small pores within the eight-membered SAPO rings, the formation of olefins higher than $C_4^=$ and even of isobutene was suppressed. During the measurements, the authors observed significant amounts of coke, which is why a closure of the carbon balances was not possible. Consequently, the kinetic measurements were recorded after 1 min TOS. The coke was analyzed by introducing air subsequent to the kinetic measurements and by monitoring the CO and CO_2 evolution with a TCD.

Setup and Conditions

Ethene, propene, 1-butene and 2-butene were separately investigated and therefore fed as gas. For this, a continuous fixed bed reactor made of quartz glass with an inner diameter of 6 mm was used. Only in three cases, the feed was diluted with nitrogen, but these data points are not considered during parameter estimation. For the remaining measurements, the partial pressure of the feed equaled the total pressure. In contrast to that, the catalyst bed was diluted each time with silica so that the ratio of bed height and bed diameter was approximately two. For product analysis, the authors applied a GC with one column and an FID.

Reaction Network

The kinetic description is conducted with eight responses during parameter estimation: $C_2^=$, $C_3^=$, $C_4^=$, $C_5^=$, C_1, C_2, C_3 and C_4. An additional lump C_x^+ is introduced which should resemble a higher protonated intermediate with arbitrary carbon number. For this, the pseudo-steady state approximation (PSSA) is applied since no experimental data for comparison are available. The authors justify this C_x^+ lump by referring to the measurements which yield an olefin composition close to the calculated thermodynamic equilibrium on this specific catalyst regardless of the feed olefin used; thus, similar intermediates should be present. The experiments with either 1-butene or 2-butene as feed showed that the linear butenes can be summarized to one lump because isomerization is fast. However, isobutene is excluded from the reaction network due to steric hindrance. For the same reason, no higher olefins and no aromatics are included. This strong molecular sieving effect could be seen as a hint that the majority of acid sites is located within the micropores. In the resulting general reaction pathways, the feed olefin is converted to C_x^+ (k_1–k_6) and then further cracked to olefins (k_7–k_{10}), as can be seen in Scheme 15. Through side reactions (k_{11}–k_{14}), the olefins can also react to the respective paraffin, whereas the lump C_x^+ is transformed to methane, respectively. This model considers the dimerization of ethene, both with itself and with propene or butenes. Only the paraffin formation is assumed to be irreversible. From the equations in Scheme 15, it can be seen that neither adsorption nor a mechanistic scheme are implemented. Furthermore, stoichiometry is neglected and the reactions are assumed to be of elementary type. Nevertheless, the order of both $C_4^=$ and $C_5^=$ dimerization is set to one.

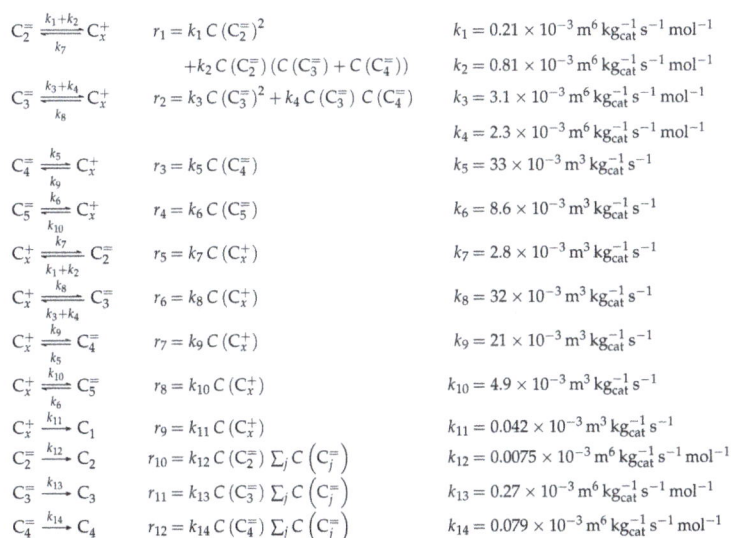

$$C_2^= \xrightleftharpoons[k_7]{k_1+k_2} C_x^+ \qquad r_1 = k_1 C \left(C_2^=\right)^2 \qquad\qquad k_1 = 0.21 \times 10^{-3}\,\mathrm{m^6\,kg_{cat}^{-1}\,s^{-1}\,mol^{-1}}$$

$$+ k_2 C \left(C_2^=\right)\left(C\left(C_3^=\right)+C\left(C_4^=\right)\right) \qquad k_2 = 0.81 \times 10^{-3}\,\mathrm{m^6\,kg_{cat}^{-1}\,s^{-1}\,mol^{-1}}$$

$$C_3^= \xrightleftharpoons[k_8]{k_3+k_4} C_x^+ \qquad r_2 = k_3 C \left(C_3^=\right)^2 + k_4 C \left(C_3^=\right) C \left(C_4^=\right) \qquad k_3 = 3.1 \times 10^{-3}\,\mathrm{m^6\,kg_{cat}^{-1}\,s^{-1}\,mol^{-1}}$$

$$k_4 = 2.3 \times 10^{-3}\,\mathrm{m^6\,kg_{cat}^{-1}\,s^{-1}\,mol^{-1}}$$

$$C_4^= \xrightleftharpoons[k_9]{k_5} C_x^+ \qquad r_3 = k_5 C \left(C_4^=\right) \qquad\qquad k_5 = 33 \times 10^{-3}\,\mathrm{m^3\,kg_{cat}^{-1}\,s^{-1}}$$

$$C_5^= \xrightleftharpoons[k_{10}]{k_6} C_x^+ \qquad r_4 = k_6 C \left(C_5^=\right) \qquad\qquad k_6 = 8.6 \times 10^{-3}\,\mathrm{m^3\,kg_{cat}^{-1}\,s^{-1}}$$

$$C_x^+ \xrightleftharpoons[k_1+k_2]{k_7} C_2^= \qquad r_5 = k_7 C \left(C_x^+\right) \qquad\qquad k_7 = 2.8 \times 10^{-3}\,\mathrm{m^3\,kg_{cat}^{-1}\,s^{-1}}$$

$$C_x^+ \xrightleftharpoons[k_3+k_4]{k_8} C_3^= \qquad r_6 = k_8 C \left(C_x^+\right) \qquad\qquad k_8 = 32 \times 10^{-3}\,\mathrm{m^3\,kg_{cat}^{-1}\,s^{-1}}$$

$$C_x^+ \xrightleftharpoons[k_5]{k_9} C_4^= \qquad r_7 = k_9 C \left(C_x^+\right) \qquad\qquad k_9 = 21 \times 10^{-3}\,\mathrm{m^3\,kg_{cat}^{-1}\,s^{-1}}$$

$$C_x^+ \xrightleftharpoons[k_6]{k_{10}} C_5^= \qquad r_8 = k_{10} C \left(C_x^+\right) \qquad\qquad k_{10} = 4.9 \times 10^{-3}\,\mathrm{m^3\,kg_{cat}^{-1}\,s^{-1}}$$

$$C_x^+ \xrightarrow{k_{11}} C_1 \qquad r_9 = k_{11} C \left(C_x^+\right) \qquad\qquad k_{11} = 0.042 \times 10^{-3}\,\mathrm{m^3\,kg_{cat}^{-1}\,s^{-1}}$$

$$C_2^= \xrightarrow{k_{12}} C_2 \qquad r_{10} = k_{12} C \left(C_2^=\right) \Sigma_j C \left(C_j^=\right) \qquad k_{12} = 0.0075 \times 10^{-3}\,\mathrm{m^6\,kg_{cat}^{-1}\,s^{-1}\,mol^{-1}}$$

$$C_3^= \xrightarrow{k_{13}} C_3 \qquad r_{11} = k_{13} C \left(C_3^=\right) \Sigma_j C \left(C_j^=\right) \qquad k_{13} = 0.27 \times 10^{-3}\,\mathrm{m^6\,kg_{cat}^{-1}\,s^{-1}\,mol^{-1}}$$

$$C_4^= \xrightarrow{k_{14}} C_4 \qquad r_{12} = k_{14} C \left(C_4^=\right) \Sigma_j C \left(C_j^=\right) \qquad k_{14} = 0.079 \times 10^{-3}\,\mathrm{m^6\,kg_{cat}^{-1}\,s^{-1}\,mol^{-1}}$$

Scheme 15. Reaction network, rate equations and estimated parameters for the model by Zhou et al. [77] with *j* ranging from 2 to 5.

The resulting net rates of production can be seen in Scheme 16. Here, the lump C_x^+ is also shown. In the original publication [77], the concentrations of all olefins are summed up for the side product formation (see Steps 12–14). It is assumed here that the consumption through this summarized value is not included in the net rates of production of the respective olefins.

$$R\left(C_2^=\right) = k_7 C\left(C_x^+\right) - k_1 C\left(C_2^=\right)^2 - k_2 C\left(C_2^=\right)\left(C\left(C_3^=\right) + C\left(C_4^=\right)\right) - k_{12} C\left(C_2^=\right) \sum_j C\left(C_j^=\right)$$

$$R\left(C_3^=\right) = k_8 C\left(C_x^+\right) - k_2 C\left(C_2^=\right)\left(C\left(C_3^=\right) + C\left(C_4^=\right)\right) - k_3 C\left(C_3^=\right)^2 - k_4 C\left(C_3^=\right) C\left(C_4^=\right) - \\ - k_{13} C\left(C_3^=\right) \sum_j C\left(C_j^=\right)$$

$$R\left(C_4^=\right) = k_9 C\left(C_x^+\right) - k_2 C\left(C_2^=\right)\left(C\left(C_3^=\right) + C\left(C_4^=\right)\right) - k_4 C\left(C_3^=\right) C\left(C_4^=\right) - k_5 C\left(C_4^=\right) - \\ - k_{14} C\left(C_4^=\right) \sum_j C\left(C_j^=\right)$$

$$R\left(C_5^=\right) = k_{10} C\left(C_x^+\right) - k_6 C\left(C_5^=\right)$$

$$R\left(C_1\right) = k_{11} C\left(C_x^+\right)$$

$$R\left(C_2\right) = k_{12} C\left(C_2^=\right) \sum_j C\left(C_j^=\right)$$

$$R\left(C_3\right) = k_{13} C\left(C_3^=\right) \sum_j C\left(C_j^=\right)$$

$$R\left(C_4\right) = k_{14} C\left(C_4^=\right) \sum_j C\left(C_j^=\right)$$

$$R\left(C_x^+\right) = 0 = k_1 C\left(C_2^=\right)^2 + k_2 C\left(C_2^=\right)\left(C\left(C_3^=\right) + C\left(C_4^=\right)\right) + k_3 C\left(C_3^=\right)^2 + k_4 C\left(C_3^=\right) C\left(C_4^=\right) + k_5 C\left(C_4^=\right) \\ + k_6 C\left(C_5^=\right) - k_7 C\left(C_x^+\right) - k_8 C\left(C_x^+\right) - k_9 C\left(C_x^+\right) - k_{10} C\left(C_x^+\right) - k_{11} C\left(C_x^+\right)$$

Scheme 16. Net rates of production of the different lumps for the model by Zhou et al. [77] with j ranging from 2 to 5.

Parameter Estimation

All mass fractions in this study are defined on a carbon basis, whereas, in the rate equations, molar concentrations per volume have to be used. No information about the actual fitting routine can be found. Only rate constants are estimated because all data points were collected at constant temperature. The reaction network is restricted to the most important dimerization reactions in order to have not too many unknown parameters. This is why the dimerization reactions of ethene with propene and with butenes are assumed to have the same rate constant. For butenes and pentenes, only the self-dimerization is considered. Finally, 14 unknown parameters are obtained. Although only the undiluted measurements are used for parameter estimation, extrapolation to lower feed partial pressures is also possible according to the authors.

3.4.2. Summary

Compared to the other examples presented in this section, the maximum carbon number is significantly lower for Zhou et al. [77] because of the smaller zeolite pores. This is why a transfer of ZSM-5 models to SAPO-34 or the other way round is difficult. The approach via reactive intermediates chosen here leads to decent agreement with experimental data and covers also side product formation; however, mechanistic insight is difficult because the rate equations seem to be rather artificial. Furthermore, no adsorption or mechanistic effects are considered. Finally, application of this model is limited to 723 K.

3.5. *Other Studies*

Chen et al. [247] performed cracking experiments on a commercial ZSM-5 zeolite (Si/Al = 42.6) with single butene, pentene and hexene feeds between 773 and 813 K. Short contact times and low conversions were applied, so dimerization could be neglected. The corresponding model focuses on different cracking steps under these differential conditions, which means it does not describe the evolution along the reactor. However, insight into the energetics of the cracking pathways is provided. By making use of group additivity and correction methods, the formation of an alkoxide as intermediate is calculated. Moreover, the theoretical evaluation of the kinetic experiments yields intrinsic activation energies of the different cracking modes. It is shown that tertiary alkoxides have the lowest stability and therefore very small concentrations. Thus, the contribution of highly branched olefins to the overall cracking performance is smaller than expected although the activation energies starting from tertiary intermediates are in a similar range than for a secondary alkoxide reactant. These results are consistent with an earlier dispersion-corrected DFT study [248]. This model allows describing the cracking products of $C_4^=$ to $C_6^=$ olefin feeds with high accuracy. Moreover, it yields detailed insight into preferred reaction pathways; however, application is limited to differential conditions which excludes consecutive and side reactions. Furthermore, model build-up is comparably complex.

In a recent study, Li et al. [249] performed experiments over a commercial ZSM-5 zeolite with a Si/Al ratio of 50. After modification, the catalyst contained 4%$_{wt}$ of phosphorus and 2%$_{wt}$ of iron. Measurements were performed at temperatures between 763 and 883 K with butenes and pentenes as co-feed. The kinetic data are described with a six-lump model, which requires 24 parameters. The model does not consider any mechanistic approaches or adsorption effects, but covers a broad picture of olefin interconversion including side product formation.

The model by Meng et al. [250] is beyond the focus of this review because of liquid products.

4. Kinetic Models for Methanol-to-Olefins without Olefin Co-Feed

Firstly, both catalyst properties and experimental conditions as well as modeling details are presented in Tables 3 and 4, respectively. Subsequently, the models are grouped into subsections, explained and compared in a short summary paragraph. This section presents all kinetic models for a feed of pure oxygenates, i.e., methanol or DME, which means that an initiation phase should be visible for short contact times (see Section 2.4.2). The first subsection contains the models by Menges and Kraushaar-Czarnetzki [139] and Jiang et al. [251] over ZSM-5 where methanol and DME are summarized to one lump, which means no differentiation of their reactivity is possible. The next subsection contains the models by Gayubo et al. [114], Aguayo et al. [252] and Pérez-Uriarte et al. [253], which are all created by the same research group. The one by Gayubo et al. [114] is the first MTO model published by this group, meaning that many elements from this approach can be found in the subsequent models and also in the one by Epelde et al. [235]. Nevertheless, all three models in this subsection have a different focus. An important similarity of them is the differentiation of methanol and DME. In the following subsection, the two microkinetic studies over ZSM-5 by Park and Froment [132,254] and Kumar et al. [19] are discussed. Whereas the former evaluates different possible mechanisms for the formation of the first C-C bond, the latter is a subsequent work which uses the same reaction network except for the mentioned C-C bond formation steps. Instead, these are replaced by reactions of the aromatic hydrocarbon pool. The last subsection involves different zeolites: Gayubo et al. [255], Ying et al. [256], Chen et al. [257] and Alwahabi and Froment [258] describe MTO over SAPO-34, whereas another model by Gayubo et al. [259] and another one by Kumar et al. [146] are valid over SAPO-18 and over ZSM-22, respectively. On all these zeolite types, deactivation is significant which is why the different approaches accounting for this fact should be compared. Both models by Gayubo et al. [255,259] are comparable to the ZSM-5 case, whereas the microkinetic studies of Alwahabi and Froment [258] and Kumar et al. [146] are subsequent models to Park and Froment [132,254] and Kumar et al. [19] over ZSM-5, respectively.

Table 3. Properties of the different catalysts which were used for the kinetic models of methanol-to-olefins without olefin co-feed; besides the zeolite type, its silicon-to-aluminum ratio (Si/Al), its total number of acid sites as well as determination method, its ratio of Brønsted to Lewis acid sites (BAS/LAS) and its surface area according to the method by Brunauer–Emmett–Teller (BET) are shown. Furthermore, the time-on-stream (TOS) after which the kinetic data were taken, the particle size (d_P) and information about whether an extrudate or pure powder was used are presented. The line separates the different subsections. A hyphen represents missing information.

Model	Zeolite Type	Si/Al	Total Acidity	BAS/LAS	BET	TOS	d_P	Extrudate
Menges [139]	ZSM-5	250 [1]	-	-	-	0–3 h	2 mm [2]	50/50%$_{wt}$ [3] (Zeolite/AlPO$_4$)
Jiang [251]	ZSM-5	200	-	-	-	2 h [4]	600–900 µm	No
Gayubo [114]	ZSM-5	24 [1]	0.51 mmol g$_{cat}^{-1}$ (NH$_3$) [1]	2.9 [1]	124 m² g$_{cat}^{-1}$	6 h [5]	150–300 µm	25/30/45%$_{wt}$ (Zeolite/Bentonite/Alumina)
Aguayo [252]	ZSM-5	30 [1]	0.23 mmol g$_{cat}^{-1}$ (NH$_3$)	1.5 at 423 K	220 m² g$_{cat}^{-1}$	5 h [5]	150–300 µm	25/30/45%$_{wt}$ (Zeolite/Bentonite/Alumina)
Pérez [253]	ZSM-5	280 [1]	0.33 mmol g$_{cat}^{-1}$ (t-BA)	-	301 m² g$_{cat}^{-1}$	0.17 h [5]	125–300 µm	50/30/20%$_{wt}$ (Zeolite/Boehmite/Alumina)
Park [132,254]	ZSM-5	200	-	-	400 m² g$_{cat}^{-1}$	0–5 h	500–1000 µm	No
Kumar [19]	ZSM-5	200	0.083 mmol g$_{cat}^{-1}$ [6]	-	400 m² g$_{cat}^{-1}$	0–5 h	500–1000 µm	No
Gayubo [255]	SAPO-34	0.16	0.135 mmol g$_{cat}^{-1}$ (NH$_3$) [7]	-	875 m² g$_{cat}^{-1}$ [1]	1 h [5]	100–300 µm	25/30/45%$_{wt}$ (Zeolite/Bentonite/Alumina)
Ying [256]	SAPO-34	-	-	-	264 m² g$_{cat}^{-1}$	0 h	250–400 µm	Yes
Chen [257]	SAPO-34	0.16	-	-	-	>0 h	105–290 µm	No
Alwahabi [258]	SAPO-34	-	-	-	-	0.25 h	1.1 µm	No
Gayubo [259,260]	SAPO-18	0.3 [8]	0.12 mmol g$_{cat}^{-1}$ (NH$_3$)	-	171 m² g$_{cat}^{-1}$	0–1.5 h	150–250 µm	25/45/30%$_{wt}$ (Zeolite/Bentonite/Alumina)
Kumar [146]	ZSM-23	26	0.62 mmol g$_{cat}^{-1}$ [6]	-	-	0–7 h [9]	250–420 µm	No

(1) Value of the zeolite, i.e., without binder. (2) Diameter of the cylindrical pellets used in this study (length = 5 mm). (3) Dry basis. (4) Additional TOS of 50 h to reach a plateau with constant propene yields. (5) Results extrapolated to 0 h TOS. (6) Calculated via the Si/Al ratio. (7) Extracted from an earlier publication [261]. (8) Value of the gel [151]. (9) Converted to an effective space time to describe a non-deactivated catalyst.

Table 4. Experimental conditions and modeling details for the kinetic models of methanol-to-olefins without olefin co-feed; the temperature range (T), the total pressure (p_t), the partial pressure range of the feed oxygenates (p_{Ox}) and the maximum contact time ($(W/F^{in})_{max}$) with resulting oxygenates conversion (X_{max}) are listed; concerning the model, the number of fitted responses (N_{Res}), the number of estimated parameters (N_{Par}), the number of experiments (N_{Exp}) and the degree of freedom (dof) are shown; and, finally, it is noted whether the model follows a type of a mechanistical scheme (Mech.), whether adsorption is considered (Ads.) and which side products are included (Side prod.). The line separates the different subsections. A hyphen represents missing information.

Model	Feed	T	p_t	p_{Ox}	$(W/F^{in})_{max}$	X_{max}	N_{Res}	N_{Par}	N_{Exp}	dof	Mech.	Ads.	Side prod.
Menges [139]	MeOH, N$_2$	673–723 K	1.65 bar	0.170–0.556 bar	280 kg$_{zeo}$ s m$_t^{-3}$ (1)	1	6	16	78 (2)	452	No	No	No
Jiang [251]	MeOH	673–773 K	-	-	384 g$_{cat}$ min mol$_C^{-1}$	-	8	20	24	172	No	No	C$_{1-3}$
Gayubo [114]	MW(3,4)	573–723 K	1.013 bar	-	0.37 g$_{cat}$ h g$_{MeOH}^{-1}$	0.8 (5)	3 (6)	15	-	-	HW	H$_2$O	-
Aguayo [252]	MeOH	673–823 K	1.013 bar	1.013 bar (7)	2.50 g$_{cat}$ h mol$_C^{-1}$	1	7	26	18	100	No	No	C$_{1-4}$, C$_{6-8}^{car}$
Pérez [253]	DME	598–673 K	1.10 bar	1.10 bar	6 g$_{cat}$ h mol$_C^{-1}$	1	10	30	39	360	HW	MW (4)	C$_{1-4}$, C$_{6-8}^{car}$
	DME, He	648 K		0.28–1.10 bar	1 g$_{cat}$ h mol$_C^{-1}$	0.5							
	DME, H$_2$O	648 K		0.99–1.10 bar	1 g$_{cat}$ h mol$_C^{-1}$	0.5							
	DME, MeOH	623–673 K		1.10 bar	1 g$_{cat}$ h mol$_C^{-1}$	0.5							
Park [132,254]	MNW(4)	633–753 K	1.04 bar	-	2 g$_{cat}$ h mol$_C^{-1}$	0.7	9	33	31	246	LLEH (8)	28 MD(4)	No
Kumar [19]	MNW(4)	633–753 K	1.04 bar	-	6.5 kg$_{cat}$ s mol$_C^{-1}$	0.7	8 (9)	29	-	-	LLEH (8)	27 MD(4)	No
Gayubo [255]	MW(3,4)	623–748 K	-	-	0.44 g$_{cat}$ h g$_{MeOH}^{-1}$	1	5 (10)	17	-	-	HW	H$_2$O	C$_{1-4}$
							4 (10)	8	-	-	HW	H$_2$O	C$_{1-4}$
Ying [256]	MW(3,4)	723–763 K	1.013 bar	0.203–1.013 bar	0.03 g$_{cat}$ h g$_{MeOH}^{-1}$	1	7	13	43	288	HW	H$_2$O	C$_{1-6}$
Chen [257]	HM(4,11)	673–823 K	-	0.072–0.830 bar	0.02 h (12)	0.95	7	12 (4x)	-	-	No	No	C$_{2-3}$
Alwahabi [258]	MeOH, H$_2$O	673–723 K	1.04 bar	0.208 bar	2.95 g$_{cat}$ h mol$_C^{-1}$	0.88	6	30	9	24	LLEH (8)	25 MD (4)	No
Gayubo [259,260]	MW(3,4)	598–748 K	-	-	0.68 g$_{cat}$ h g$_{MeOH}^{-1}$	0.9 (5)	4	11	-	-	HW	H$_2$O	C$_1$
Kumar [146]	MeOH, He	673 K	1.013 bar	-	57.7 kg$_{cat}$ s mol$_C^{-1}$	0.95	7	8	12	76	LLEH (8)	27 MD (4)	No

(1) Value based on outlet volumetric flow rate where volume expansion is considered. (2) Extracted from [262]. (3) Experiments both with and without water dilution. (4) D = DME, H = He, M = MeOH, N = N$_2$, W = H$_2$O, 27 = C$_{2-7}$, and 28 = C$_{2-8}^=$. (5) Calculated with the lowest oxygenates fraction shown; conversion might be higher for measurements not presented. (6) Another lump (higher hydrocarbons) calculated via conservation of mass. (7) No feed dilution mentioned. (8) LLEH = L, LH, ER, HW. (9) Two additional responses (methanol and water) from carbon and hydrogen balance. (10) Another lump (methanol plus DME) calculated via conservation of mass. (11) Extracted from [263]. (12) Inverse value of the minimum WHSV.

4.1. Studies with Lumped Oxygenates over ZSM-5

4.1.1. Menges and Kraushaar-Czarnetzki: Six-Lump Approach Focusing on Lower Olefins Production

Catalyst

In this study [139], a self-extruded catalyst was applied. It consisted of a commercial zeolite from Zeochem and aluminium phosphate from Riedel-de Haen. In earlier studies [264–266], both the high Si/Al ratio of 250 and the binder were shown to be advantageous for high propene yields. With regular binders such as alumina, the Si/Al ratio could decrease during extrusion because of alumination which means the migration of extra aluminium from the binder into the zeolite. Moreover, alumina is known to produce both methane and coke, whereas aluminium phosphate is non-reactive, leading to a catalyst which has the advantageous macropores, but no changed reactivity. For the kinetic experiments, the catalyst was shaped into cylinders. The measurements were performed with fresh catalyst up to a TOS of 3 h [262] to avoid deactivation effects. For the same reason, the combination of the highest reaction temperature with the highest methanol partial pressure was ignored.

Setup and Conditions

The measurements were performed in an electrically heated continuous stainless steel fixed bed reactor [262] with an inner diameter of 16 mm. The methanol feed was introduced via a saturator configuration with nitrogen as carrier and dilution gas. Isothermality and plug flow conditions were assured by having SiC particles in front of, behind and also within the catalyst bed. At the latter position, the temperature was controlled via a thermocouple. The setup also contained a pre-reactor with the same dimensions as the main reactor, but filled with 10 g alumina. At a temperature of 573 K, the equilibrated state between methanol, DME and water was reached when leaving the pre-reactor in order to be closer to industrial conditions. The GC for product analysis had an FID and one column, but could not separate side products. An internal standard was used and with the combination of an afterburner and an infrared (IR) spectroscopy, the amount of CO and CO_2 was analyzed to screen the carbon balance. More details about the setup can be found elsewhere [262]. In addition to the experiments with a pure methanol feed, $C_2^=$ to $C_4^=$ olefins were separately co-fed with methanol for mechanistic analyses, but not for extending into the model.

Reaction Network

During preliminary studies, the authors observed individual reactivities of the olefins with different carbon numbers, which is why they divided them into separate lumps. Moreover, the experiments with different methanol partial pressures showed a behavior which was not necessarily first order. Finally, the methylation reactions revealed a dependency on both methanol and olefin partial pressure. This leads to six lumps: Ox (methanol plus DME), $C_2^=$, $C_3^=$, $C_4^=$, $C_5^=$ and $C_{6+}^=$. As it is obvious, no reaction between methanol and DME is considered. Some side products such as aromatics were measured, but could not be separated from the higher olefins and are thus not included in the reaction network. Because of the relatively high minimum conversion values, no initiation phase can be observed. Scheme 17 contains three different types of reactions: conversion of oxygenates to olefins (k_1–k_4), methylation of olefins (k_5–k_7) and cracking of higher olefins to $C_3^=$ (k_8). The latter step is the only one representing olefin interconversion which means no dimerization is implemented. Ethene is a final product arising only from the oxygenates as methylation is restricted to $C_3^=$, $C_4^=$ and $C_5^=$. All steps are assumed to be irreversible. The influence of water is neglected as well as adsorption. The rate equations represent power law kinetics without any mechanistic background. From Scheme 18, it can be seen that stoichiometry is neglected for the net rates of production. The reaction orders result from a preliminary fitting, where these were also adjustable parameters and thus have no physical meaning.

In [262], an alternate reaction network can be found which includes the dimerization of $C_3^=$, $C_4^=$ and $C_5^=$ to higher olefins, but which has no improvement in describing the experimental data.

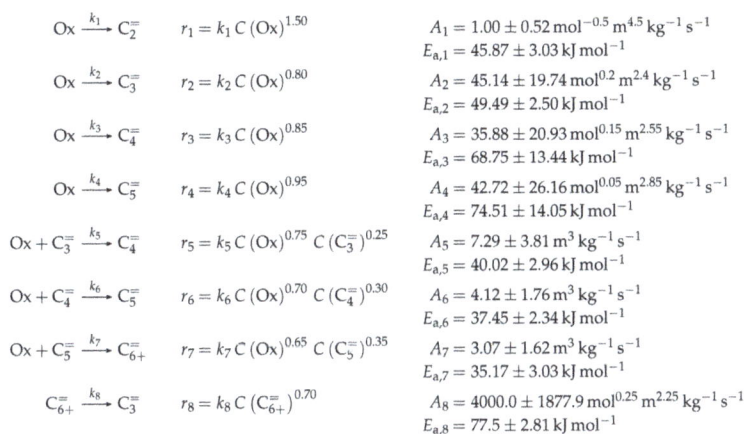

$$Ox \xrightarrow{k_1} C_2^= \qquad r_1 = k_1 C (Ox)^{1.50}$$
$$A_1 = 1.00 \pm 0.52 \, mol^{-0.5} \, m^{4.5} \, kg^{-1} \, s^{-1}$$
$$E_{a,1} = 45.87 \pm 3.03 \, kJ \, mol^{-1}$$

$$Ox \xrightarrow{k_2} C_3^= \qquad r_2 = k_2 C (Ox)^{0.80}$$
$$A_2 = 45.14 \pm 19.74 \, mol^{0.2} \, m^{2.4} \, kg^{-1} \, s^{-1}$$
$$E_{a,2} = 49.49 \pm 2.50 \, kJ \, mol^{-1}$$

$$Ox \xrightarrow{k_3} C_4^= \qquad r_3 = k_3 C (Ox)^{0.85}$$
$$A_3 = 35.88 \pm 20.93 \, mol^{0.15} \, m^{2.55} \, kg^{-1} \, s^{-1}$$
$$E_{a,3} = 68.75 \pm 13.44 \, kJ \, mol^{-1}$$

$$Ox \xrightarrow{k_4} C_5^= \qquad r_4 = k_4 C (Ox)^{0.95}$$
$$A_4 = 42.72 \pm 26.16 \, mol^{0.05} \, m^{2.85} \, kg^{-1} \, s^{-1}$$
$$E_{a,4} = 74.51 \pm 14.05 \, kJ \, mol^{-1}$$

$$Ox + C_3^= \xrightarrow{k_5} C_4^= \qquad r_5 = k_5 C (Ox)^{0.75} C (C_3^=)^{0.25}$$
$$A_5 = 7.29 \pm 3.81 \, m^3 \, kg^{-1} \, s^{-1}$$
$$E_{a,5} = 40.02 \pm 2.96 \, kJ \, mol^{-1}$$

$$Ox + C_4^= \xrightarrow{k_6} C_5^= \qquad r_6 = k_6 C (Ox)^{0.70} C (C_4^=)^{0.30}$$
$$A_6 = 4.12 \pm 1.76 \, m^3 \, kg^{-1} \, s^{-1}$$
$$E_{a,6} = 37.45 \pm 2.34 \, kJ \, mol^{-1}$$

$$Ox + C_5^= \xrightarrow{k_7} C_{6+}^= \qquad r_7 = k_7 C (Ox)^{0.65} C (C_5^=)^{0.35}$$
$$A_7 = 3.07 \pm 1.62 \, m^3 \, kg^{-1} \, s^{-1}$$
$$E_{a,7} = 35.17 \pm 3.03 \, kJ \, mol^{-1}$$

$$C_{6+}^= \xrightarrow{k_8} C_3^= \qquad r_8 = k_8 C (C_{6+}^=)^{0.70}$$
$$A_8 = 4000.0 \pm 1877.9 \, mol^{0.25} \, m^{2.25} \, kg^{-1} \, s^{-1}$$
$$E_{a,8} = 77.5 \pm 2.81 \, kJ \, mol^{-1}$$

Scheme 17. Reaction network, rate equations and estimated parameters for the model by Menges and Kraushaar-Czarnetzki [139].

The net rates of production are listed in Scheme 18; the stoichiometric coefficients are extracted from [262].

$$R (Ox) = -k_1 C (Ox)^{1.50} - k_2 C (Ox)^{0.80} - k_3 C (Ox)^{0.85} - k_4 C (Ox)^{0.95} - k_5 C (Ox)^{0.75} C (C_3^=)^{0.25}$$
$$- k_6 C (Ox)^{0.70} C (C_4^=)^{0.30} - k_7 C (Ox)^{0.65} C (C_5^=)^{0.35}$$
$$R (C_2^=) = k_1 C (Ox)^{1.50}$$
$$R (C_3^=) = k_2 C (Ox)^{0.80} + k_8 C (C_{6+}^=)^{0.70} - k_5 C (Ox)^{0.75} C (C_3^=)^{0.25}$$
$$R (C_4^=) = k_3 C (Ox)^{0.85} + k_5 C (Ox)^{0.75} C (C_3^=)^{0.25} - k_6 C (Ox)^{0.70} C (C_4^=)^{0.30}$$
$$R (C_5^=) = k_4 C (Ox)^{0.95} + k_6 C (Ox)^{0.70} C (C_4^=)^{0.30} - k_7 C (Ox)^{0.65} C (C_5^=)^{0.35}$$
$$R (C_{6+}^=) = k_7 C (Ox)^{0.65} C (C_5^=)^{0.35} - k_8 C (C_{6+}^=)^{0.70}$$

Scheme 18. Net rates of production of the different lumps for the model by Menges and Kraushaar-Czarnetzki [139].

Parameter Estimation

For the rate equations in Scheme 17, molar concentrations per volume are necessary. These are obtained via the molar flow rate of the respective compound and the current total volumetric flow rate. The differential equations are integrated with the solver *ode23s* in MATLAB, whereas *lsqnonlin* minimizes the unweighted sum of squared residuals between the molar concentrations in model and experiment with the trust-region-algorithm. For the objective function, each residual is normalized by dividing it by the respective experimental value. An Arrhenius equation that is not reparameterized is used for the rate constants, which causes 16 unknown parameters: eight preexponential factors and eight activation energies.

4.1.2. Jiang et al.: Eight-Lump Model Including Side Product Formation

Catalyst

A commercial ZSM-5 zeolite by SINOPEC with a high Si/Al ratio of 200 was used [251]. The authors specify a TOS of 2 h, however, another 50 h prereaction phase was applied to reach a stable plateau of propene yield. The authors state this should avoid any deactivation effects impeding the kinetic measurements.

Setup and Conditions

The experiments were performed in an electrically heated continuous fixed bed reactor with an inner diameter of 20 mm. Here, relatively large particles (600–900 μm) could be investigated. Four thermocouples were installed to control the temperature: three outside of the tube at the top, the middle and the bottom, and one within the catalyst bed. The methanol was provided in liquid state and pumped through a vaporizer before entering the reactor. Neither feed nor catalyst dilution is mentioned. A GC equipped with one column and an FID enabled product analysis.

Reaction Network

The model is composed of eight lumps: Ox (methanol plus DME), $C_2^=$, $C_3^=$, $C_4^=$, $C_{5+}^=$, C_1, C_2 and C_3. It should characterize the reactivity in a moving bed reactor where the catalyst slowly settles down to be regenerated at the end. Such a setup would allow the use of a methanol feed of significantly less purity and an adjustment of the catalyst to have optimum performance. The reactions in Scheme 19 can be classified into five types: conversion of oxygenates to olefins (k_2 and k_5–k_7) or to paraffins (k_1, k_3 and k_4), methylation of olefins (k_8), cracking (k_{10}) and a simplified hybrid reaction for $C_4^=$ which should resemble both methylation and dimerization (k_9). The two types of oxygenates are not differentiated. Ethene is a final product in the reaction network, thus not acting as reactant and not being methylated. Furthermore, because its formation is mechanistically separated from the other olefins, the authors omitted cracking reactions leading to $C_2^=$. As side products, small paraffins from C_1 to C_3 are included. Because of missing data points for short contact times, no initiation phase is detected. All steps are formulated as irreversible elementary reactions without any stoichiometry or adsorption effects.

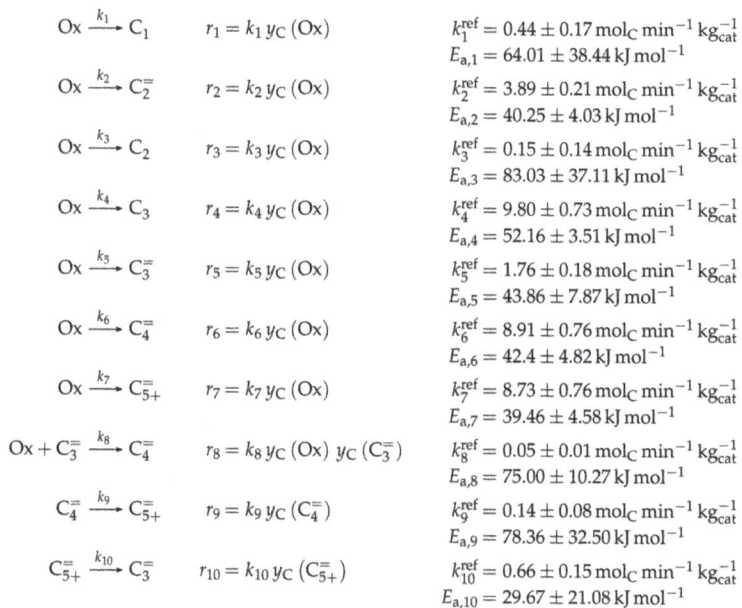

$$\text{Ox} \xrightarrow{k_1} C_1 \qquad r_1 = k_1\, y_C\,(\text{Ox}) \qquad k_1^{\text{ref}} = 0.44 \pm 0.17\, \text{mol}_C\,\text{min}^{-1}\,\text{kg}_{\text{cat}}^{-1}$$
$$E_{a,1} = 64.01 \pm 38.44\, \text{kJ}\,\text{mol}^{-1}$$

$$\text{Ox} \xrightarrow{k_2} C_2^= \qquad r_2 = k_2\, y_C\,(\text{Ox}) \qquad k_2^{\text{ref}} = 3.89 \pm 0.21\, \text{mol}_C\,\text{min}^{-1}\,\text{kg}_{\text{cat}}^{-1}$$
$$E_{a,2} = 40.25 \pm 4.03\, \text{kJ}\,\text{mol}^{-1}$$

$$\text{Ox} \xrightarrow{k_3} C_2 \qquad r_3 = k_3\, y_C\,(\text{Ox}) \qquad k_3^{\text{ref}} = 0.15 \pm 0.14\, \text{mol}_C\,\text{min}^{-1}\,\text{kg}_{\text{cat}}^{-1}$$
$$E_{a,3} = 83.03 \pm 37.11\, \text{kJ}\,\text{mol}^{-1}$$

$$\text{Ox} \xrightarrow{k_4} C_3 \qquad r_4 = k_4\, y_C\,(\text{Ox}) \qquad k_4^{\text{ref}} = 9.80 \pm 0.73\, \text{mol}_C\,\text{min}^{-1}\,\text{kg}_{\text{cat}}^{-1}$$
$$E_{a,4} = 52.16 \pm 3.51\, \text{kJ}\,\text{mol}^{-1}$$

$$\text{Ox} \xrightarrow{k_5} C_3^= \qquad r_5 = k_5\, y_C\,(\text{Ox}) \qquad k_5^{\text{ref}} = 1.76 \pm 0.18\, \text{mol}_C\,\text{min}^{-1}\,\text{kg}_{\text{cat}}^{-1}$$
$$E_{a,5} = 43.86 \pm 7.87\, \text{kJ}\,\text{mol}^{-1}$$

$$\text{Ox} \xrightarrow{k_6} C_4^= \qquad r_6 = k_6\, y_C\,(\text{Ox}) \qquad k_6^{\text{ref}} = 8.91 \pm 0.76\, \text{mol}_C\,\text{min}^{-1}\,\text{kg}_{\text{cat}}^{-1}$$
$$E_{a,6} = 42.4 \pm 4.82\, \text{kJ}\,\text{mol}^{-1}$$

$$\text{Ox} \xrightarrow{k_7} C_{5+}^= \qquad r_7 = k_7\, y_C\,(\text{Ox}) \qquad k_7^{\text{ref}} = 8.73 \pm 0.76\, \text{mol}_C\,\text{min}^{-1}\,\text{kg}_{\text{cat}}^{-1}$$
$$E_{a,7} = 39.46 \pm 4.58\, \text{kJ}\,\text{mol}^{-1}$$

$$\text{Ox} + C_3^= \xrightarrow{k_8} C_4^= \qquad r_8 = k_8\, y_C\,(\text{Ox})\, y_C\,(C_3^=) \qquad k_8^{\text{ref}} = 0.05 \pm 0.01\, \text{mol}_C\,\text{min}^{-1}\,\text{kg}_{\text{cat}}^{-1}$$
$$E_{a,8} = 75.00 \pm 10.27\, \text{kJ}\,\text{mol}^{-1}$$

$$C_4^= \xrightarrow{k_9} C_{5+}^= \qquad r_9 = k_9\, y_C\,(C_4^=) \qquad k_9^{\text{ref}} = 0.14 \pm 0.08\, \text{mol}_C\,\text{min}^{-1}\,\text{kg}_{\text{cat}}^{-1}$$
$$E_{a,9} = 78.36 \pm 32.50\, \text{kJ}\,\text{mol}^{-1}$$

$$C_{5+}^= \xrightarrow{k_{10}} C_3^= \qquad r_{10} = k_{10}\, y_C\,(C_{5+}^=) \qquad k_{10}^{\text{ref}} = 0.66 \pm 0.15\, \text{mol}_C\,\text{min}^{-1}\,\text{kg}_{\text{cat}}^{-1}$$
$$E_{a,10} = 29.67 \pm 21.08\, \text{kJ}\,\text{mol}^{-1}$$

Scheme 19. Reaction network, rate equations and estimated parameters for the model by Jiang et al. [251].

This leads to the net rates of production presented in Scheme 20.

$$R\,(\text{Ox}) = -k_1\,y_C\,(\text{Ox}) - k_2\,y_C\,(\text{Ox}) - k_3\,y_C\,(\text{Ox}) - k_4\,y_C\,(\text{Ox}) - k_5\,y_C\,(\text{Ox}) - k_6\,y_C\,(\text{Ox})$$
$$- k_7\,y_C\,(\text{Ox}) - k_8\,y_C\,(\text{Ox})\,y_C\,(C_3^=)$$
$$R\,(C_2^=) = k_2\,y_C\,(\text{Ox})$$
$$R\,(C_3^=) = k_5\,y_C\,(\text{Ox}) + k_{10}\,y_C\,(C_{5+}^=) - k_8\,y_C\,(\text{Ox})\,y_C\,(C_3^=)$$
$$R\,(C_4^=) = k_6\,y_C\,(\text{Ox}) + k_8\,y_C\,(\text{Ox})\,y_C\,(C_3^=) - k_9\,y_C\,(C_4^=)$$
$$R\,(C_{5+}^=) = k_7\,y_C\,(\text{Ox}) + k_9\,y_C\,(C_4^=) - k_{10}\,y_C\,(C_{5+}^=)$$
$$R\,(C_1) = k_1\,y_C\,(\text{Ox})$$
$$R\,(C_2) = k_3\,y_C\,(\text{Ox})$$
$$R\,(C_3) = k_4\,y_C\,(\text{Ox})$$

Scheme 20. Net rates of production of the different lumps for the model by Jiang et al. [251].

The reaction rates in Scheme 19 require mole fractions based on carbon. The objective function equals the unweighted sum of squared residuals between calculated and measured mole fractions and is minimized using *lsqnonlin* within MATLAB. The solver *ode45* is applied to the differential equations. For parameter estimation, the reparameterized Arrhenius approach of Equation (21) is used with a reference temperature of 733 K, which is the upper limit of the investigated range. Twenty parameters have to be fitted: ten reference rate constants and ten activation energies.

Parameter Estimation

4.1.3. Summary

The underlying reaction networks of both models show several similarities. Manifold pathways converting oxygenates to lower olefins and describing methylation reactions are considered. These models are fast and simple because no oxygenates interconversion is regarded. However, the different reactivity of methanol and DME cannot be expressed. Furthermore, in both studies, olefin interconversion reactions are implemented in a simplified way, meaning that dimerization reactions are missing and cracking is limited to one step. Extrapolation is difficult because of missing adsorption and mechanistic assumptions; furthermore, stoichiometry is not retained throughout the whole models. The model by Jiang et al. [251] allows a description of side product formation.

4.2. Studies with Differentiated Reactivity of Methanol and Dimethyl Ether over ZSM-5

4.2.1. Gayubo et al.: Four-Lump Approach Analyzing the Inhibiting Effect of Water Adsorption

Catalyst

In this study [114], a self-synthesized ZSM-5 zeolite with a low Si/Al ratio of 24 was mixed with bentonite and alumina. The measurements were started after a TOS of 6 h, but, similar to Epelde et al. [235], the results are extrapolated to a TOS of 0 h. This routine should yield the performance of a fresh catalyst although deactivation is significant for this system according to the authors. In a subsequent study [267], the deactivation through coke is modeled based on the kinetics presented in this section. Furthermore, in another publication [268], the authors derived a kinetic description of the irreversible deactivation caused by dealumination.

Setup and Conditions

The automated reaction equipment is described in detail in an earlier contribution [269]. It consisted of a continuous stainless steel fixed bed reactor with an inner diameter of 7 mm. This unit was surrounded by an oven and allowed measuring the temperature at three locations within the catalyst bed: close to the reactor wall, in the center and at the end of the bed. Methanol was provided in liquid state and evaporated, whereas the setup enabled the feeding of both liquid and gaseous compounds. For product analysis, the authors used a GC which had three columns and both an FID and a TCD. Additionally, the GC was coupled with a Fourier transform infrared (GC-FTIR) and a mass spectrometer (GC-MS). The catalyst bed was diluted with inert alumina. For the kinetics, the authors applied pure methanol without any dilution as feed. However, the influence of water co-feeding could be investigated in parts of the measurements: the water to methanol ratio based on weight was either zero or one.

Reaction Network

Four lumps are defined: MeOH, DME, $C_{2-3}^=$ and C_{4+}^{HC}. However, the latter is not fitted to experimental data, but calculated with the results of the remaining lumps and the conservation of mass. The assumed network can be divided into four parts (see Scheme 21): the reaction between methanol and DME (k_1), oxygenates transformation to olefins (k_2 and k_3), methylation reactions (k_5 and k_6) and olefin interconversion (k_4, k_7 and k_8). The concentrations of oxygenates are not implemented as equilibrated: the authors observed a DME amount that is much lower than the theoretical equilibrium value and attributed this to the higher reactivity of DME. This is why the reaction to DME is implemented as kinetic step with the backward reaction being expressed via a thermodynamic equilibrium constant. For the latter, the empiric correlation by Spivey [117] is used, which itself is a citation from Hayashi and Moffat [110]. However, in Gayubo et al.'s publication [114], different numeric values for this correlation are used (see Section 2.1). All other steps in the reaction network are assumed to be irreversible. No initiation phase can be observed during the measurements. The methylation is implemented both via methanol and via DME to account for the different reactivities. In addition, the conversion to olefins can start from both types of oxygenates according to the model. The olefin interconversion is limited to one cracking step and two reactions to higher compounds. It can be seen in Scheme 21 that the reaction rates are expressed as HW type of mechanism. Only the adsorption of water is considered because this step is significantly slower compared to the hydrocarbons adsorption and not *quasi*-equilibrated according to the authors. Side products are not explicitly mentioned, but especially aromatics should be included in the lump C_{4+}^{HC} as it was done in an earlier model [270]. The reactions are defined as elementary steps, but no stoichiometry is considered.

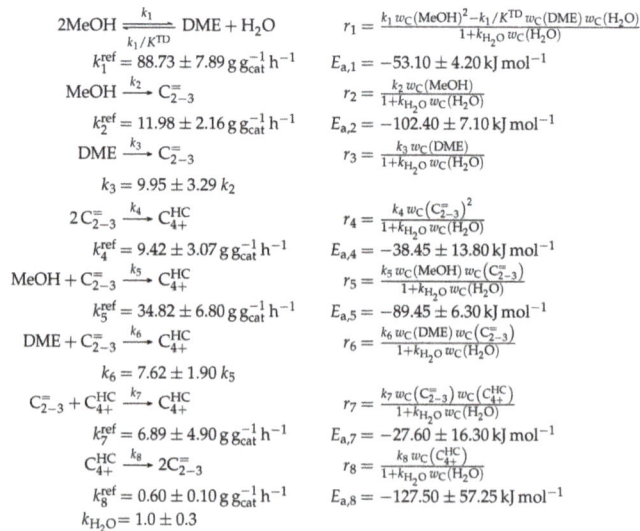

$$2\text{MeOH} \underset{k_1/K^{TD}}{\overset{k_1}{\rightleftharpoons}} \text{DME} + \text{H}_2\text{O} \qquad r_1 = \frac{k_1 w_C(\text{MeOH})^2 - k_1/K^{TD} w_C(\text{DME}) w_C(\text{H}_2\text{O})}{1 + k_{\text{H}_2\text{O}} w_C(\text{H}_2\text{O})}$$

$$k_1^{\text{ref}} = 88.73 \pm 7.89 \,\text{g}\,\text{g}_{\text{cat}}^{-1}\,\text{h}^{-1} \qquad E_{a,1} = -53.10 \pm 4.20 \,\text{kJ}\,\text{mol}^{-1}$$

$$\text{MeOH} \xrightarrow{k_2} C_{2-3}^= \qquad r_2 = \frac{k_2 w_C(\text{MeOH})}{1 + k_{\text{H}_2\text{O}} w_C(\text{H}_2\text{O})}$$

$$k_2^{\text{ref}} = 11.98 \pm 2.16 \,\text{g}\,\text{g}_{\text{cat}}^{-1}\,\text{h}^{-1} \qquad E_{a,2} = -102.40 \pm 7.10 \,\text{kJ}\,\text{mol}^{-1}$$

$$\text{DME} \xrightarrow{k_3} C_{2-3}^= \qquad r_3 = \frac{k_3 w_C(\text{DME})}{1 + k_{\text{H}_2\text{O}} w_C(\text{H}_2\text{O})}$$

$$k_3 = 9.95 \pm 3.29 \,k_2$$

$$2 C_{2-3}^= \xrightarrow{k_4} C_{4+}^{HC} \qquad r_4 = \frac{k_4 w_C(C_{2-3}^=)^2}{1 + k_{\text{H}_2\text{O}} w_C(\text{H}_2\text{O})}$$

$$k_4^{\text{ref}} = 9.42 \pm 3.07 \,\text{g}\,\text{g}_{\text{cat}}^{-1}\,\text{h}^{-1} \qquad E_{a,4} = -38.45 \pm 13.80 \,\text{kJ}\,\text{mol}^{-1}$$

$$\text{MeOH} + C_{2-3}^= \xrightarrow{k_5} C_{4+}^{HC} \qquad r_5 = \frac{k_5 w_C(\text{MeOH}) w_C(C_{2-3}^=)}{1 + k_{\text{H}_2\text{O}} w_C(\text{H}_2\text{O})}$$

$$k_5^{\text{ref}} = 34.82 \pm 6.80 \,\text{g}\,\text{g}_{\text{cat}}^{-1}\,\text{h}^{-1} \qquad E_{a,5} = -89.45 \pm 6.30 \,\text{kJ}\,\text{mol}^{-1}$$

$$\text{DME} + C_{2-3}^= \xrightarrow{k_6} C_{4+}^{HC} \qquad r_6 = \frac{k_6 w_C(\text{DME}) w_C(C_{2-3}^=)}{1 + k_{\text{H}_2\text{O}} w_C(\text{H}_2\text{O})}$$

$$k_6 = 7.62 \pm 1.90 \,k_5$$

$$C_{2-3}^= + C_{4+}^{HC} \xrightarrow{k_7} C_{4+}^{HC} \qquad r_7 = \frac{k_7 w_C(C_{2-3}^=) w_C(C_{4+}^{HC})}{1 + k_{\text{H}_2\text{O}} w_C(\text{H}_2\text{O})}$$

$$k_7^{\text{ref}} = 6.89 \pm 4.90 \,\text{g}\,\text{g}_{\text{cat}}^{-1}\,\text{h}^{-1} \qquad E_{a,7} = -27.60 \pm 16.30 \,\text{kJ}\,\text{mol}^{-1}$$

$$C_{4+}^{HC} \xrightarrow{k_8} 2 C_{2-3}^= \qquad r_8 = \frac{k_8 w_C(C_{4+}^{HC})}{1 + k_{\text{H}_2\text{O}} w_C(\text{H}_2\text{O})}$$

$$k_8^{\text{ref}} = 0.60 \pm 0.10 \,\text{g}\,\text{g}_{\text{cat}}^{-1}\,\text{h}^{-1} \qquad E_{a,8} = -127.50 \pm 57.25 \,\text{kJ}\,\text{mol}^{-1}$$

$$k_{\text{H}_2\text{O}} = 1.0 \pm 0.3$$

Scheme 21. Reaction network, rate equations and estimated parameters for the model by Gayubo et al. [114] over ZSM-5; K^{TD} is calculated with a modified version of Hayashi and Moffat's correlation [110] (see Section 2.1).

The net rates of production for all four lumps are presented in Scheme 22.

$$R\,(\text{MeOH}) = \frac{k_1/K^{\text{TD}} w_C\,(\text{DME})\,w_C\,(\text{H}_2\text{O}) - k_1\,w_C\,(\text{MeOH})^2 - k_2\,w_C\,(\text{MeOH})}{1 + k_{\text{H}_2\text{O}}\,w_C\,(\text{H}_2\text{O})}$$

$$\frac{-k_5\,w_C\,(\text{MeOH})\,w_C\,(\text{C}_{2-3}^{=})}{1 + k_{\text{H}_2\text{O}}\,w_C\,(\text{H}_2\text{O})}$$

$$R\,(\text{DME}) = \frac{k_1\,w_C\,(\text{MeOH})^2 - k_1/K^{\text{TD}}\,w_C\,(\text{DME})\,w_C\,(\text{H}_2\text{O}) - k_3\,w_C\,(\text{DME})}{1 + k_{\text{H}_2\text{O}}\,w_C\,(\text{H}_2\text{O})}$$

$$\frac{-k_6\,w_C\,(\text{DME})\,w_C\,(\text{C}_{2-3}^{=})}{1 + k_{\text{H}_2\text{O}}\,w_C\,(\text{H}_2\text{O})}$$

$$R\,(\text{C}_{2-3}^{=}) = \frac{k_2\,w_C\,(\text{MeOH}) + k_3\,w_C\,(\text{DME}) - k_4\,w_C\,(\text{C}_{2-3}^{=})^2 - k_5\,w_C\,(\text{MeOH})\,w_C\,(\text{C}_{2-3}^{=})}{1 + k_{\text{H}_2\text{O}}\,w_C\,(\text{H}_2\text{O})}$$

$$\frac{-k_6\,w_C\,(\text{DME})\,w_C\,(\text{C}_{2-3}^{=}) - k_7\,w_C\,(\text{C}_{2-3}^{=})\,w_C\,(\text{C}_{4+}^{\text{HC}}) + k_8\,w_C\,(\text{C}_{4+}^{\text{HC}})}{1 + k_{\text{H}_2\text{O}}\,w_C\,(\text{H}_2\text{O})}$$

$$R\left(\text{C}_{4+}^{\text{HC}}\right) = \frac{k_4\,w_C\,(\text{C}_{2-3}^{=})^2 + k_5\,w_C\,(\text{MeOH})\,w_C\,(\text{C}_{2-3}^{=}) + k_6\,w_C\,(\text{DME})\,w_C\,(\text{C}_{2-3}^{=})}{1 + k_{\text{H}_2\text{O}}\,w_C\,(\text{H}_2\text{O})}$$

$$\frac{+k_7\,w_C\,(\text{C}_{2-3}^{=})\,w_C\,(\text{C}_{4+}^{\text{HC}}) - k_8\,w_C\,(\text{C}_{4+}^{\text{HC}})}{1 + k_{\text{H}_2\text{O}}\,w_C\,(\text{H}_2\text{O})}$$

Scheme 22. Net rates of production of the different lumps for the model by Gayubo et al. [114] over ZSM-5; the $\text{C}_{4+}^{\text{HC}}$ lump is calculated via conservation of mass within the model.

Parameter Estimation

The reaction rates in Scheme 21 are defined with mass fractions of organic compounds where water is explicitly excluded. Even the mass fraction of water is related to the water-free composition. Integration of the differential equations is performed with a code written in FORTRAN which makes use of the DGEAR subroutine of the IMSL library. The objective function returns the sum of squared residuals between modeled and experimental organic mass fractions and is additionally divided by the number of lumps and experiments. This average value is then minimized with the Complex algorithm, as explained in earlier work [271]. After obtaining the parameters of best description, another fitting is performed with the Marquardt algorithm. Reparameterization according to Equation (21) is applied with a reference temperature of 673 K, which is in the upper third of the investigated range. For the DME transformation to olefins and the methylation via DME, the same activation energies as for the respective methanol related steps are assumed, which only requires the fitting of a separate preexponential factor. In total, 15 parameters are estimated: eight reference rate constants, six activation energies and one rate constant for water adsorption.

4.2.2. Aguayo et al.: Seven-Lump Model for Significant Side Product Formation and Resulting Interconversion Reactions

Catalyst

The authors [252] used a commercial ZSM-5 zeolite from Zeolyst International with a low Si/Al ratio of 30 which they further processed to an extrudate. The resulting catalyst showed sufficient activity at 0 h TOS during earlier studies [272], combined with high olefin selectivity, low coke amounts and increased hydrothermal stability. It could be shown that up to ten reaction–regeneration cycles without irreversible deactivation were possible with this catalytic system. The measurements were performed at a TOS of 5 h, but, although it is not explicitly mentioned, the results might be extrapolated to 0 h.

Setup and Conditions

A continuous fixed bed reactor made of stainless steel with an inner diameter of 9 mm and surrounded by a heated steel chamber with a ceramic cover was applied. The temperature was measured inside the catalyst bed and also within the chamber and at the transfer line connecting the GC. The setup allowed providing methanol in liquid state which was evaporated before being fed to the reactor. More details about the reaction equipment which has many similarities to Epelde et al. [235] are shown in earlier work [272]. The catalyst bed was diluted with SiC in a way that the bed height remained almost constant. The authors analyzed the products with a micro GC equipped with three columns and a TCD.

Reaction Network

In this model, seven lumps are defined: MeOH, DME, $C_{2-4}^=$, C_{5+}^{HC}, C_1, C_{2-4} and C_{n4}. The lump C_{5+}^{HC} comprises C_{6-8} aromatics as well as C_{5-10} aliphatics. In Scheme 23, eight different types of reactions can be identified: the one between methanol and DME (k_1), the conversion of oxygenates to olefins (k_2 and k_3), methylations (k_5 and k_6), olefin interconversion (k_9), olefin–paraffin interconversion (k_7 and k_8), paraffin formation through oxygenates (k_4) or olefins (k_{10} and k_{11}) and aromatization steps (k_{12} and k_{13}). A separate consideration of *n*-butane is performed because it was co-fed with methanol in an earlier study [272]. All reaction steps except the one between methanol and DME are treated as irreversible. Because the authors performed measurements at relatively short contact times, the initiation phase is clearly visible which means that no detectable conversion to hydrocarbons and only oxygenates equilibration takes place. Nevertheless, the methanol dehydration is implemented as step of kinetic relevance. Its backward reaction is expressed via a thermodynamic equilibrium constant which is calculated with an own correlation derived in another publication [109] (see Section 2.1). As in Gayubo et al.'s model [114] (Section 4.2.1), both methanol and DME can perform methylation reactions, which yields not only higher hydrocarbons, but also lower olefins. Both types of oxygenates can be converted to olefins or to methane; in the latter reaction, no differentiation between the reactivity of methanol and DME is performed. The olefin interconversion is restricted to one cracking step, whereas dimerization is neglected. Instead, several reactions starting with or leading to paraffins are implemented. The interaction between the formed water and the zeolite is not considered. Neither a mechanistic model nor any adsorption effects are included. The reactions are assumed to be elementary, except for steps k_{10} and k_{11} which are arbitrarily set to second order because of a better agreement with experimental data. Stoichiometry is not considered for the net rates of production; as for the second order reactions in Steps 10 and 11, the reaction rates of Steps 12 and 13 are arbitrarily multiplied by 2.

Scheme 24 contains the net rates of production of the different lumps.

Parameter Estimation

Whereas the mole fractions shown in the figures of Aguayo et al. [252] are defined with carbon units, regular mole fractions including water have to be inserted for the rate equations in Scheme 23. The kinetic parameters are obtained via multivariable nonlinear regression using MATLAB. The objective function returns the weighted sum of squared residuals between modeled and measured output, see Section 3.1.1 and an earlier publication [273] for details. Reparameterization according to Equation (21) is performed with a reference temperature of 773 K which is in the upper third of the investigated range. Finally, 26 parameters have to be fitted: 13 reference rate constants and 13 activation energies.

$$2\text{MeOH} \xrightleftharpoons[k_1/K^{\text{TD}}]{k_1} \text{DME} + \text{H}_2\text{O} \qquad r_1 = k_1\,y\,(\text{MeOH})^2 - k_1/K^{\text{TD}}\,y\,(\text{DME})\,y\,(\text{H}_2\text{O})$$

$$k_1^{\text{ref}} = 24.5 \pm 2.4\,\text{mol}_\text{C}\,\text{g}_\text{cat}^{-1}\,\text{h}^{-1} \qquad E_{\text{a},1} = 54.4 \pm 5.4\,\text{kJ}\,\text{mol}^{-1}$$

$$\text{MeOH} \xrightarrow{k_2} \text{C}_{2-4}^{=} \qquad r_2 = k_2\,y\,(\text{MeOH})$$

$$k_2^{\text{ref}} = (1.2 \pm 0.1) \times 10^{-6}\,\text{mol}_\text{C}\,\text{g}_\text{cat}^{-1}\,\text{h}^{-1} \qquad E_{\text{a},2} = 4.60 \pm 0.41\,\text{kJ}\,\text{mol}^{-1}$$

$$\text{DME} \xrightarrow{k_3} \text{C}_{2-4}^{=} \qquad r_3 = k_3\,y\,(\text{DME})$$

$$k_3^{\text{ref}} = (1.9 \pm 0.2) \times 10^{-6}\,\text{mol}_\text{C}\,\text{g}_\text{cat}^{-1}\,\text{h}^{-1} \qquad E_{\text{a},3} = 186 \pm 22\,\text{kJ}\,\text{mol}^{-1}$$

$$\text{MeOH/DME} \xrightarrow{k_4} \text{C}_1 \qquad r_4 = k_4\,y\,(\text{MeOH})\,/\,r_4 = k_4\,y\,(\text{DME})$$

$$k_4^{\text{ref}} = (19.9 \pm 2.6) \times 10^{-2}\,\text{mol}_\text{C}\,\text{g}_\text{cat}^{-1}\,\text{h}^{-1} \qquad E_{\text{a},4} = 111 \pm 14\,\text{kJ}\,\text{mol}^{-1}$$

$$\text{MeOH} + \text{C}_{2-4}^{=} \xrightarrow{k_5} \text{C}_{2-4}^{=} \qquad r_5 = k_5\,y\,(\text{MeOH})\,y\,(\text{C}_{2-4}^{=})$$

$$k_5^{\text{ref}} = 88.3 \pm 8.8\,\text{mol}_\text{C}\,\text{g}_\text{cat}^{-1}\,\text{h}^{-1} \qquad E_{\text{a},5} = 4.44 \pm 0.45\,\text{kJ}\,\text{mol}^{-1}$$

$$\text{DME} + \text{C}_{2-4}^{=} \xrightarrow{k_6} \text{C}_{2-4}^{=} \qquad r_6 = k_6\,y\,(\text{DME})\,y\,(\text{C}_{2-4}^{=})$$

$$k_6^{\text{ref}} = 331 \pm 23\,\text{mol}_\text{C}\,\text{g}_\text{cat}^{-1}\,\text{h}^{-1} \qquad E_{\text{a},6} = 53.6 \pm 3.7\,\text{kJ}\,\text{mol}^{-1}$$

$$\text{C}_{n4} + \text{C}_{2-4}^{=} \xrightarrow{k_7} \text{C}_{2-4}^{=} \qquad r_7 = k_7\,y\,(\text{C}_{n4})\,y\,(\text{C}_{2-4}^{=})$$

$$k_7^{\text{ref}} = 342 \pm 17\,\text{mol}_\text{C}\,\text{g}_\text{cat}^{-1}\,\text{h}^{-1} \qquad E_{\text{a},7} = 14.1 \pm 0.7\,\text{kJ}\,\text{mol}^{-1}$$

$$\text{C}_{2-4} + \text{C}_{2-4}^{=} \xrightarrow{k_8} \text{C}_{2-4}^{=} \qquad r_8 = k_8\,y\,(\text{C}_{2-4})\,y\,(\text{C}_{2-4}^{=})$$

$$k_8^{\text{ref}} = 228 \pm 23\,\text{mol}_\text{C}\,\text{g}_\text{cat}^{-1}\,\text{h}^{-1} \qquad E_{\text{a},8} = 56.5 \pm 5.6\,\text{kJ}\,\text{mol}^{-1}$$

$$\text{C}_{5+}^{\text{HC}} \xrightarrow{k_9} \text{C}_{2-4}^{=} \qquad r_9 = k_9\,y\,(\text{C}_{5+}^{\text{HC}})$$

$$k_9^{\text{ref}} = (7.2 \pm 0.5) \times 10^{-2}\,\text{mol}_\text{C}\,\text{g}_\text{cat}^{-1}\,\text{h}^{-1} \qquad E_{\text{a},9} = 25.0 \pm 2.1\,\text{kJ}\,\text{mol}^{-1}$$

$$\text{C}_{2-4}^{=} \xrightarrow{k_{10}} \text{C}_{n4} \qquad r_{10} = k_{10}\,y\,(\text{C}_{2-4}^{=})^2$$

$$k_{10}^{\text{ref}} = 35.5 \pm 3.6\,\text{mol}_\text{C}\,\text{g}_\text{cat}^{-1}\,\text{h}^{-1} \qquad E_{\text{a},10} = 4.35 \pm 0.43\,\text{kJ}\,\text{mol}^{-1}$$

$$\text{C}_{2-4}^{=} \xrightarrow{k_{11}} \text{C}_{2-4} \qquad r_{11} = k_{11}\,y\,(\text{C}_{2-4}^{=})^2$$

$$k_{11}^{\text{ref}} = 111 \pm 12\,\text{mol}_\text{C}\,\text{g}_\text{cat}^{-1}\,\text{h}^{-1} \qquad E_{\text{a},11} = 5.02 \pm 0.55\,\text{kJ}\,\text{mol}^{-1}$$

$$\text{MeOH} + \text{C}_{2-4}^{=} \xrightarrow{k_{12}} \text{C}_{5+}^{\text{HC}} \qquad r_{12} = k_{12}\,y\,(\text{MeOH})\,y\,(\text{C}_{2-4}^{=})$$

$$k_{12}^{\text{ref}} = 10.8 \pm 1.6\,\text{mol}_\text{C}\,\text{g}_\text{cat}^{-1}\,\text{h}^{-1} \qquad E_{\text{a},12} = 16.2 \pm 2.4\,\text{kJ}\,\text{mol}^{-1}$$

$$\text{DME} + \text{C}_{2-4}^{=} \xrightarrow{k_{13}} \text{C}_{5+}^{\text{HC}} \qquad r_{13} = k_{13}\,y\,(\text{DME})\,y\,(\text{C}_{2-4}^{=})$$

$$k_{13}^{\text{ref}} = 60.2 \pm 8.4\,\text{mol}_\text{C}\,\text{g}_\text{cat}^{-1}\,\text{h}^{-1} \qquad E_{\text{a},13} = 36.4 \pm 5.1\,\text{kJ}\,\text{mol}^{-1}$$

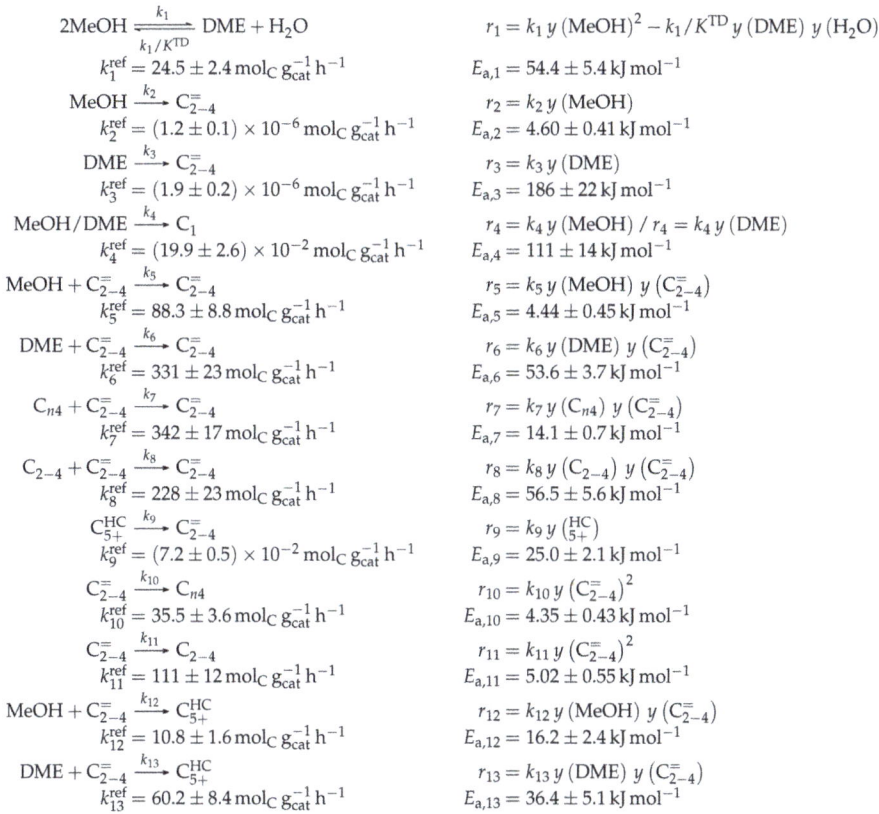

Scheme 23. Reaction network, rate equations and estimated parameters for the model by Aguayo et al. [252]; K^{TD} is calculated with an own correlation [109] (see Section 2.1).

$$R\,(\text{MeOH}) = k_1/K^{\text{TD}}\,y\,(\text{DME})\,y\,(\text{H}_2\text{O}) - k_1\,y\,(\text{MeOH})^2 - k_2\,y\,(\text{MeOH}) - k_4\,y\,(\text{MeOH})$$
$$- k_5\,y\,(\text{MeOH})\,y\,(\text{C}_{2-4}^{=}) - k_{12}\,y\,(\text{MeOH})\,y\,(\text{C}_{2-4}^{=})$$

$$R\,(\text{DME}) = k_1\,y\,(\text{MeOH})^2 - k_1/K^{\text{TD}}\,y\,(\text{DME})\,y\,(\text{H}_2\text{O}) - k_3\,y\,(\text{DME}) - k_4\,y\,(\text{DME})$$
$$- k_6\,y\,(\text{DME})\,y\,(\text{C}_{2-4}^{=}) - k_{13}\,y\,(\text{DME})\,y\,(\text{C}_{2-4}^{=})$$

$$R\,(\text{C}_{2-4}^{=}) = k_2\,y\,(\text{MeOH}) + k_3\,y\,(\text{DME}) + k_5\,y\,(\text{MeOH})\,y\,(\text{C}_{2-4}^{=}) + k_6\,y\,(\text{DME})\,y\,(\text{C}_{2-4}^{=})$$
$$+ k_7\,y\,(\text{C}_{n4})\,y\,(\text{C}_{2-4}^{=}) + k_8\,y\,(\text{C}_{2-4})\,y\,(\text{C}_{2-4}^{=}) + k_9\,y\,(\text{C}_{5+}^{\text{HC}}) - k_{10}\,y\,(\text{C}_{2-4}^{=})^2$$
$$- k_{11}\,y\,(\text{C}_{2-4}^{=})^2 - k_{12}\,y\,(\text{MeOH})\,y\,(\text{C}_{2-4}^{=}) - k_{13}\,y\,(\text{DME})\,y\,(\text{C}_{2-4}^{=})$$

$$R\left(\text{C}_{5+}^{\text{HC}}\right) = 2\,k_{12}\,y\,(\text{MeOH})\,y\,(\text{C}_{2-4}^{=}) + 2\,k_{13}\,y\,(\text{DME})\,y\,(\text{C}_{2-4}^{=}) - k_9\,y\left(\text{C}_{5+}^{\text{HC}}\right)$$

$$R\,(\text{C}_1) = k_4\,y\,(\text{MeOH}) + k_4\,y\,(\text{DME})$$

$$R\,(\text{C}_{2-4}) = k_{11}\,y\,(\text{C}_{2-4}^{=})^2 - k_8\,y\,(\text{C}_{2-4})\,y\,(\text{C}_{2-4}^{=})$$

$$R\,(\text{C}_{n4}) = k_{10}\,y\,(\text{C}_{2-4}^{=})^2 - k_7\,y\,(\text{C}_{n4})\,y\,(\text{C}_{2-4}^{=})$$

Scheme 24. Net rates of production of the different lumps for the model by Aguayo et al. [252].

4.2.3. Pérez-Uriarte et al.: Eleven-Lump Approach for Dimethyl Ether Feeds

Catalyst

A central difference of this study [253] is the use of DME as feed which changes oxygenates conversion, product selectivities and deactivation rates and therefore requires different conditions and kinetic models [274]. The authors extruded a commercial high-silica (Si/Al = 280) ZSM-5 catalyst from Zeolyst International with boehmite from Sasol as binder and with α-alumina as inert filler. This composition showed a satisfying compromise between activity, stability and mechanical resistance in earlier work [275] through moderate acidity, a mesoporous structure and additional acid sites through γ-alumina, which is a calcination product of boehmite. The measurements were performed at 0.17 h TOS, but extrapolated to 0 h in order to represent the fresh catalyst. Deactivation through coke should be higher compared to methanol feeds according to the authors because of the lower water content. These effects are ignored in this study, but considered in a subsequent kinetic model of deactivation [276].

Setup and Conditions

The setup was almost identical to the studies by Aguayo et al. [252] and Epelde et al. [235] (see Sections 3.1.1 and 4.2.2 as well as [275] for more details). Additionally, liquid components could be fed by pumping them through an evaporator. Besides the experiments with pure DME as feed, some other combinations with DME/helium, DME/methanol and DME/water mixtures were investigated and are also included in the model. The catalyst bed was diluted with SiC particles to reach a uniform height of 50 mm. A GC with four columns and a TCD enabled product analysis.

Reaction Network

Eleven lumps are defined: MeOH, DME, $C_2^=$, $C_3^=$, $C_4^=$, $C_{5+}^=$, C_1, C_{2-4}, C_{6-8}^{ar}, CO and H_2O. For parameter estimation, the amount of water is not fitted to the experimental data, thereby reducing the number of responses in Table 4 to ten. The steps in Scheme 25 can be summarized to nine sections: DME to lower olefins (k_1), the reaction between DME and methanol (k_2), methanol to lower olefins (k_3), reactions of lower olefins with DME (k_4–k_6) or with methanol (k_7–k_9), conversion of lower to higher olefins (k_{10}, $k_{10'}$ and $k_{10''}$), formation of BTX aromatics and lower paraffins out of higher (k_{11}) or lower olefins (k_{12}, $k_{12'}$ and $k_{12''}$) and DME cracking to CO and methane (k_{13}). It follows that the differing reactivity of methanol and DME is considered; both can react either to or with lower olefins. The mechanism of the latter step is not resolved, but it is different to the methylations postulated in other models. For the oxygenates interconversion, no instant equilibrium is assumed, causing a kinetic rate constant; for the backward reaction, the same equilibrium constant as for Aguayo et al. [252] is assumed. In the experimental data, no initiation phase can be observed, which might be either due to relatively high minimum contact times or due to the higher reactivity of DME. The olefin interconversion includes only dimerization, no cracking is included. The different reactivity for ethene and butenes compared to propene is accounted for via multiplying the rate constant for propene with a specific factor. The same is done for the side product formation out of these olefins. Besides C_{2-4} paraffins and BTX aromatics, methane and CO are implemented, both being directly produced out of DME. All steps are formulated as elementary reactions, but with partially deviating reaction orders and except for the reaction between the oxygenates, all steps are irreversible. For the HW type of mechanism, the adsorption of methanol and water is considered with a common equilibrium constant K_{MW}^{ads}. Arbitrary values are used for the stoichiometric coefficients (see Scheme 26). The reaction network is compared with two other versions where Steps 4–9 are summarized to two reactions or where more olefin interconversion steps are implemented [253]; however, statistical evaluation proves that no improvement is achieved with these two variations.

$$\text{DME} \xrightarrow{k_1} C_2^= + C_3^= + C_4^= + H_2O$$

$$k_1^{\text{ref}} = (4.99 \pm 0.01) \times 10^{-2}\,\text{mol}_C\,\text{g}_{\text{cat}}^{-1}\,\text{h}^{-1}\,\text{atm}^{-1}$$

$$\text{DME} + H_2O \underset{k_2/K^{\text{TD}}}{\overset{k_2}{\rightleftharpoons}} 2\text{MeOH}$$

$$num = k_2\, p_C(\text{DME})\, p_C(H_2O) - k_2/K^{\text{TD}}\, p_C(\text{MeOH})^2$$

$$k_2^{\text{ref}} = (7.88 \pm 0.50) \times 10^1\,\text{mol}_C\,\text{g}_{\text{cat}}^{-1}\,\text{h}^{-1}\,\text{atm}^{-2}$$

$$\text{MeOH} \xrightarrow{k_3} C_2^= + C_3^= + C_4^= + H_2O$$

$$k_3^{\text{ref}} = (2.42 \pm 0.15) \times 10^{-3}\,\text{mol}_C\,\text{g}_{\text{cat}}^{-1}\,\text{h}^{-1}\,\text{atm}^{-1}$$

$$C_2^= + \text{DME} \xrightarrow{k_4} 2C_2^= + H_2O$$

$$k_4^{\text{ref}} = (2.48 \pm 0.34) \times 10^{-1}\,\text{mol}_C\,\text{g}_{\text{cat}}^{-1}\,\text{h}^{-1}\,\text{atm}^{-2}$$

$$2C_3^= + 3\text{DME} \xrightarrow{k_5} 4C_3^= + 3H_2O$$

$$k_5^{\text{ref}} = 2.54 \pm 0.01\,\text{mol}_C\,\text{g}_{\text{cat}}^{-1}\,\text{h}^{-1}\,\text{atm}^{-2}$$

$$C_4^= + 2\text{DME} \xrightarrow{k_6} 2C_4^= + 4H_2O$$

$$k_6^{\text{ref}} = (1.44 \pm 0.88) \times 10^{-1}\,\text{mol}_C\,\text{g}_{\text{cat}}^{-1}\,\text{h}^{-1}\,\text{atm}^{-2}$$

$$C_2^= + 2\text{MeOH} \xrightarrow{k_7} 2C_2^= + 2H_2O$$

$$k_7^{\text{ref}} = (3.02 \pm 0.05) \times 10^1\,\text{mol}_C\,\text{g}_{\text{cat}}^{-1}\,\text{h}^{-1}\,\text{atm}^{-2}$$

$$C_3^= + 3\text{MeOH} \xrightarrow{k_8} 2C_3^= + 3H_2O$$

$$k_8^{\text{ref}} = 2.63 \pm 0.09\,\text{mol}_C\,\text{g}_{\text{cat}}^{-1}\,\text{h}^{-1}\,\text{atm}^{-2}$$

$$C_4^= + 4\text{MeOH} \xrightarrow{k_9} 2C_4^= + 4H_2O$$

$$k_9^{\text{ref}} = (4.24 \pm 1.14) \times 10^{-1}\,\text{mol}_C\,\text{g}_{\text{cat}}^{-1}\,\text{h}^{-1}\,\text{atm}^{-2}$$

$$C_3^= \xrightarrow{k_{10}} C_{5+}^=$$

$$k_{10}^{\text{ref}} = 1.03 \pm 0.01\,\text{mol}_C\,\text{g}_{\text{cat}}^{-1}\,\text{h}^{-1}\,\text{atm}^{-1}$$

$$C_2^= \xrightarrow{k_{10'}} C_{5+}^=$$

$$k_{10'} = 1.78 \pm 0.02\, k_{10}$$

$$C_4^= \xrightarrow{k_{10''}} C_{5+}^=$$

$$k_{10''} = (6.92 \pm 0.20) \times 10^{-1}\, k_{10}$$

$$C_{5+}^= \xrightarrow{k_{11}} C_{2-4}^= + C_{6-8}^{\text{ar}}$$

$$k_{11}^{\text{ref}} = (1.16 \pm 2.54) \times 10^{-2}\,\text{mol}_C\,\text{g}_{\text{cat}}^{-1}\,\text{h}^{-1}\,\text{atm}^{-1}$$

$$C_3^= \xrightarrow{k_{12}} C_{2-4}^= + C_{6-8}^{\text{ar}}$$

$$k_{12}^{\text{ref}} = (3.10 \pm 0.02) \times 10^{-1}\,\text{mol}_C\,\text{g}_{\text{cat}}^{-1}\,\text{h}^{-1}\,\text{atm}^{-1}$$

$$C_2^= \xrightarrow{k_{12'}} C_{2-4}^= + C_{6-8}^{\text{ar}}$$

$$k_{12'} = 1.78 \pm 0.02\, k_{12}$$

$$C_4^= \xrightarrow{k_{12''}} C_{2-4}^= + C_{6-8}^{\text{ar}}$$

$$k_{12''} = (6.92 \pm 0.20) \times 10^{-1}\, k_{15}$$

$$\text{DME} \xrightarrow{k_{13}} C_1 + CO$$

$$k_{13}^{\text{ref}} = (5.63 \pm 1.10) \times 10^{-4}\,\text{mol}_C\,\text{g}_{\text{cat}}^{-1}\,\text{h}^{-1}\,\text{atm}^{-1}$$

$$K_{\text{MW}}^{\text{ads,ref}} = (1.27 \pm 0.01) \times 10^1\,\text{atm}^{-1}$$

$$r_1 = \frac{k_1\, p_C(\text{DME})}{1 + K_{\text{MW}}^{\text{ads}}\,(p_C(\text{MeOH}) + p_C(H_2O))}$$

$$E_{a,1} = 41.5 \pm 0.3\,\text{kJ}\,\text{mol}^{-1}$$

$$r_2 = \frac{num}{1 + K_{\text{MW}}^{\text{ads}}\,(p_C(\text{MeOH}) + p_C(H_2O))}$$

$$E_{a,2} = 11.9 \pm 0.7\,\text{kJ}\,\text{mol}^{-1}$$

$$r_3 = \frac{k_3\, p_C(\text{MeOH})}{1 + K_{\text{MW}}^{\text{ads}}\,(p_C(\text{MeOH}) + p_C(H_2O))}$$

$$E_{a,3} = 33.8 \pm 0.2\,\text{kJ}\,\text{mol}^{-1}$$

$$r_4 = \frac{k_4\, p_C(\text{DME})\, p_C(C_2^=)}{1 + K_{\text{MW}}^{\text{ads}}\,(p_C(\text{MeOH}) + p_C(H_2O))}$$

$$E_{a,4} = 17.2 \pm 0.2\,\text{kJ}\,\text{mol}^{-1}$$

$$r_5 = \frac{k_5\, p_C(\text{DME})\, p_C(C_3^=)}{1 + K_{\text{MW}}^{\text{ads}}\,(p_C(\text{MeOH}) + p_C(H_2O))}$$

$$E_{a,5} = 25.7 \pm 0.3\,\text{kJ}\,\text{mol}^{-1}$$

$$r_6 = \frac{k_6\, p_C(\text{DME})\, p_C(C_4^=)}{1 + K_{\text{MW}}^{\text{ads}}\,(p_C(\text{MeOH}) + p_C(H_2O))}$$

$$E_{a,6} = 9.8 \pm 0.7\,\text{kJ}\,\text{mol}^{-1}$$

$$r_7 = \frac{k_7\, p_C(\text{MeOH})\, p_C(C_2^=)}{1 + K_{\text{MW}}^{\text{ads}}\,(p_C(\text{MeOH}) + p_C(H_2O))}$$

$$E_{a,7} = 16.3 \pm 0.1\,\text{kJ}\,\text{mol}^{-1}$$

$$r_8 = \frac{k_8\, p_C(\text{MeOH})\, p_C(C_3^=)}{1 + K_{\text{MW}}^{\text{ads}}\,(p_C(\text{MeOH}) + p_C(H_2O))}$$

$$E_{a,8} = 16.9 \pm 0.3\,\text{kJ}\,\text{mol}^{-1}$$

$$r_9 = \frac{k_9\, p_C(\text{MeOH})\, p_C(C_4^=)}{1 + K_{\text{MW}}^{\text{ads}}\,(p_C(\text{MeOH}) + p_C(H_2O))}$$

$$E_{a,9} = 69.0 \pm 0.3\,\text{kJ}\,\text{mol}^{-1}$$

$$r_{10} = \frac{k_{10}\, p_C(C_3^=)}{1 + K_{\text{MW}}^{\text{ads}}\,(p_C(\text{MeOH}) + p_C(H_2O))}$$

$$E_{a,10} = 21.2 \pm 0.1\,\text{kJ}\,\text{mol}^{-1}$$

$$r_{10'} = \frac{k_{10'}\, p_C(C_2^=)}{1 + K_{\text{MW}}^{\text{ads}}\,(p_C(\text{MeOH}) + p_C(H_2O))}$$

$$r_{10''} = \frac{k_{10''}\, p_C(C_4^=)}{1 + K_{\text{MW}}^{\text{ads}}\,(p_C(\text{MeOH}) + p_C(H_2O))}$$

$$r_{11} = \frac{k_{11}\, p_C(C_{5+}^=)}{1 + K_{\text{MW}}^{\text{ads}}\,(p_C(\text{MeOH}) + p_C(H_2O))}$$

$$E_{a,11} = 0.601 \pm 0.135\,\text{kJ}\,\text{mol}^{-1}$$

$$r_{12} = \frac{k_{12}\, p_C(C_3^=)}{1 + K_{\text{MW}}^{\text{ads}}\,(p_C(\text{MeOH}) + p_C(H_2O))}$$

$$E_{a,12} = 20.5 \pm 0.1\,\text{kJ}\,\text{mol}^{-1}$$

$$r_{12'} = \frac{k_{12'}\, p_C(C_2^=)}{1 + K_{\text{MW}}^{\text{ads}}\,(p_C(\text{MeOH}) + p_C(H_2O))}$$

$$r_{12''} = \frac{k_{12''}\, p_C(C_4^=)}{1 + K_{\text{MW}}^{\text{ads}}\,(p_C(\text{MeOH}) + p_C(H_2O))}$$

$$r_{13} = \frac{k_{13}\, p_C(\text{DME})}{1 + K_{\text{MW}}^{\text{ads}}\,(p_C(\text{MeOH}) + p_C(H_2O))}$$

$$E_{a,13} = 33.7 \pm 0.4\,\text{kJ}\,\text{mol}^{-1}$$

$$\Delta_{\text{ads}}H_{\text{MW}}^{\circ} = 0.2 \pm 0.1\,\text{kJ}\,\text{mol}^{-1}$$

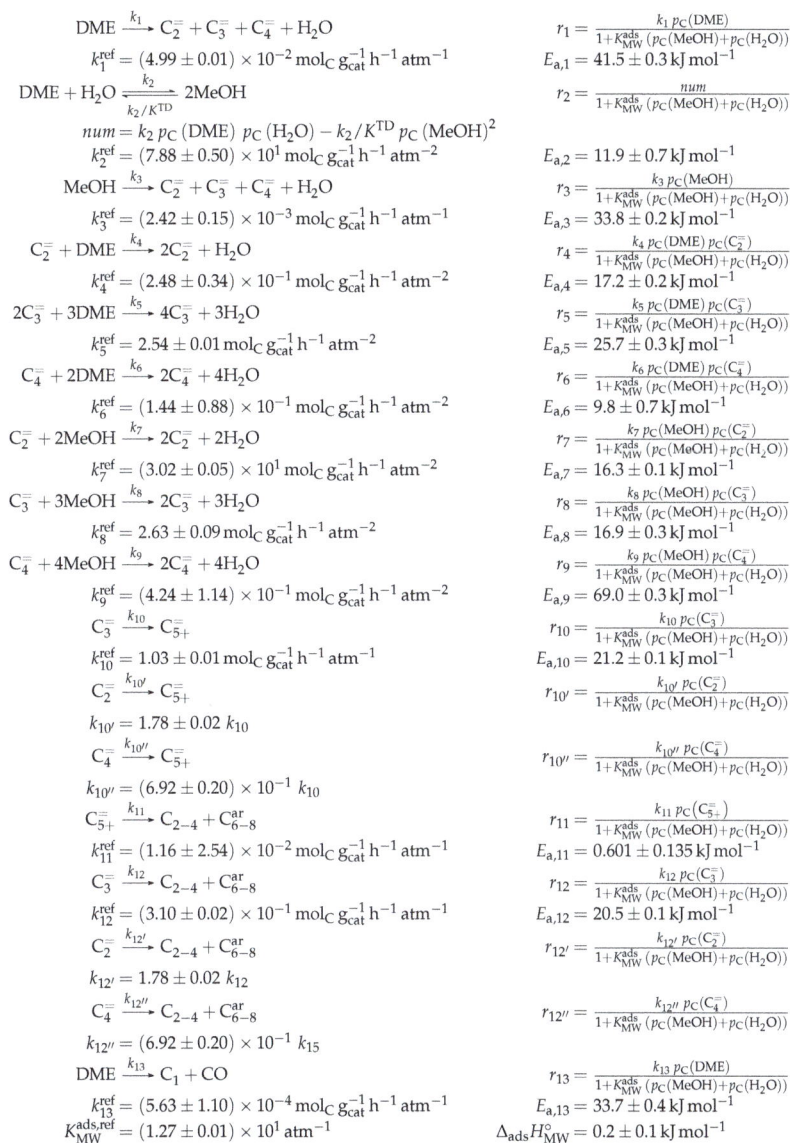

Scheme 25. Reaction network, rate equations and estimated parameters for the model by Pérez-Uriarte et al. [253]; K^{TD} is calculated with an own correlation [109] (see Section 2.1); some values are from [276].

The net rates of production are presented in Scheme 26.

$$R\,(\text{MeOH}) = \frac{2\,k_2\,p_C\,(\text{DME})\,p_C\,(\text{H}_2\text{O}) - 2\,k_2/K^{\text{TD}}\,p_C\,(\text{MeOH})^2 - 3\,k_3\,p_C\,(\text{MeOH})}{1 + K_{\text{MW}}^{\text{ads}}\,(p_C\,(\text{MeOH}) + p_C\,(\text{H}_2\text{O}))}$$
$$\frac{-k_7\,p_C\,(\text{MeOH})\,p_C\,(\text{C}_2^=) - k_8\,p_C\,(\text{MeOH})\,p_C\,(\text{C}_3^=) - k_9\,p_C\,(\text{MeOH})\,p_C\,(\text{C}_4^=)}{1 + K_{\text{MW}}^{\text{ads}}\,(p_C\,(\text{MeOH}) + p_C\,(\text{H}_2\text{O}))}$$

$$R\,(\text{DME}) = \frac{2\,k_2/K^{\text{TD}}\,p_C\,(\text{MeOH})^2 - 6\,k_1\,p_C\,(\text{DME}) - 2\,k_2\,p_C\,(\text{DME})\,p_C\,(\text{H}_2\text{O})}{1 + K_{\text{MW}}^{\text{ads}}\,(p_C\,(\text{MeOH}) + p_C\,(\text{H}_2\text{O}))}$$
$$\frac{-2\,k_4\,p_C\,(\text{DME})\,p_C\,(\text{C}_2^=) - 2\,k_5\,p_C\,(\text{DME})\,p_C\,(\text{C}_3^=) - 2\,k_6\,p_C\,(\text{DME})\,p_C\,(\text{C}_4^=)}{1 + K_{\text{MW}}^{\text{ads}}\,(p_C\,(\text{MeOH}) + p_C\,(\text{H}_2\text{O}))}$$
$$\frac{-6\,k_{13}\,p_C\,(\text{DME})}{1 + K_{\text{MW}}^{\text{ads}}\,(p_C\,(\text{MeOH}) + p_C\,(\text{H}_2\text{O}))}$$

$$R\,(\text{C}_2^=) = \frac{2\,k_1\,p_C\,(\text{DME}) + k_3\,p_C\,(\text{MeOH}) + 2\,k_4\,p_C\,(\text{DME})\,p_C\,(\text{C}_2^=) + k_7\,p_C\,(\text{MeOH})\,p_C\,(\text{C}_2^=)}{1 + K_{\text{MW}}^{\text{ads}}\,(p_C\,(\text{MeOH}) + p_C\,(\text{H}_2\text{O}))}$$
$$\frac{-k_{10'}\,p_C\,(\text{C}_2^=) - 2.5\,k_{12'}\,p_C\,(\text{C}_2^=)}{1 + K_{\text{MW}}^{\text{ads}}\,(p_C\,(\text{MeOH}) + p_C\,(\text{H}_2\text{O}))}$$

$$R\,(\text{C}_3^=) = \frac{2\,k_1\,p_C\,(\text{DME}) + k_3\,p_C\,(\text{MeOH}) + 2\,k_5\,p_C\,(\text{DME})\,p_C\,(\text{C}_3^=) + k_8\,p_C\,(\text{MeOH})\,p_C\,(\text{C}_3^=)}{1 + K_{\text{MW}}^{\text{ads}}\,(p_C\,(\text{MeOH}) + p_C\,(\text{H}_2\text{O}))}$$
$$\frac{-k_{10}\,p_C\,(\text{C}_3^=) - 2.5\,k_{12}\,p_C\,(\text{C}_3^=)}{1 + K_{\text{MW}}^{\text{ads}}\,(p_C\,(\text{MeOH}) + p_C\,(\text{H}_2\text{O}))}$$

$$R\,(\text{C}_4^=) = \frac{2\,k_1\,p_C\,(\text{DME}) + k_3\,p_C\,(\text{MeOH}) + 2\,k_6\,p_C\,(\text{DME})\,p_C\,(\text{C}_4^=) + k_9\,p_C\,(\text{MeOH})\,p_C\,(\text{C}_4^=)}{1 + K_{\text{MW}}^{\text{ads}}\,(p_C\,(\text{MeOH}) + p_C\,(\text{H}_2\text{O}))}$$
$$\frac{-k_{10''}\,p_C\,(\text{C}_4^=) - 2.5\,k_{12''}\,p_C\,(\text{C}_4^=)}{1 + K_{\text{MW}}^{\text{ads}}\,(p_C\,(\text{MeOH}) + p_C\,(\text{H}_2\text{O}))}$$

$$R\,(\text{C}_{5+}^=) = \frac{k_{10}\,p_C\,(\text{C}_3^=) + k_{10'}\,p_C\,(\text{C}_2^=) + k_{10''}\,p_C\,(\text{C}_4^=) - 2.5\,k_{11}\,p_C\,(\text{C}_{5+}^=)}{1 + K_{\text{MW}}^{\text{ads}}\,(p_C\,(\text{MeOH}) + p_C\,(\text{H}_2\text{O}))}$$

$$R\,(\text{C}_1) = \frac{4\,k_{13}\,p_C\,(\text{DME})}{1 + K_{\text{MW}}^{\text{ads}}\,(p_C\,(\text{MeOH}) + p_C\,(\text{H}_2\text{O}))}$$

$$R\,(\text{C}_{2-4}) = \frac{k_{11}\,p_C\,(\text{C}_{5+}^=) + k_{12}\,p_C\,(\text{C}_3^=) + k_{12'}\,p_C\,(\text{C}_2^=) + k_{12''}\,p_C\,(\text{C}_4^=)}{1 + K_{\text{MW}}^{\text{ads}}\,(p_C\,(\text{MeOH}) + p_C\,(\text{H}_2\text{O}))}$$

$$R\,(\text{C}_{6-8}^{\text{ar}}) = \frac{1.5\,k_{11}\,p_C\,(\text{C}_{5+}^=) + 1.5\,k_{12'}\,p_C\,(\text{C}_2^=) + 1.5\,k_{12}\,p_C\,(\text{C}_3^=) + 1.5\,k_{12''}\,p_C\,(\text{C}_4^=)}{1 + K_{\text{MW}}^{\text{ads}}\,(p_C\,(\text{MeOH}) + p_C\,(\text{H}_2\text{O}))}$$

$$R\,(\text{CO}) = \frac{2\,k_{13}\,p_C\,(\text{DME})}{1 + K_{\text{MW}}^{\text{ads}}\,(p_C\,(\text{MeOH}) + p_C\,(\text{H}_2\text{O}))}$$

$$R\,(\text{H}_2\text{O}) = \frac{k_1\,p_C\,(\text{DME}) + k_2/K^{\text{TD}}\,p_C\,(\text{MeOH})^2 + k_3\,p_C\,(\text{MeOH}) + k_4\,p_C\,(\text{DME})\,p_C\,(\text{C}_2^=)}{1 + K_{\text{MW}}^{\text{ads}}\,(p_C\,(\text{MeOH}) + p_C\,(\text{H}_2\text{O}))}$$
$$\frac{+k_5\,p_C\,(\text{DME})\,p_C\,(\text{C}_3^=) + k_6\,p_C\,(\text{DME})\,p_C\,(\text{C}_4^=) + k_7\,p_C\,(\text{MeOH})\,p_C\,(\text{C}_2^=)}{1 + K_{\text{MW}}^{\text{ads}}\,(p_C\,(\text{MeOH}) + p_C\,(\text{H}_2\text{O}))}$$
$$\frac{+k_8\,p_C\,(\text{MeOH})\,p_C\,(\text{C}_3^=) + k_9\,p_C\,(\text{MeOH})\,p_C\,(\text{C}_4^=) - k_2\,p_C\,(\text{DME})\,p_C\,(\text{H}_2\text{O})}{1 + K_{\text{MW}}^{\text{ads}}\,(p_C\,(\text{MeOH}) + p_C\,(\text{H}_2\text{O}))}$$

Scheme 26. Net rates of production of the different lumps for the model by Pérez-Uriarte et al. [253].

Parameter Estimation

According to the authors, the mole fractions are defined with carbon units and therefore only for carbon containing species which pertains for the partial pressures. The numeric routine is similar to Epelde et al. [235] and is therefore described in Section 3.1.1. As reference temperature, 623 K is chosen, which is within the lower third of the pure DME experiments. Thirty parameters have to be estimated: 13 reference rate constants, 13 activation energies, 1 reference equilibrium constant, 1 adsorption enthalpy and 2 factors relating the rate constants for ethene and butenes with that of propene.

4.2.4. Summary

All models in this subsection differentiate the methylation via methanol and via DME. However, this causes several additional parameters, not only for the methylations themselves, but also for the interconversion of the oxygenates. In all three examples, the latter reaction is implemented as step of kinetic relevance with the backward reaction being described by the thermodynamic equilibrium constant. For this value, no reasonable results are obtained with the equation shown in [114], whereas the outcome for the other two models is close to thermodynamics. When a detailed description of lower olefins is desired, the combined ethene and propene lump of Gayubo et al. [114] might be problematic. In addition, no side products are described here. On the other hand, this model explicitly includes the effect of water adsorption, similar to Pérez-Uriarte et al. [253]. Moreover, Gayubo et al. [114] described olefin interconversion in a simple, but effective way. In contrast, more reactions were considered by Aguayo et al. [252]. Here, the separate description of *n*-butane is noteworthy. In general, this model is useful for significant side product formation: these evolve to such an extent that they interact with olefins. In this model, both adsorption and mechanistic effects are missing. These were considered by Pérez-Uriarte et al. [253] who, besides water, implemented methanol adsorption. Further improvement would be possible through extending this with olefin and DME adsorption. Their model is the only one in the review that is explicitly created for DME feeds. Nevertheless, it should be also valid at least for oxygenates mixtures as feed. A vast reactivity including side products is covered by the reaction network of Pérez-Uriarte et al. [253]; however, some reactions and especially their stoichiometry seem to be arbitrarily chosen. Furthermore, the number of estimated parameters is comparably high.

4.3. Microkinetic Studies over ZSM-5

4.3.1. Park and Froment: Analysis of First C-C Bond Formation Mechanisms

Catalyst

A self-synthesized ZSM-5 zeolite powder having a high Si/Al ratio of 200 was used without any binder [132,254]. The particle size of 500–1000 μm was small enough to exclude any effects of heat and mass transfer limitations according to the authors. Until 5 h TOS, no deactivation effects could be observed. Complete catalyst regeneration in air was possible up to 50 times.

Setup and Conditions

The measurements were performed in a continuous fixed bed stainless steel reactor with an inner diameter of 21.4 mm. A tube made of titanium was chosen for experiments at temperatures higher than 723 K. The reactor could be positioned within a molten salt bath during reaction. For product analysis, a GC with several columns, an FID and a TCD was applied with nitrogen as internal standard. Moreover, the authors evaluated the C_{6+} fraction with a GC-MS. The catalyst bed was diluted 5:1 in a volumetric ratio using glass beads; these were also stacked in front of the catalyst bed. The latter was located on a stainless steel sieve and both glass wool and beads were placed in between. For dilution of the gaseous methanol feed, both water and nitrogen were used. The temperature within the bed could be controlled with a thermocouple sliding inside a well. More details about the setup can be found in [277].

Reaction Network

In the original publications [132,254], a detailed overview of all included reactions, rate equations and net rates of formation can be found. This is why only references and no schemes are shown here. In this microkinetic study, formation of the first C-C bond is modeled with the oxonium ylide mechanism [278]. This route comprises the formation of an oxonium methylide (OM) out of a surface methyl group (R_1^+) and a basic site (Step (iii.a''.1) in Table 1 in [254]), the reaction of OM and a protonated DME species (DMO^+) to protonated ethene and methanol (Step (iii.a".2) in Table 1 in [254], LH type of mechanism) as well as to protonated propene and water (Step (iii.a".4) in Table 1 in [254]). Similar to these three steps, the deprotonation of ethene is also considered as reaction of kinetic relevance (Step (iii.a".3) in Table 1 in [254]). Both formation steps of the DMO^+ species are assumed to be irreversible. The surface methyl group is formed by methanol protonation (Step (i.1) in Table 1 in [254]) and subsequent water release (Step (i.2) in Table 1 in [254]). This surface group can perform methylation reactions; when methanol is the other reactant, DMO^+ is the product (Step (i.3) in Table 1 in [254]) whose desorption releases DME (reverse reaction of DME protonation, Step (i.4) in Table 1 in [254]). Another pathway starting from these two reactants is the formation of methane and formaldehyde (Step (ii.1) in Table 1 in [254]). Whereas the two protonations are assumed to be *quasi*-equilibrated, the remaining steps are kinetically relevant; except for the methane formation, all reactions are further assumed to be reversible. The surface methyl group can also methylate gas phase olefins (ER type of mechanism); all possible steps for C_2–C_7 as reactants are included (Table 3 in [254]). In the olefin interconversion network, all cracking (Table 5 in [254]), dimerization (Table 4 in [254]) and isomerization steps are implemented for a maximum carbon number of eight. Only methyl side groups are allowed because of the small ZSM-5 pores. On the other hand, quaternary carbon atoms are considered. All steps starting from or leading to a primary intermediate are excluded except for ethene methylation and ethene self-dimerization to butene. It follows that methylation reactions are assumed to be irreversible in contrast to cracking/dimerization. The isomerization, protonation and deprotonation steps are assumed to be *quasi*-equilibrated. Cracking and dimerization are expressed as L and ER types of mechanism, respectively, where protonation of the gas phase olefins leads to the surface intermediates. Finally, 172 pathways of kinetic relevance remain in the whole network. No side product formation is covered by this model because only measurements at methanol conversions less than 0.7 are included. In this regime, side product formation is negligible. Only 31 of the original 222 data points are evaluated by the model. For comparison with the experimental results, the isomers of same carbon number are lumped, leading to the following fitting responses: DME, $C_2^=$, $C_3^=$, $C_4^=$, $C_5^=$, $C_6^=$, $C_7^=$, $C_8^=$ and C_1. The corresponding net rates of formation, in the same order, are formulated in Equations (8), (9), (11), (31) (exemplary equation for $C_4^=$–$C_8^=$) and 7 in [254]. The reaction rates for methylation, dimerization and cracking are shown in Equation (24) in [254]. The concentration of reactive intermediates, i.e., surface methyl group and OM, is accessed via the PSSA (Equations (14) and (16) in [254]) and site balances are applied both for acid and for basic centers (Equations (21) and (23) in [254]) according to a HW type of mechanism.

Parameter Estimation

Results in [132] are shown with weight based yields, whereas the rate equations require partial pressures. The objective function compares the weighted squared differences of measured and modeled yields; the weighting factors are not obtained via replicate experiments, but calculated via Equation (30) where -1 as exponent is replaced by 0.3 for $C_7^=$, $C_8^=$ and C_1. Integration of the differential equations of stiff character is performed using Gear's method. The deviation between model and experiment is minimized via a hybrid genetic algorithm (GA) approach [279]: at first, the GA searches for parameters satisfying the constraints and lowering the objective function to a significant extent. When these conditions are fulfilled, two local optimizers use the initial guesses obtained by the GA: a sequential quadratic program called FFSQP, which also considers constraints and the Levenberg–Marquardt algorithm for an unconstrained final parameter estimation. Even when a suitable solution is found, new initial guesses are tried up to the maximum number of GA iterations. Through this routine, finding of the global minimum should be ensured. For the GA, values of 0.10 and 0.005 are chosen for the crossover and mutation probability, respectively. The mentioned constraints should avoid physically unreasonable values, i.e., negative activation energies or positive reaction enthalpies for methylation and dimerization; furthermore, the protonation values have to match Boudart's criteria [280]. Finally, the higher the carbon number, the lower the protonation enthalpy should be. All the linearized constraints can be found in Table 2 in [132]. Except for the methylation and dimerization rates, reparameterization is performed according to Equations (22) and (25); Table 1 in [132] contains all dimensionless fitting parameters. Their number is drastically reduced by using the single-event methodology [20,21] in combination with the Evans–Polanyi relation [281] and the concept of thermodynamic consistency [246]. The required thermodynamic data are calculated via Benson's group contribution method [96] as well as via quantum chemical approaches. In total, 33 values have to be estimated: eight preexponential factors, eight activation energies, nine protonation enthalpies, three protonation entropies, one hydration enthalpy, one hydration entropy and one combination of preexponential factor, activation energy and transfer coefficient for the Evans–Polanyi relation (see Table 3 in [132] for the values). This model is used in a subsequent study to simulate isothermal fixed bed and adiabatic multi stage reactors [282].

4.3.2. Kumar et al.: Implementation of Aromatic Hydrocarbon Pool

Catalyst

The authors [19] used the same catalyst as Park and Froment [132,254], see Section 4.3.1.

Setup and Conditions

The setup is already explained in Section 4.3.1.

Reaction Network

In contrast to the previous microkinetic implementation [132,254], the conversion of oxygenates to lower olefins is implemented via the side-chain mechanism of the aromatic hydrocarbon pool according to Arstad et al. [195] (see Figure 1 in [19]). The formation of these polymethylated aromatics is not described by the model, but their contribution to the overall MTO reactivity is explicitly considered. Starting from *para*-xylene, a sequence of methylations (Steps (i), (iii) and (vi) in Figure 1 in [19]), deprotonations (Steps (ii), (v) and (viii) in Figure 1 in [19]) and dealkylations (Steps (iv) and (vii) in Figure 1 in [19]) releases ethene and propene. All these steps are assumed to be reversible and of kinetic relevance. The DME and methane formation (Steps (i)–(v)) is implemented in a similar way to Park and Froment [132,254] (see Section 4.3.1). Finally, olefin interconversion is accounted for up to a maximum carbon number of seven. This excludes all transformations of a tertiary reactant to a tertiary product intermediate. Moreover, the ethene self-dimerization is not considered. The other assumptions of Park and Froment [132,254] are retained, including the negligence of side products. The protonation is extended with a physisorption step before, the data for which are taken from Denayer et al. [283,284]. The reaction network leads to 64 pathways of kinetic relevance. The experimental data are fitted to eight responses: DME, $C_2^=$, $C_3^=$, $C_4^=$, $C_5^=$, $C_6^=$, $C_7^=$ and C_1. The net rates of formation for DME, ethene, propene and methane can be found in Equations (12), (13), (16) and (17) in [19]. For the higher olefins, this coherence is not shown, but it follows from a summation of all methylation (Equation (18)) and alkylation rates (Equation (19)) where these species are involved. The amount of methanol and water is calculated using a carbon and hydrogen balance, respectively. The concentration of surface methyl groups and the seven intermediates in the aromatic hydrocarbon pool follows from applying the PSSA (Equations (15) and (14) in [19]). In addition, the total concentration of all aromatic hydrocarbon pool species is fitted (Equation (26) in [19]); the value is comparable to a concentration of active catalyst sites. The balance for the acid sites is found in Equation (29) in [19].

Parameter Estimation

All figures in [19] use weight based yields. For the reaction rates, partial pressures have to be used. The objective function evaluates the weighted squared residuals between calculated and measured molar flow rates and is minimized by a combination of a Rosenbrock and a Levenberg–Marquardt algorithm. For the former, an in-house code is used, whereas the latter is provided by the ordinary least-squares option of *ODRPACK*, version 2.01, from Netlib. Integration of the differential equations is performed by *DDASPK*, which is also part of Netlib. Here, a consistent set of boundary conditions is required, which is accessible for the gas-phase species, but not for the reactive intermediates. The latter is obtained by applying the numerical routine *DNSQE* which solves the PSSA conditions via a hybrid Powell method. The weighting factors are calculated according to Equation (30). For parameter reduction, the single-event methodology [20,21] is applied as well as the thermodynamic consistency [246]. In addition, all protonation entropies and preexponential factors are calculated before fitting based on statistical thermodynamics and the principle of microscopic reversibility [18] (see Table 3 in [19]). Here, the necessary values for entropy changes are extracted from databases [80], calculated via group contribution methods [96] or obtained via DFT. The aromatic hydrocarbon pool is characterized by only one average concentration and, finally, two deprotonation, two methylation and two dealkylation steps within this catalytic cycle are assumed to have similar activation energies. During fitting, Boudart's criteria [280] and the ordering according to carbon number are introduced as constraints for the protonation enthalpies. The rate constants are reparameterized according to Equation (21). Finally, besides the total concentration of aromatic hydrocarbon pool species, 29 parameters are fitted: 21 activation energies and eight protonation enthalpies.

4.3.3. Summary

Both models show high complexity and cause much computational effort therefore. On the other hand, an almost complete picture of reactivity is obtained here as the reaction network covers oxygenates interaction, olefin production out of oxygenates, methylation and olefin interconversion reactions. Only side product formation is left out because the maximum methanol conversion is limited to 0.7. In contrast to all other models in this review, the formation of lower olefins out of the oxygenates is not simply characterized by an arbitrary rate equation; in contrast, the pathways and intermediates are included. Whereas Park and Froment [132,254] focused on the formation of the first C-C bond on a direct way, Kumar et al. [19] considered the indirect formation via the aromatic hydrobarbon pool. In both models, the interaction between water and the zeolite is assumed to be negligible.

4.4. Studies with Significant Deactivation Effects over SAPO-34, SAPO-18 and ZSM-22

4.4.1. Gayubo et al.: Six- and Five-Lump Approach with and without Differentiation in Side Products over SAPO-34

Catalyst

The authors [255] synthesized a SAPO-34 zeolite with moderate acid strength that consisted mainly of Brønsted acid sites [261]. After that, the final catalyst was obtained by mixing the zeolite with bentonite and inert alumina. The deactivation through coke is considerably fast over SAPO-34 systems: it was observed that, under harsh conditions, 3–4% of the methanol fed was deposited on the catalyst even during the first minute TOS. However, this coking rate could be decreased through higher water contents, temperatures or contact times. The maximum coking rate under water co-feeding was 1%. Nevertheless, the measurements were performed at a TOS of 1 h with the results being extrapolated to the fresh catalyst.

Setup and Conditions

The experimental setup was the same as for Gayubo et al.'s model over ZSM-5 [114]; for more details, refer to Section 4.2.1 or [269]. For the SAPO-34 experiments, the catalyst was diluted 1:3 on a weight base with alumina. Furthermore, dilution of the methanol feed with water was performed with the following weight ratios: 0, 1 and 3.

Reaction Network

In this model, six lumps are applied: Ox (methanol plus DME), $C_2^=$, $C_3^=$, $C_4^=$, $C_5^=$ and C_{1-4}. Consequently, no differentiation and no reactions between the oxygenates are considered. Their amount is not fitted to experimental data, but calculated via conservation of mass. No methylation reactions are included; the network in Scheme 27 is restricted to either oxygenates conversion to olefins (k_1–k_4) or olefin interconversion (k_6–k_8). The latter comprises conversion of $C_3^=$ to $C_5^=$ olefins to ethene and propene. In this model, ethene is seen as final product because it cannot act as reactant. Similar to Zhou et al. [77], no compounds with more than five carbon atoms are detected because of the shape selectivity. For the same reason, the amount of isobutene is lower than expected. Side product formation is considered as oxygenates conversion to C_{1-4} paraffins (k_5). All steps are assumed to be irreversible and elementary, whereas stoichiometry is neglected; the consumption of pentenes in Scheme 28 is arbitrarily set to two. Water adsorption is taken into account according to the ZSM-5 model by the same group [114], meaning that the reaction rates in Scheme 27 are formulated as HW type of mechanism where the adsorption of all hydrocarbons is not considered.

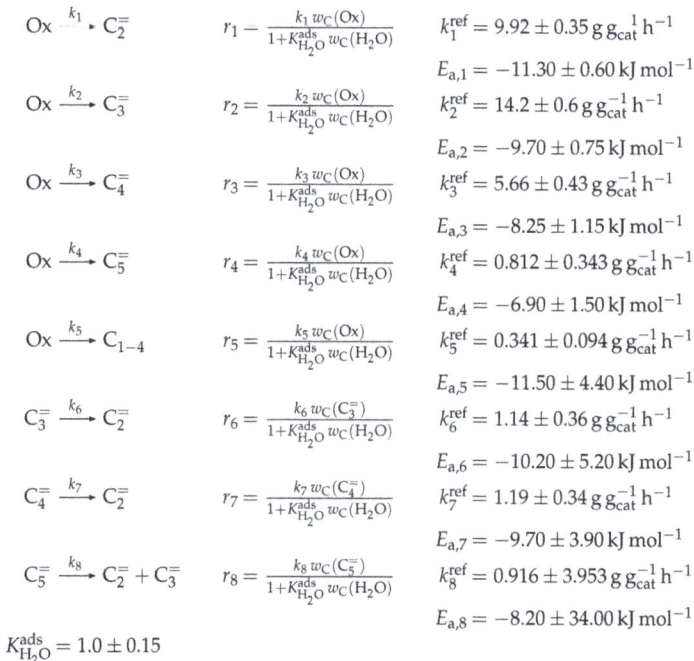

$$Ox \xrightarrow{k_1} C_2^= \qquad r_1 = \frac{k_1\, w_C(Ox)}{1+K_{H_2O}^{ads}\, w_C(H_2O)} \qquad k_1^{ref} = 9.92 \pm 0.35\, \mathrm{g\, g_{cat}^{-1}\, h^{-1}}$$
$$E_{a,1} = -11.30 \pm 0.60\, \mathrm{kJ\, mol^{-1}}$$

$$Ox \xrightarrow{k_2} C_3^= \qquad r_2 = \frac{k_2\, w_C(Ox)}{1+K_{H_2O}^{ads}\, w_C(H_2O)} \qquad k_2^{ref} = 14.2 \pm 0.6\, \mathrm{g\, g_{cat}^{-1}\, h^{-1}}$$
$$E_{a,2} = -9.70 \pm 0.75\, \mathrm{kJ\, mol^{-1}}$$

$$Ox \xrightarrow{k_3} C_4^= \qquad r_3 = \frac{k_3\, w_C(Ox)}{1+K_{H_2O}^{ads}\, w_C(H_2O)} \qquad k_3^{ref} = 5.66 \pm 0.43\, \mathrm{g\, g_{cat}^{-1}\, h^{-1}}$$
$$E_{a,3} = -8.25 \pm 1.15\, \mathrm{kJ\, mol^{-1}}$$

$$Ox \xrightarrow{k_4} C_5^= \qquad r_4 = \frac{k_4\, w_C(Ox)}{1+K_{H_2O}^{ads}\, w_C(H_2O)} \qquad k_4^{ref} = 0.812 \pm 0.343\, \mathrm{g\, g_{cat}^{-1}\, h^{-1}}$$
$$E_{a,4} = -6.90 \pm 1.50\, \mathrm{kJ\, mol^{-1}}$$

$$Ox \xrightarrow{k_5} C_{1-4} \qquad r_5 = \frac{k_5\, w_C(Ox)}{1+K_{H_2O}^{ads}\, w_C(H_2O)} \qquad k_5^{ref} = 0.341 \pm 0.094\, \mathrm{g\, g_{cat}^{-1}\, h^{-1}}$$
$$E_{a,5} = -11.50 \pm 4.40\, \mathrm{kJ\, mol^{-1}}$$

$$C_3^= \xrightarrow{k_6} C_2^= \qquad r_6 = \frac{k_6\, w_C(C_3^=)}{1+K_{H_2O}^{ads}\, w_C(H_2O)} \qquad k_6^{ref} = 1.14 \pm 0.36\, \mathrm{g\, g_{cat}^{-1}\, h^{-1}}$$
$$E_{a,6} = -10.20 \pm 5.20\, \mathrm{kJ\, mol^{-1}}$$

$$C_4^= \xrightarrow{k_7} C_2^= \qquad r_7 = \frac{k_7\, w_C(C_4^=)}{1+K_{H_2O}^{ads}\, w_C(H_2O)} \qquad k_7^{ref} = 1.19 \pm 0.34\, \mathrm{g\, g_{cat}^{-1}\, h^{-1}}$$
$$E_{a,7} = -9.70 \pm 3.90\, \mathrm{kJ\, mol^{-1}}$$

$$C_5^= \xrightarrow{k_8} C_2^= + C_3^= \qquad r_8 = \frac{k_8\, w_C(C_5^=)}{1+K_{H_2O}^{ads}\, w_C(H_2O)} \qquad k_8^{ref} = 0.916 \pm 3.953\, \mathrm{g\, g_{cat}^{-1}\, h^{-1}}$$
$$E_{a,8} = -8.20 \pm 34.00\, \mathrm{kJ\, mol^{-1}}$$

$$K_{H_2O}^{ads} = 1.0 \pm 0.15$$

Scheme 27. Reaction network, rate equations and estimated parameters for the model by Gayubo et al. [255] over SAPO-34 (six lumps).

The resulting net rates of production can be found in Scheme 28.

$$R\left(Ox\right) = \frac{-k_1\,w_C\left(Ox\right) - k_2\,w_C\left(Ox\right) - k_3\,w_C\left(Ox\right) - k_4\,w_C\left(Ox\right) - k_5\,w_C\left(Ox\right)}{1 + K_{H_2O}^{ads}\,w_C\left(H_2O\right)}$$

$$R\left(C_2^=\right) = \frac{k_1\,w_C\left(Ox\right) + k_6\,w_C\left(C_3^=\right) + k_7\,w_C\left(C_4^=\right) + k_8\,w_C\left(C_5^=\right)}{1 + K_{H_2O}^{ads}\,w_C\left(H_2O\right)}$$

$$R\left(C_3^=\right) = \frac{k_2\,w_C\left(Ox\right) + k_8\,w_C\left(C_5^=\right) - k_6\,w_C\left(C_3^=\right)}{1 + K_{H_2O}^{ads}\,w_C\left(H_2O\right)}$$

$$R\left(C_4^=\right) = \frac{k_3\,w_C\left(Ox\right) - k_7\,w_C\left(C_4^=\right)}{1 + K_{H_2O}^{ads}\,w_C\left(H_2O\right)}$$

$$R\left(C_5^=\right) = \frac{k_4\,w_C\left(Ox\right) - 2k_8\,w_C\left(C_5^=\right)}{1 + K_{H_2O}^{ads}\,w_C\left(H_2O\right)}$$

$$R\left(C_{1-4}\right) = \frac{k_5\,w_C\left(Ox\right)}{1 + K_{H_2O}^{ads}\,w_C\left(H_2O\right)}$$

Scheme 28. Net rates of production of the different lumps for the model by Gayubo et al. [255] over SAPO-34 (six lumps); the Ox lump (methanol plus DME) is calculated via conservation of mass within the model.

Compared to the model with six lumps, $C_5^=$ and C_{1-4} are summarized to the lump Rest. Furthermore, all olefin interconversion reactions are neglected (see Scheme 29). The resulting network thus only considers oxygenates conversion to olefins and to the Rest lump.

$$Ox \xrightarrow{k_1} C_2^= \qquad r_1 = \frac{k_1\,w_C(Ox)}{1 + K_{H_2O}^{ads}\,w_C(H_2O)} \qquad k_1^{ref} = 10.5 \pm 0.4\,\mathrm{g\,g_{cat}^{-1}\,h^{-1}}$$

$$E_{a,1} = -11.8 \pm 0.6\,\mathrm{kJ\,mol^{-1}}$$

$$Ox \xrightarrow{k_2} C_3^= \qquad r_2 = \frac{k_2\,w_C(Ox)}{1 + K_{H_2O}^{ads}\,w_C(H_2O)} \qquad k_2^{ref} = 13.7 \pm 0.5\,\mathrm{g\,g_{cat}^{-1}\,h^{-1}}$$

$$E_{a,2} = -9.5 \pm 0.6\,\mathrm{kJ\,mol^{-1}}$$

$$Ox \xrightarrow{k_3} C_4^= \qquad r_3 = \frac{k_3\,w_C(Ox)}{1 + K_{H_2O}^{ads}\,w_C(H_2O)} \qquad k_3^{ref} = 5.49 \pm 0.37\,\mathrm{g\,g_{cat}^{-1}\,h^{-1}}$$

$$E_{a,3} = -7.1 \pm 1.0\,\mathrm{kJ\,mol^{-1}}$$

$$Ox \xrightarrow{k_4} Rest \qquad r_4 = \frac{k_4\,w_C(Ox)}{1 + K_{H_2O}^{ads}\,w_C(H_2O)} \qquad k_4^{ref} = 1.00 \pm 0.22\,\mathrm{g\,g_{cat}^{-1}\,h^{-1}}$$

$$E_{a,4} = -9.0 \pm 3.6\,\mathrm{kJ\,mol^{-1}}$$

Scheme 29. Reaction network, rate equations and estimated parameters for the model by Gayubo et al. [255] over SAPO-34 (five lumps).

The network reduces the net rates of formation to the ones in Scheme 30.

$$R\left(Ox\right) = \frac{-k_1\,w_C\left(Ox\right) - k_2\,w_C\left(Ox\right) - k_3\,w_C\left(Ox\right) - k_4\,w_C\left(Ox\right)}{1 + K_{H_2O}^{ads}\,w_C\left(H_2O\right)}$$

$$R\left(C_2^=\right) = \frac{k_1\,w_C\left(Ox\right)}{1 + K_{H_2O}^{ads}\,w_C\left(H_2O\right)}$$

$$R\left(C_3^=\right) = \frac{k_2\,w_C\left(Ox\right)}{1 + K_{H_2O}^{ads}\,w_C\left(H_2O\right)}$$

$$R\left(C_4^=\right) = \frac{k_3\,w_C\left(Ox\right)}{1 + K_{H_2O}^{ads}\,w_C\left(H_2O\right)}$$

$$R\left(Rest\right) = \frac{k_4\,w_C\left(Ox\right)}{1 + K_{H_2O}^{ads}\,w_C\left(H_2O\right)}$$

Scheme 30. Net rates of production of the different lumps for the model by Gayubo et al. [255] over SAPO-34 (five lumps); the Ox lump (methanol plus DME) is calculated via conservation of mass within the model.

Parameter Estimation

The numeric routine is similar to the ZSM-5 model by Gayubo et al. [114] (see Section 4.2.1). The reference temperature for reparameterization is set to 698 K which is close to the mean value of the experimentally covered range. The reaction scheme causes 17 unknown parameters: eight reference rate constants, eight activation energies and one equilibrium constant for water adsorption. As for the study over ZSM-5, the latter value is determined to unity, which converts the organic mass fractions to total mass fractions. Because of the poor numeric significance of the estimated parameters in Scheme 27, the network is reduced to five lumps in the following. For the version with five lumps, only eight parameters have to be estimated due to the simplified network: four reference rate constants and four activation energies. The equilibrium constant of water adsorption is kept fixed at a value of 1.

4.4.2. Ying et al.: Seven-Lump Model with Subsequent Fitting of Deactivation Parameters over SAPO-34

Catalyst

The authors [256] used the commercial DMTO catalyst by Chia Tai Energy Materials in order to ensure transferability of their model to an industrial plant. It is based over SAPO-34 crystals. Because of confidentiality, not many details about the final catalyst extrudate are given. Concerning the kinetic model, the measured data were obtained with a fresh catalyst. In addition to this, the authors analyzed deactivation effects using TOS values up to 1.67 h. For this, they applied a fluidized bed reactor because coke evolution in a fixed bed is prone to zoning effects, which cause a non-uniform coke distribution. The authors found that the reactor type has no influence on the final coke content. Furthermore, the coke growth rate was comparably high at the beginning and leveled off towards the maximum. At a certain deactivation level, the catalyst showed a maximum olefin production rate. This is why the authors include a seventh step to their kinetic scheme which accounts for coke formation out of methanol. The other six reaction rates are multiplied by a deactivation value φ_l. This stems from an exponential approach consisting of several constants and a rate-specific value α_l. The deactivation model as well as the resulting reactor model is beyond the scope of this review, so the reader is referred to the original contribution [256].

Setup and Conditions

The experiments were performed in a continuous fixed bed quartz glass reactor with an inner diameter of 4 mm. The liquid feed consisted of either pure methanol or a water–methanol stream with a molar ratio of 2:1 or 4:1; it was vaporized before entering the reactor. A GC with one column and an FID was used for product analysis. For the coking experiments, the authors applied a fluidized bed reactor with an inner diameter of 19 mm where the evolved coke could be evaluated via TG.

Reaction Network

The reaction network in Scheme 31 considers no DME formation; all reactions start from methanol as reactant. These steps are assumed to be irreversible and lead to both olefins (k_2, k_3, k_5 and k_6) and paraffins (k_1, k_4 and k_6). Neither methylations nor olefin interconversion reactions are implemented. The following lumps are defined: MeOH, $C_2^=$, $C_3^=$, C_4^{HC}, C_1, C_3 and $C_{2,5,6}^{HC}$. The latter comprises both olefins and paraffins with five and six carbon atoms as well as ethane. In preliminary experiments, no further side products and no higher compounds could be detected. Water attenuates the overall reaction rates, which is why its adsorption is included via a HW type of mechanism. The interaction of the hydrocarbons with the acid sites is neglected. All reactions are formulated as first-order with respect to methanol, which is the result of an experimental observation. Stoichiometry is retained for the net rates of formation.

$$\text{MeOH} \xrightarrow{k_1} C_1 \qquad r_1 = \frac{k_1\, C(\text{MeOH})\, M_w\, (C_1)}{1+K^{ads}_{H_2O}\, w(H_2O)} \qquad k^{ref}_1 = 0.10\,\text{L}\,g^{-1}_{cat}\,\text{min}^{-1}$$

$$E_{a,1} = 117.7\,\text{kJ}\,\text{mol}^{-1}$$

$$\text{MeOH} \xrightarrow{k_2} C_2^= \qquad r_2 = \frac{k_2\, C(\text{MeOH})\, M_w\, (C_2^=)}{1+K^{ads}_{H_2O}\, w(H_2O)} \qquad k^{ref}_2 = 4.93\,\text{L}\,g^{-1}_{cat}\,\text{min}^{-1}$$

$$E_{a,2} = 56.9\,\text{kJ}\,\text{mol}^{-1}$$

$$\text{MeOH} \xrightarrow{k_3} C_3^= \qquad r_3 = \frac{k_3\, C(\text{MeOH})\, M_w\, (C_3^=)}{1+K^{ads}_{H_2O}\, w(H_2O)} \qquad k^{ref}_3 = 7.32\,\text{L}\,g^{-1}_{cat}\,\text{min}^{-1}$$

$$E_{a,3} = 41.9\,\text{kJ}\,\text{mol}^{-1}$$

$$\text{MeOH} \xrightarrow{k_4} C_3 \qquad r_4 = \frac{k_4\, C(\text{MeOH})\, M_w\, (C_3)}{1+K^{ads}_{H_2O}\, w(H_2O)} \qquad k^{ref}_4 = 0.52\,\text{L}\,g^{-1}_{cat}\,\text{min}^{-1}$$

$$E_{a,4} = 13.4\,\text{kJ}\,\text{mol}^{-1}$$

$$\text{MeOH} \xrightarrow{k_5} C_4^{HC} \qquad r_5 = \frac{k_5\, C(\text{MeOH})\, M_w\, (C_4^{HC})}{1+K^{ads}_{H_2O}\, w(H_2O)} \qquad k^{ref}_5 = 2.60\,\text{L}\,g^{-1}_{cat}\,\text{min}^{-1}$$

$$E_{a,5} = 31.2\,\text{kJ}\,\text{mol}^{-1}$$

$$\text{MeOH} \xrightarrow{k_6} C_{2,5,6}^{HC} \qquad r_6 = \frac{k_6\, C(\text{MeOH})\, M_w\, (C_{2,5,6}^{HC})}{1+K^{ads}_{H_2O}\, w(H_2O)} \qquad k^{ref}_6 = 1.02\,\text{L}\,g^{-1}_{cat}\,\text{min}^{-1}$$

$$E_{a,6} = 45.8\,\text{kJ}\,\text{mol}^{-1}$$

$$K^{ads}_{H_2O} = 3.05$$

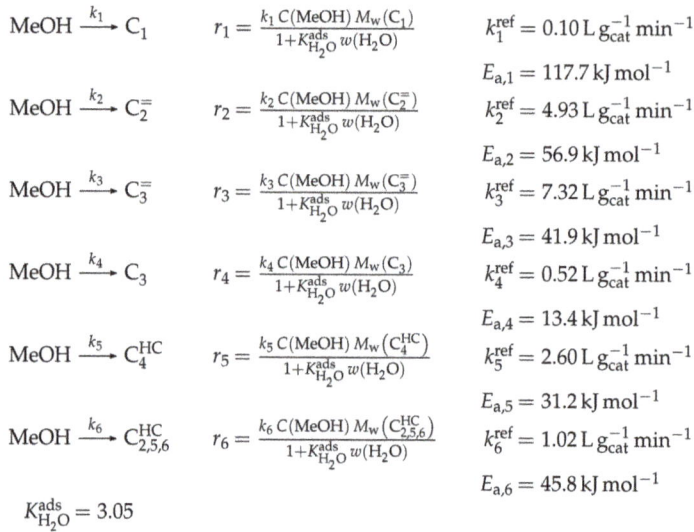

Scheme 31. Reaction network, rate equations and estimated parameters for the model by Ying et al. [256].

An overview of the net rates of formation can be found in Scheme 32.

$$R(\text{MeOH}) = \frac{-k_1\, C(\text{MeOH})\, M_w\,(\text{MeOH}) - k_2\, C(\text{MeOH})\, M_w\,(\text{MeOH}) - k_3\, C(\text{MeOH})\, M_w\,(\text{MeOH})}{1+K^{ads}_{H_2O}\, w(H_2O)}$$
$$\frac{-k_4\, C(\text{MeOH})\, M_w\,(\text{MeOH}) - k_5\, C(\text{MeOH})\, M_w\,(\text{MeOH}) - k_6\, C(\text{MeOH})\, M_w\,(\text{MeOH})}{1+K^{ads}_{H_2O}\, w(H_2O)}$$

$$R(C_2^=) = \frac{1}{2}\frac{k_2\, C(\text{MeOH})\, M_w\,(C_2^=)}{1+K^{ads}_{H_2O}\, w(H_2O)}$$

$$R(C_3^=) = \frac{1}{3}\frac{k_3\, C(\text{MeOH})\, M_w\,(C_3^=)}{1+K^{ads}_{H_2O}\, w(H_2O)}$$

$$R\left(C_4^{HC}\right) = \frac{1}{4}\frac{k_5\, C(\text{MeOH})\, M_w\,\left(C_4^{HC}\right)}{1+K^{ads}_{H_2O}\, w(H_2O)}$$

$$R(C_1) = \frac{k_1\, C(\text{MeOH})\, M_w\,(C_1)}{1+K^{ads}_{H_2O}\, w(H_2O)}$$

$$R(C_3) = \frac{1}{3}\frac{k_3\, C(\text{MeOH})\, M_w\,(C_3)}{1+K^{ads}_{H_2O}\, w(H_2O)}$$

$$R\left(C_{2,5,6}^{HC}\right) = \frac{1}{5}\frac{k_6\, C(\text{MeOH})\, M_w\,\left(C_{2,5,6}^{HC}\right)}{1+K^{ads}_{H_2O}\, w(H_2O)}$$

Scheme 32. Net rates of production of the different lumps for the model by Ying et al. [256].

Parameter Estimation

In Scheme 31, molar concentrations per volume have to be used for organic compounds while the water content is expressed as mass fraction. Whereas water is excluded for the figures shown in [256], the integrated rate expressions lead to mass fractions where water is included. Parameter estimation is performed via the Levenberg–Marquardt algorithm which minimizes the objective function. The latter returns the weighted sum of squared residuals between the modeled and the experimental mass fractions, but the calculation of the weighting factors is not shown. The adsorption equilibrium constant of water is assumed to be the same for all steps. Reparameterization according to Equation (21) is performed with a reference temperature of 723 K, the lowest experimentally investigated value. In total, without the coking values, 13 parameters are obtained: six reference rate constants, six activation energies and one equilibrium constant for water adsorption.

4.4.3. Chen et al.: Seven-Lump Model with Simultaneous Fitting of Deactivation Parameters over SAPO-34

Catalyst

A commercial SAPO-34 powder from SINTEF was used [257]. As it is known for this zeolite type, the coking rate was high and significant deactivation could be observed from the beginning on. In an earlier contribution [285], a detailed kinetic model of the coke evolution was derived. For [257], a simpler approach via a linear function is chosen: a deactivation constant α_l is multiplied by the weight percent of coke on the catalyst; subtracting the result from 1 yields the corresponding deactivation function φ_l. It depends on the reaction step l because of a selective deactivation, which means the higher the carbon number, the more selectivity loss through coke deposition can be observed. The authors supposed changes in shape selectivity for this behavior.

Setup and Conditions

The experiments were performed in a tapered element oscillating microbalance reactor which is described elsewhere [263,286]. This allowed for measuring mass changes without bypass effects, making it a useful tool to measure product evolution and coke formation, equivalent to main and deactivation kinetics, simultaneously. The setup exhibited fixed-bed characteristics with almost gradientless operation. Temperature control was ensured by two thermocouples, one at the outside and one below the outlet of the reactor. The latter consisted of proprietary glass. Liquid feeds were provided from a storage cylinder and evaporated. The catalyst bed was diluted with quartz particles and the feed stream with helium. Because of the rapid deactivation, methanol was fed in pulses of 3 min at mild and of 1 min at harsh conditions. It could be shown that such a procedure does not affect conversion and selectivity [285]. For the same reason, not all combinations of conditions shown in Table 4 were performed (see [257]). The products were analyzed via a GC using one column and an FID.

Reaction Network

The reaction network is derived from preliminary measurements evaluated via yield-conversion plots. For one specific condition, several pulse amounts are applied. When connecting all data points of the first pulse, an optimum performance envelope is obtained, which gives further insight. The authors concluded that all olefins are stable secondary products forming in parallel out of DME. The effect of side reactions is low because of high WHSV values; only the stable tertiary products ethane and propane are produced at high oxygenates conversions. Methane is also detected as stable primary and secondary product, but it is excluded from modeling because of very small mole fractions. As coke deposition is significant throughout all experiments, its formation has to be included in the reaction network. It is classified as stable secondary and tertiary product. These observations lead to seven lumps: Ox (methanol plus DME), $C_2^=$, $C_3^=$, $C_4^=$, $C_5^=$, $C_6^=$ and C_{2-3}. Scheme 33 includes two different types of reactions, the conversion of oxygenates to olefins (k_1–k_5) and the subsequent reaction of olefins to paraffins (k_6). Consequently, no methylation reactions are considered and the reactivity of methanol is restricted to the step converting it to DME. Both oxygenates are summarized to one lump because of intracrystalline diffusion effects, which impede the reliable modeling of DME evolution. As mentioned above, although being lumped together with methanol, the olefin formation is assumed to originate only from DME. No olefin interconversion reactions are considered which is justified with their comparably low reactivity. The reaction rates are formulated as irreversible elementary steps without any stoichiometry. Neither the effect of water nor adsorption phenomena are implemented. An initiation phase is not observed, but the autocatalytic effect should be significantly lower over SAPO-34 according to the authors.

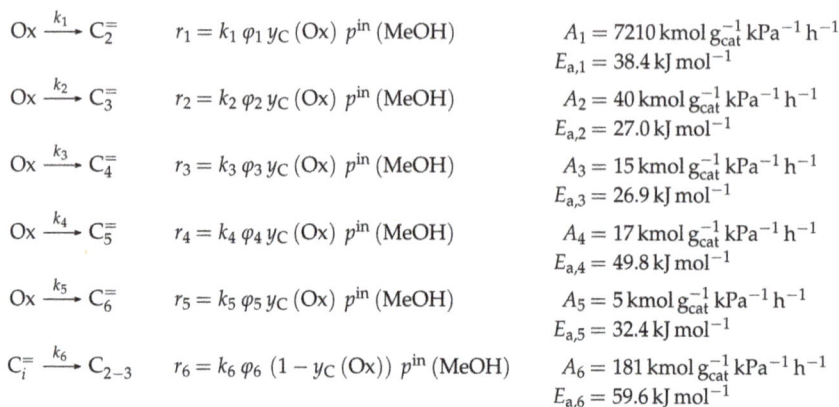

$$\text{Ox} \xrightarrow{k_1} C_2^= \qquad r_1 = k_1\, \varphi_1\, y_C\,(\text{Ox})\, p^{\text{in}}\,(\text{MeOH}) \qquad \begin{aligned} A_1 &= 7210\,\text{kmol}\,g_{\text{cat}}^{-1}\,\text{kPa}^{-1}\,\text{h}^{-1} \\ E_{a,1} &= 38.4\,\text{kJ}\,\text{mol}^{-1} \end{aligned}$$

$$\text{Ox} \xrightarrow{k_2} C_3^= \qquad r_2 = k_2\, \varphi_2\, y_C\,(\text{Ox})\, p^{\text{in}}\,(\text{MeOH}) \qquad \begin{aligned} A_2 &= 40\,\text{kmol}\,g_{\text{cat}}^{-1}\,\text{kPa}^{-1}\,\text{h}^{-1} \\ E_{a,2} &= 27.0\,\text{kJ}\,\text{mol}^{-1} \end{aligned}$$

$$\text{Ox} \xrightarrow{k_3} C_4^= \qquad r_3 = k_3\, \varphi_3\, y_C\,(\text{Ox})\, p^{\text{in}}\,(\text{MeOH}) \qquad \begin{aligned} A_3 &= 15\,\text{kmol}\,g_{\text{cat}}^{-1}\,\text{kPa}^{-1}\,\text{h}^{-1} \\ E_{a,3} &= 26.9\,\text{kJ}\,\text{mol}^{-1} \end{aligned}$$

$$\text{Ox} \xrightarrow{k_4} C_5^= \qquad r_4 = k_4\, \varphi_4\, y_C\,(\text{Ox})\, p^{\text{in}}\,(\text{MeOH}) \qquad \begin{aligned} A_4 &= 17\,\text{kmol}\,g_{\text{cat}}^{-1}\,\text{kPa}^{-1}\,\text{h}^{-1} \\ E_{a,4} &= 49.8\,\text{kJ}\,\text{mol}^{-1} \end{aligned}$$

$$\text{Ox} \xrightarrow{k_5} C_6^= \qquad r_5 = k_5\, \varphi_5\, y_C\,(\text{Ox})\, p^{\text{in}}\,(\text{MeOH}) \qquad \begin{aligned} A_5 &= 5\,\text{kmol}\,g_{\text{cat}}^{-1}\,\text{kPa}^{-1}\,\text{h}^{-1} \\ E_{a,5} &= 32.4\,\text{kJ}\,\text{mol}^{-1} \end{aligned}$$

$$C_i^= \xrightarrow{k_6} C_{2-3} \qquad r_6 = k_6\, \varphi_6\,(1 - y_C\,(\text{Ox}))\, p^{\text{in}}\,(\text{MeOH}) \qquad \begin{aligned} A_6 &= 181\,\text{kmol}\,g_{\text{cat}}^{-1}\,\text{kPa}^{-1}\,\text{h}^{-1} \\ E_{a,6} &= 59.6\,\text{kJ}\,\text{mol}^{-1} \end{aligned}$$

Scheme 33. Reaction network, rate equations and estimated parameters for the model by Chen et al. [257] with i ranging from 2 to 6; see [257] for the deactivation parameters φ_l.

The resulting net rates of formation are listed in Scheme 34.

$$R\,(\text{Ox}) = -k_1\, \varphi_1\, y_C\,(\text{Ox})\, p^{\text{in}}\,(\text{MeOH}) - k_2\, \varphi_2\, y_C\,(\text{Ox})\, p^{\text{in}}\,(\text{MeOH}) - k_3\, \varphi_3\, y_C\,(\text{Ox})\, p^{\text{in}}\,(\text{MeOH})$$
$$- k_4\, \varphi_4\, y_C\,(\text{Ox})\, p^{\text{in}}\,(\text{MeOH}) - k_5\, \varphi_5\, y_C\,(\text{Ox})\, p^{\text{in}}\,(\text{MeOH})$$
$$R\,(C_2^=) = k_1\, \varphi_1\, y_C\,(\text{Ox})\, p^{\text{in}}\,(\text{MeOH})$$
$$R\,(C_3^=) = k_2\, \varphi_2\, y_C\,(\text{Ox})\, p^{\text{in}}\,(\text{MeOH})$$
$$R\,(C_4^=) = k_3\, \varphi_3\, y_C\,(\text{Ox})\, p^{\text{in}}\,(\text{MeOH})$$
$$R\,(C_5^=) = k_4\, \varphi_4\, y_C\,(\text{Ox})\, p^{\text{in}}\,(\text{MeOH})$$
$$R\,(C_6^=) = k_5\, \varphi_5\, y_C\,(\text{Ox})\, p^{\text{in}}\,(\text{MeOH})$$
$$R\,(C_{2-3}) = k_6\, \varphi_6\,(1 - y_C\,(\text{Ox}))\, p^{\text{in}}\,(\text{MeOH})$$

Scheme 34. Net rates of production of the different lumps for the model by Chen et al. [257].

Parameter Estimation

Conversions and selectivities are based on carbon units, as is the mole fraction of oxygenates in Scheme 33. Here, the inlet partial pressure of methanol is also necessary. The reaction rates depend on the coke content wherefore a uniform distribution is assumed. The objective function which equals the weighted sum of squared residuals between predicted and measured mole fractions is minimized using *lsqnonlin* in MATLAB with the Levenberg–Marquardt algorithm. No information about the calculation of the weighting factors is given. The differential equations are integrated via a fourth-order Runge–Kutta method. The parameters of best description shown in Scheme 33 are obtained via isothermal regression at the four different temperatures and a subsequent Arrhenius plot. This causes twelve unknown values during one fitting run: six rate constants and six deactivation constants.

4.4.4. Alwahabi and Froment: Microkinetic Implementation over SAPO-34

Catalyst

The investigated SAPO-34 zeolite [258] powder had a small particle size of 1.1 μm. Measurements were performed after 0.25 h TOS where neither deactivation effects nor coke could be observed. In the final section of [258], TOS values of up to 3 h were achieved to model deactivation effects.

Setup and Conditions

For the measurements, a continuous fixed bed reactor was used. The feed consisted of 80%$_{mol}$ water to suppress deactivation effects. The catalyst bed was diluted 1:4 on a weight base with α-alumina in three layers. All experimental data points are shown in [287].

Reaction Network

The same network as for the work by Park and Froment [254] is applied. Thus, 172 pathways of kinetic relevance are included. However, because of the smaller catalyst pores, fitting is only performed for the following responses: DME, $C_2^=$, $C_3^=$, C_4^-, $C_5^=$ and C_1.

Parameter Estimation

The numerical method is identical to Park and Froment [254]. Due to the lack of higher olefins in the product stream, three parameters are missing here, i.e, protonation enthalpies of $C_6^=$, $C_7^=$ and $C_8^=$.

4.4.5. Gayubo et al.: Four- and Five-Lump Approach Including Deactivation Parameters over SAPO-18

Catalyst

A self-synthesized SAPO-18 zeolite was further processed to an extrudate [259]. The total number of acid sites was smaller compared to SAPO-34 and the acid strength was lower with a more uniform distribution which caused less deactivation. The measurements were performed up to a TOS of 1.5 h.

Setup and Conditions

In this study, a fluidized bed reactor with an internal diameter of 20 mm was applied. The catalyst bed was placed on a porous plate at a height of 285 mm from the bottom (total height of 465 mm). A ceramic chamber with a heating surrounded the whole reactor where the feed reactants were provided in liquid state. The temperature was measured both within the catalyst bed and in the vaporization chamber. A GC equipped with one column and an FID was used for product analysis. The whole setup is explained in detail in another publication [288]. For the experiments, alumina as diluent was mixed with the catalyst using a ratio of 1:4 on a weight base. Feed compositions with different gravimetric water/methanol ratios from 0 to 3 were analyzed.

Reaction Network

As it is obvious from Scheme 35, this model describes the reaction system with four lumps: Ox (methanol plus DME), $C_{2-5}^=$, C_1 and Int. The latter considers the initiation phase during which the oxygenates build up the first compounds of the hydrocarbon pool which themselves react with further oxygenates to higher intermediates (see Section 2.4.2). This lump is not further classified, but both the formation out of oxygenates (k_2) and the autocatalytic behaviour (k_3) are taken into account. The two remaining steps describe the olefin (k_4) and methane (k_5) evolution, the latter being the only side product detected. Because of the small pores, no species with a carbon number higher than five are detected. Although the reaction between methanol and DME is shown with a kinetic rate constant (k_1) in the original publication, which is similar to the ZSM-5 model by the same authors [114] (see

Section 4.2.1), both oxygenates are summarized to one lump in the model [260]. In another study [288], the authors observed that the amount of intermediates is almost independent of contact time. Thus, their evolution is only evaluated as time-dependent variable (see Scheme 35). The adsorption of water is assumed to attenuate the other reaction rates which is why the equations are written as HW type of mechanism without the adsorption of all other compounds. The steps are implemented as elementary reactions and no stoichiometry is retained. All steps are defined as irreversible. The models also describes the deactivation through coke deposition. For this, a rate constant for deactivation is introduced. Furthermore, all reaction rates except for methane production are multiplied with the activity a. This value expresses the ratio of the olefin production rate at a certain TOS to the one when activity would be unity, i.e., the fresh catalyst. For the deactivation rate, a different equilibrium constant and a different exponent of water adsorption are assumed.

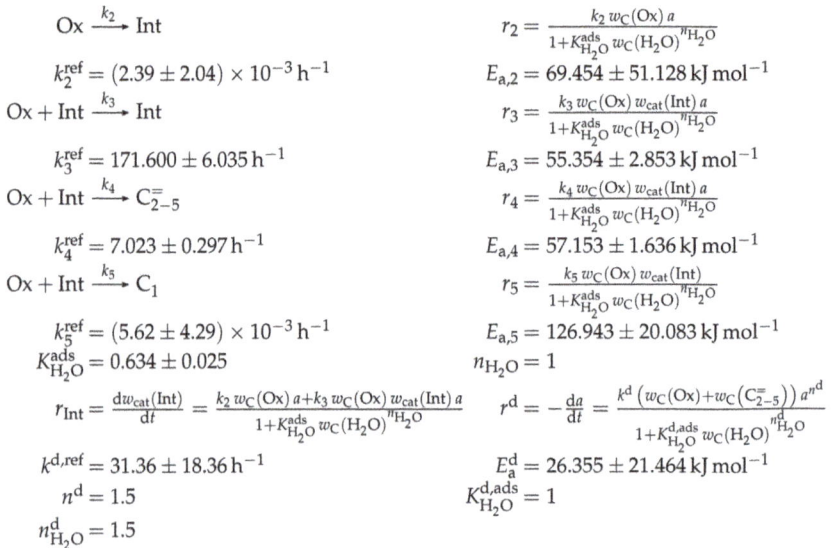

$$Ox \xrightarrow{k_2} Int$$

$$k_2^{ref} = (2.39 \pm 2.04) \times 10^{-3}\,h^{-1}$$

$$Ox + Int \xrightarrow{k_3} Int$$

$$k_3^{ref} = 171.600 \pm 6.035\,h^{-1}$$

$$Ox + Int \xrightarrow{k_4} C_{2-5}^=$$

$$k_4^{ref} = 7.023 \pm 0.297\,h^{-1}$$

$$Ox + Int \xrightarrow{k_5} C_1$$

$$k_5^{ref} = (5.62 \pm 4.29) \times 10^{-3}\,h^{-1}$$
$$K_{H_2O}^{ads} = 0.634 \pm 0.025$$

$$r_{Int} = \frac{dw_{cat}(Int)}{dt} = \frac{k_2\,w_C(Ox)\,a + k_3\,w_C(Ox)\,w_{cat}(Int)\,a}{1 + K_{H_2O}^{ads}\,w_C(H_2O)^{n_{H_2O}}}$$

$$k^{d,ref} = 31.36 \pm 18.36\,h^{-1}$$
$$n^d = 1.5$$
$$n_{H_2O}^d = 1.5$$

$$r_2 = \frac{k_2\,w_C(Ox)\,a}{1 + K_{H_2O}^{ads}\,w_C(H_2O)^{n_{H_2O}}}$$
$$E_{a,2} = 69.454 \pm 51.128\,kJ\,mol^{-1}$$

$$r_3 = \frac{k_3\,w_C(Ox)\,w_{cat}(Int)\,a}{1 + K_{H_2O}^{ads}\,w_C(H_2O)^{n_{H_2O}}}$$
$$E_{a,3} = 55.354 \pm 2.853\,kJ\,mol^{-1}$$

$$r_4 = \frac{k_4\,w_C(Ox)\,w_{cat}(Int)\,a}{1 + K_{H_2O}^{ads}\,w_C(H_2O)^{n_{H_2O}}}$$
$$E_{a,4} = 57.153 \pm 1.636\,kJ\,mol^{-1}$$

$$r_5 = \frac{k_5\,w_C(Ox)\,w_{cat}(Int)}{1 + K_{H_2O}^{ads}\,w_C(H_2O)^{n_{H_2O}}}$$
$$E_{a,5} = 126.943 \pm 20.083\,kJ\,mol^{-1}$$
$$n_{H_2O} = 1$$

$$r^d = -\frac{da}{dt} = \frac{k^d\left(w_C(Ox) + w_C(C_{2-5}^=)\right)a^{n^d}}{1 + K_{H_2O}^{d,ads}\,w_C(H_2O)^{n_{H_2O}^d}}$$
$$E_a^d = 26.355 \pm 21.464\,kJ\,mol^{-1}$$
$$K_{H_2O}^{d,ads} = 1$$

Scheme 35. Reaction network, rate equations and estimated parameters for the model by Gayubo et al. [259] over SAPO-18 (four lumps).

Scheme 36 contains the resulting net rates of production.

$$R\,(Ox) = \frac{-k_4\,w_C\,(Ox)\,w_{cat}\,(Int)\,a - k_5\,w_C\,(Ox)\,w_{cat}\,(Int)}{1 + K_{H_2O}^{ads}\,w_C\,(H_2O)^{n_{H_2O}}}$$

$$R\,(C_{2-5}^=) = \frac{k_4\,w_C\,(Ox)\,w_{cat}\,(Int)\,a}{1 + K_{H_2O}^{ads}\,w_C\,(H_2O)^{n_{H_2O}}}$$

$$R\,(C_1) = \frac{k_5\,w_C\,(Ox)\,w_{cat}\,(Int)}{1 + K_{H_2O}^{ads}\,w_C\,(H_2O)^{n_{H_2O}}}$$

Scheme 36. Net rates of production of the different lumps for the model by Gayubo et al. [259] over SAPO-18 (four lumps).

The reaction network for the five-lump model is based on the previous one for SAPO-18, but the lump $C_{2-5}^=$ is replaced with three separate olefin lumps $C_2^=$, $C_3^=$ and $C_{4+}^=$ in order to account for their different reactivities and evolutions depending on the reaction conditions. In the publication, five different networks with varying complexity for olefin interconversion are introduced and the one presented in Scheme 37 is chosen after an evaluation with the Fisher test. It should be noted that, except for the replacement of Step k_4 with Steps k_{4a}, k_{4b} and k_{4c}, the same reaction network as in Scheme 35 applies. The earlier rate constant k_4 should yield the sum of k_{4a}, k_{4b} and k_{4c}. The deactivation approach is unselective: the activity a relates the production rate after a certain TOS to the value at $a = 1$ for ethene, propene and higher olefins.

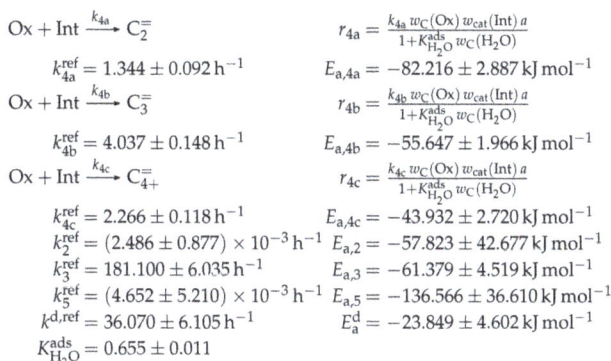

$$Ox + Int \xrightarrow{k_{4a}} C_2^=$$
$$r_{4a} = \frac{k_{4a}\, w_C(Ox)\, w_{cat}(Int)\, a}{1 + K_{H_2O}^{ads}\, w_C(H_2O)}$$

$$k_{4a}^{ref} = 1.344 \pm 0.092\,\text{h}^{-1} \qquad E_{a,4a} = -82.216 \pm 2.887\,\text{kJ}\,\text{mol}^{-1}$$

$$Ox + Int \xrightarrow{k_{4b}} C_3^=$$
$$r_{4b} = \frac{k_{4b}\, w_C(Ox)\, w_{cat}(Int)\, a}{1 + K_{H_2O}^{ads}\, w_C(H_2O)}$$

$$k_{4b}^{ref} = 4.037 \pm 0.148\,\text{h}^{-1} \qquad E_{a,4b} = -55.647 \pm 1.966\,\text{kJ}\,\text{mol}^{-1}$$

$$Ox + Int \xrightarrow{k_{4c}} C_{4+}^=$$
$$r_{4c} = \frac{k_{4c}\, w_C(Ox)\, w_{cat}(Int)\, a}{1 + K_{H_2O}^{ads}\, w_C(H_2O)}$$

$$k_{4c}^{ref} = 2.266 \pm 0.118\,\text{h}^{-1} \qquad E_{a,4c} = -43.932 \pm 2.720\,\text{kJ}\,\text{mol}^{-1}$$
$$k_2^{ref} = (2.486 \pm 0.877) \times 10^{-3}\,\text{h}^{-1} \quad E_{a,2} = -57.823 \pm 42.677\,\text{kJ}\,\text{mol}^{-1}$$
$$k_3^{ref} = 181.100 \pm 6.035\,\text{h}^{-1} \qquad E_{a,3} = -61.379 \pm 4.519\,\text{kJ}\,\text{mol}^{-1}$$
$$k_5^{ref} = (4.652 \pm 5.210) \times 10^{-3}\,\text{h}^{-1} \quad E_{a,5} = -136.566 \pm 36.610\,\text{kJ}\,\text{mol}^{-1}$$
$$k^{d,ref} = 36.070 \pm 6.105\,\text{h}^{-1} \qquad E_a^d = -23.849 \pm 4.602\,\text{kJ}\,\text{mol}^{-1}$$
$$K_{H_2O}^{ads} = 0.655 \pm 0.011$$

Scheme 37. Reaction network, rate equations and estimated parameters for the model by Gayubo et al. [255] over SAPO-18 (five lumps); the corresponding equations for Steps 2, 3, 5 and d can be found in Scheme 35.

The net rates of formation are the same as for the four-lump model except that $R\left(C_{2-5}^=\right)$ has to be replaced with $R\left(C_2^=\right)$, $R\left(C_3^=\right)$ and $R\left(C_{4+}^=\right)$, which correspond to r_{4a}, r_{4b} and r_{4c}, respectively.

Parameter Estimation

The mass fractions in Scheme 35 are defined with carbon units except for the intermediates where y_{cat} (Int) is related to the mass of the fresh catalyst. Both the contact time dependent kinetic expressions as well as the TOS dependent equations for deactivation and intermediates have to be solved simultaneously. For this, a MATLAB script based on finite differences in combination with orthogonal collocation [269] is written. Parameter estimation is performed with the Levenberg–Marquardt algorithm where the objective function evaluates the unweighted squared differences between modeled and experimental mass fractions. Reparameterization according to Equation (21) is performed with a reference temperature of 623 K which is close to the lowest investigated value. Different values for n^d, n_{H_2O}, $n_{H_2O}^d$ and $K_{H_2O}^{d,ads}$ are tried, the results with the best fit are shown in Scheme 35. Without these, eleven unknown parameters remain: five reference rate constants, five activation energies and one equilibrium constant for water adsorption. For the version with five lumps, 15 parameters are estimated: seven reference rate constants, seven activation energies and one equilibrium constant for water adsorption. As it can be seen in Scheme 37, the values which were already included in the model with four lumps [259] are fitted another time here.

4.4.6. Kumar et al.: Microkinetic Implementation over ZSM-22

Catalyst

The authors [146] used a commercial ZSM-23 sample without any binder provided by Zeolyst International. The relatively low Si/Al ratio of 26 caused a high number of acid sites ($0.62\,\text{mol}\,\text{kg}_{cat}^{-1}$). The zeolite showed significant deactivation effects due to coke formation [147]. However, it was observed that the selectivity at a specific conversion level is independent of the coke amount [231]. Through the linear dependence between TOS and contact time until a certain conversion is achieved, an effective contact time is calculated in this study. This allows describing intrinsic kinetics free of interfering deactivation effects.

Setup and Conditions

The continuous fixed bed glass reactor had an inner diameter of 10 mm. It was fed by a saturator with helium as carrier and dilution gas. A GC equipped with an FID and one column enabled product analysis. In this study, only one temperature was analyzed; it was controlled by a thermocouple placed in the middle of the catalyst bed [289,290].

Reaction Network

The network is almost similar to the ZSM-5 model by the same authors; the only differences are caused by the use of a different catalyst [11]. Over ZSM-23, profound ethene formation out of olefins is observed which is why two additional cracking routes leading to primary intermediates are introduced, starting from either tertiary or from secondary intermediates. Because the reverse reaction takes also place, protonation to a primary intermediate has to be included; the stability difference between secondary and primary intermediates is an additional fitting parameter in this model. Physisorption is included with own experimental data of alkanes over ZSM-22 [291] which are applicable to ZSM-23 [292]. As in the ZSM-5 case, the formation of side products and especially of aromatics is negligible. Finally, 142 pathways of kinetic relevance are obtained. The following responses are fitted to the measurements: Ox (methanol plus DME), $C_2^=$, $C_3^=$, $C_4^=$, $C_5^=$, $C_{6+}^=$ and C_1. The amount of methanol within the Ox lump is calculated from a carbon balance, whereas water is obtained from a hydrogen balance.

Parameter Estimation

The numerical routine is similar to the ZSM-5 case. The kinetic descriptors determined earlier [19] are held constant, whereas the different catalyst descriptors are estimated. This leads to eight fitted parameters: two activation energies including primary intermediates, five protonation enthalpies and one stability difference between primary and secondary intermediates. In addition, the total concentration of aromatic hydrocarbon pool species is also obtained via regression as this value changes with a different catalyst type.

4.4.7. Summary

Because of the smaller pore size, deactivation is more pronounced over SAPO-34, SAPO-18 and ZSM-22 compared to ZSM-5 and cannot be ignored during kinetic evaluation. The four models over SAPO-34 show different methodologies to consider this fact. Gayubo et al. [255] chose conditions where deactivation effect are minimized and extrapolate their results to a fresh catalyst. This is why they could neglect coking effects in their model. Alwahabi and Froment [258] had a similar approach as they use kinetic measurements of an almost fresh catalyst and simulate deactivation with separate data. Ying et al. [256] estimated their parameters according to their kinetic scheme first; in a subsequent step, these are held constant, whereas rate-specific deactivation parameters are fitted. This procedure requires kinetic data free of deactivation effects for the first step. In contrast, Chen et al. [257] estimated these rate-specific deactivation values directly with the kinetic parameters. Except for the microkinetic approach of Alwahabi and Froment [258], the different reactivity of DME is ignored through lumping both oxygenates (Gayubo et al. [255] and Chen et al. [257]) or through considering only reactions starting from methanol (Ying et al. [256]). In the latter study, all olefin interconversion steps as well as methylation reactions are neglected. The same holds for the model by Chen et al. [257]. On the other hand, both approaches consider side product formation which is also included for Gayubo et al. [255]. In the latter, methylation is also missing, whereas some olefin interconversion steps are assumed. In the five lump version, lumping of final and intermediate products might impede extrapolation; this is also observed for Ying et al. [256]. Furthermore, in the approach with five lumps, the olefin interconversion steps are removed. On the other hand, this model as well as Ying et al. [256] consider water adsorption in an HW type of mechanism which is ignored for Chen et al. [257]. Finally, the approach by Alwahabi and Froment [258] depicts almost complete reactivity, but at cost of complex reaction networks and high computational effort. The SAPO-18 model by Gayubo et al. [259] is comparable to the SAPO-34 case. However, this version includes deactivation parameters which are directly fitted to the kinetic data. This model is the only one in the review that explicitly describes the evolution of the initiation phase via a lump of intermediates. The five lump version additionally has the advantage that the lower olefins are split up to separate lumps. Finally, Kumar et al. [146] took advantage of the effect that selectivity is independent of coking at a certain conversion level. Through a linear approach, they could convert data at specific TOS to the performance of a fresh catalyst. Besides this, the model is almost identical to the one over ZSM-5 with the same advantages and disadvantages. The transfer to ZSM-22 shows how a separation of kinetic and catalyst descriptors [54] allows one to move a specific model obtained on a certain catalyt to another one by holding the kinetics constant and by adapting the reaction network and catalyst specific values.

4.5. Other Studies

Another well-known model is the one by Kaarsholm et al. [293]. Here, a commercial ZSM-5 zeolite was further modified. The final catalyst contained 1.5% phosphorus. Experiments were performed at temperatures between 673 and 823 K. The feed consisted either of pure methanol or of mixtures with water or argon. Deactivation effects can be neglected for the kinetics. A fluidized bed reactor model is combined with a kinetic scheme consisting of eleven lumps. Here, 16 unknown parameters are estimated to experimental data. The model includes water adsorption in an HW type of mechanism. Furthermore, side product formation is covered. Methanol and DME are assumed to be equilibrated throughout the whole reactor. All steps producing hydrocarbons out of the oxygenates have to proceed via a protonated intermediate with ten carbon atoms.

In a recent study by Yuan et al. [294], a kinetic model is derived for converting methanol feeds on a commercial SAPO-34 catalyst. The authors conducted experiments in a fluidized bed reactor at temperatures between 698–763 K. The feed was diluted using nitrogen. The kinetic model consists of nine lumps and requires 34 parameters. The dual cycle is implemented via two virtual species, one characterizing the olefin and another one resembling the aromatic hydrocarbon pool. Deactivation is also considered to describe the product evolution as function of TOS. Several reactor modeling studies were already published by this group [295–297].

In the approach by Strizhak et al. [298], a 1:1 mixture of commercial ZSM-5 zeolize (Si/Al of 35.4) and alumina was analyzed at temperatures between 513 and 693 K. The methanol feed was diluted with argon, leading to methanol partial pressures between 0.055 and 0.236 bar. Different theoretical reaction mechanisms are compared to the experimental data. Highest agreement is achieved when the DME formation is assumed to occur on LAS, whereas the conversion of oxygenates takes place on BAS.

Other studies in this context are the ones by Sedighi et al. [299], Fatourehchi et al. [300], Taheri Najafabadi et al. [301] and Azarhoosh et al. [302].

5. Kinetic Models for Methanol-to-Olefins with Olefin Co-Feed

The properties of the catalysts are listed in Table 5 and an overview of experimental conditions as well as modeling details are found in Table 6. Then, an explanation of the different models follows, focusing on studies where olefins are co-fed with the oxygenates. Consequently, the initiation phase should disappear which leads to a direct increase of oxygenates conversion (see Section 2.4.2). No division into different subsections is performed because there are only two models. Nevertheless, a summary section is shown at the end. This section is about the models by Huang et al. [240] and Wen et al. [303]. The former is a subsequent study to the olefin interconversion work discussed above.

Table 5. Properties of the different catalysts which were used for the kinetic models of methanol-to-olefins with olefin co-feed; besides the zeolite type, its silicon-to-aluminum ratio (Si/Al), its total number of acid sites as well as the determination method, its ratio of Brønsted to Lewis acid sites (BAS/LAS) and its surface area according to the method by Brunauer–Emmett–Teller (BET) are shown. Furthermore, the time-on-stream (TOS) after which the kinetic data were taken, the particle size (d_P) and the information on whether an extrudate or pure powder was used are presented. A hyphen represents missing information.

Model	Zeolite Type	Si/Al	Total Acidity	BAS/LAS	BET	TOS	d_P	Extrudate
Huang [240]	ZSM-5	200 [1]	0.012 mmol g_{cat}^{-1} (NH_3)	1.35 at 423 K	301.1 m² g_{cat}^{-1}	0–10 h [2]	125–149 µm	70/30%wt (Zeolite/Alumina)
Wen [303]	ZSM-5 on microfibers	147	-	-	93 m² g_{cat}^{-1} [3]	-	16.1 mm [4]	No, but 19/81%wt (Zeolite/Microfiber)
	ZSM-5	155	-	-	-	-	100–300 µm	No

(1) Value of the zeolite, i.e., without binder. (2) Regeneration after 10 h TOS. (3) Extracted from [304]. (4) Diameter of the circular sample chips.

Table 6. Experimental conditions and modeling details for the kinetic models of methanol-to-olefins with olefin co-feed; the feed components, the temperature range (T), the total pressure (p_t), the partial pressure range of the feed oxygenates as well as the feed olefins (p_i, in that order) and the maximum contact time ($(W/F^{in})_{max}$) with resulting oxygenates conversion (X_{max}) are listed; concerning the model, the number of fitted responses (N_{Res}), the number of estimated parameters (N_{Par}), the number of experiments (N_{Exp}) and the degree of freedom (dof) and, finally, it is noted whether the model follows a type of a mechanistical scheme (Mech.), whether adsorption is considered (Ads.) and which side products are included (Side prod.). A hyphen represents missing information.

Model	Feed	T	p_t	p_i	$(W/F^{in})_{max}$	X_{max}	N_{Res}	N_{Par}	N_{Exp}	dof	Mech.	Ads.	Side prod.
Huang [240]	MeOH, $C_3^=$, N_2, H_2O [1]	673–763 K	1.013 bar	0.050 bar / 0.050 bar	4.3 kg_{cat} s mol_t^{-1}	0.90	8	20	79	612	LH, HW, ER	$C_{2-7}^=$, H_2O, MeOH	No [2]
	MeOH, $C_4^=$, N_2, H_2O [1]				2.9 kg_{cat} s mol_t^{-1}	0.92							
	MeOH, $C_5^=$, N_2, H_2O [1]				2.5 kg_{cat} s mol_t^{-1}	0.92							
	MeOH, $C_6^=$, N_2, H_2O [1]				2.5 kg_{cat} s mol_t^{-1}	0.92							
Wen [303][3]	MeOH, N_2 [4]	673–753 K	1.013 bar [5]	0.304 bar	32 g_{cat} h mol_C^{-1}	1	10	38	46	422	No	No	C_{1-6}
Wen [303][3]	MeOH, N_2 [4]	673–753 K	1.013 bar [5]	0.304 bar	32 g_{cat} h mol_C^{-1}	1	10	38	45	412	No	No	C_{1-6}

(1) Respective n-alcohol was fed instead of the olefin. (2) Side product formation included in subsequent publication [305]. (3) Similar kinetic model for microfibered (first line) and powdered (second line) catalyst, but different dof. (4) Pure methanol as feed, but model only works when olefins are present (comparable to co-feeding studies). (5) Extracted from [304].

5.1. Huang et al.: Eight-Lump Approach Extending the Olefin Cracking Model to Methanol-to-Olefins

5.1.1. Catalyst

The authors [240] used the same catalyst as for the olefin cracking study [166], see Section 3.1.3.

5.1.2. Setup and Conditions

The reaction equipment is already described in Section 3.1.3. However, another GC column was used here for better separating the oxygenates from the olefins. The partial pressure of water was held constant at 0.24 bar for all measurements which includes the amount released during alcohol dehydration.

5.1.3. Reaction Network

All reactions from the olefin interconversion model of the same authors [166] (see Section 3.1.3) are also included here. This network is extended with the methanol related reactions in Scheme 38, which include the conversion to DME and water (k_{11}) as well as methylation steps (k_{12}–k_{18}). Consequently, the following lumps are described: MeOH, DME, $C_2^=$, C_3^-, $C_4^=$, C_5^-, $C_6^=$ and $C_{7+}^=$. Because of the fast reaction progress under co-feeding conditions, no comparably slow conversion steps of oxygenates to hydrocarbons are implemented. For the same reason, the methanol reaction to DME and water is not treated as equilibrated: based on experiments and calculations, the authors could show that the fast methylation disturbs the equilibration of the oxygenates. Only propene to hexene are considered as possible reactants for methylations as an earlier study proved this reaction to be very slow when having ethene as co-feed [210]. A mechanistic pathway is implemented here: the methanol chemisorption on a Brønsted acid site leads to a surface methyl group, which methylates an olefin reacting out of the gas phase in a subsequent step. Besides the olefin interconversion steps implemented as combination of LH and HW types of mechanism (Section 3.1.3), the network thus comprises irreversible methylation steps expressed as ER type of mechanism. A first regression without the steps k_{16}–k_{18} showed significant deviation especially for propene and butenes, indicating the absence of an important pathway for these species. The comparison with experimental results from Svelle et al. [209] leads to the formulation of double methylation reactions. In [209], it was observed that pentenes contain marked ^{13}C methanol in an amount that cannot be explained by simple stepwise methylation reactions of the co-fed ^{12}C propene. Based on these experiments, Huang et al. [240] formulate the double methylation of propene and butenes as well as a triple methylation of propene; all these are also assumed as ER type of mechanism, meaning that two or three methanol molecules have to be chemisorbed first. As for the olefin interconversion model by the same authors [166], stoichiometry is retained and adsorption is considered for all hydrocarbons, for methanol and for water (HW type of mechanism), but not for DME. Furthermore, the methylation through DME is not considered. In this work, side products are neglected because of short contact times. The yield of aromatics and paraffins was below 0.4% in all experiments. However, in a subsequent study [305], their formation is explicitly included.

$$2\text{MeOH} \xrightleftharpoons[k_{11}/K_{11}]{k_{11}} \text{DME} + \text{H}_2\text{O} \qquad r_{11} = \frac{k_{11}}{Den^2}\left(p\left(\text{MeOH}\right)^2 - \frac{1}{K_{11}}p\left(\text{DME}\right)p\left(\text{H}_2\text{O}\right)\right)$$

$$k_{11}^{ref} = (6.78 \pm 0.23) \times 10^{-3}\,\text{mol kg}_{cat}^{-1}\,\text{s}^{-1}\,\text{kPa}^{-2} \qquad E_{a,11} = 36.29 \pm 6.94\,\text{kJ mol}^{-1}$$

$$K_{11}^{ref} = 1.19 \pm 0.30\,\text{kPa}^{-2} \qquad \Delta_r H_{11}^\circ = -15.85 \pm 3.35\,\text{kJ mol}^{-1}$$

$$\text{C}_3^= + \text{MeOH} \xrightarrow{k_{12}} \text{C}_4^= + \text{H}_2\text{O} \qquad r_{12} = \frac{k_{12}}{Den^2}p\left(\text{C}_3^=\right)p\left(\text{MeOH}\right)$$

$$k_{12}^{ref} = (2.11 \pm 0.39) \times 10^{-3}\,\text{mol kg}_{cat}^{-1}\,\text{s}^{-1}\,\text{kPa}^{-2} \qquad E_{a,12} = 49.68 \pm 6.14\,\text{kJ mol}^{-1}$$

$$\text{C}_4^= + \text{MeOH} \xrightarrow{k_{13}} \text{C}_5^= + \text{H}_2\text{O} \qquad r_{13} = \frac{k_{13}}{Den^2}p\left(\text{C}_4^=\right)p\left(\text{MeOH}\right)$$

$$k_{13}^{ref} = (5.20 \pm 0.78) \times 10^{-3}\,\text{mol kg}_{cat}^{-1}\,\text{s}^{-1}\,\text{kPa}^{-2} \qquad E_{a,13} = 69.19 \pm 4.73\,\text{kJ mol}^{-1}$$

$$\text{C}_5^= + \text{MeOH} \xrightarrow{k_{14}} \text{C}_6^= + \text{H}_2\text{O} \qquad r_{14} = \frac{k_{14}}{Den^2}p\left(\text{C}_5^=\right)p\left(\text{MeOH}\right)$$

$$k_{14}^{ref} = (7.80 \pm 0.98) \times 10^{-3}\,\text{mol kg}_{cat}^{-1}\,\text{s}^{-1}\,\text{kPa}^{-2} \qquad E_{a,14} = 53.10 \pm 5.69\,\text{kJ mol}^{-1}$$

$$\text{C}_6^= + \text{MeOH} \xrightarrow{k_{15}} \text{C}_7^= + \text{H}_2\text{O} \qquad r_{15} = \frac{k_{15}}{Den^2}p\left(\text{C}_6^=\right)p\left(\text{MeOH}\right)$$

$$k_{15}^{ref} = (7.50 \pm 1.08) \times 10^{-3}\,\text{mol kg}_{cat}^{-1}\,\text{s}^{-1}\,\text{kPa}^{-2} \qquad E_{a,15} = 41.99 \pm 5.63\,\text{kJ mol}^{-1}$$

$$\text{C}_3^= + 2\text{MeOH} \xrightarrow{k_{16}} \text{C}_5^= + 2\text{H}_2\text{O} \qquad r_{16} = \frac{k_{16}}{Den^3}p\left(\text{C}_3^=\right)p\left(\text{MeOH}\right)^2$$

$$k_{16}^{ref} = (7.80 \pm 0.33) \times 10^{-5}\,\text{mol kg}_{cat}^{-1}\,\text{s}^{-1}\,\text{kPa}^{-3} \qquad E_{a,16} = 25.27 \pm 11.27\,\text{kJ mol}^{-1}$$

$$\text{C}_4^= + 2\text{MeOH} \xrightarrow{k_{17}} \text{C}_6^= + 2\text{H}_2\text{O} \qquad r_{17} = \frac{k_{17}}{Den^3}p\left(\text{C}_4^=\right)p\left(\text{MeOH}\right)^2$$

$$k_{17}^{ref} = (9.24 \pm 1.52) \times 10^{-4}\,\text{mol kg}_{cat}^{-1}\,\text{s}^{-1}\,\text{kPa}^{-3} \qquad E_{a,17} = 11.98 \pm 3.66\,\text{kJ mol}^{-1}$$

$$\text{C}_3^= + 3\text{MeOH} \xrightarrow{k_{18}} \text{C}_6^= + 3\text{H}_2\text{O} \qquad r_{18} = \frac{k_{18}}{Den^4}p\left(\text{C}_3^=\right)p\left(\text{MeOH}\right)^3$$

$$k_{18}^{ref} = (1.83 \pm 0.42) \times 10^{-4}\,\text{mol kg}_{cat}^{-1}\,\text{s}^{-1}\,\text{kPa}^{-4} \qquad E_{a,18} = 1.10 \pm 2.87\,\text{kJ mol}^{-1}$$

$$Den = 1 + K_{C_{2-7}^=}^{ads}\sum_j p\left(\text{C}_j^=\right) + K_{\text{H}_2\text{O}}^{ads}p\left(\text{H}_2\text{O}\right) + K_{\text{MeOH}}^{ads}p\left(\text{MeOH}\right)$$

$$K_{\text{MeOH}}^{ads,ref} = (4.36 \pm 0.35) \times 10^{-2}\,\text{kPa}^{-1} \qquad \Delta_{ads}H_{\text{MeOH}}^\circ = -47.50 \pm 4.59\,\text{kJ mol}^{-1}$$

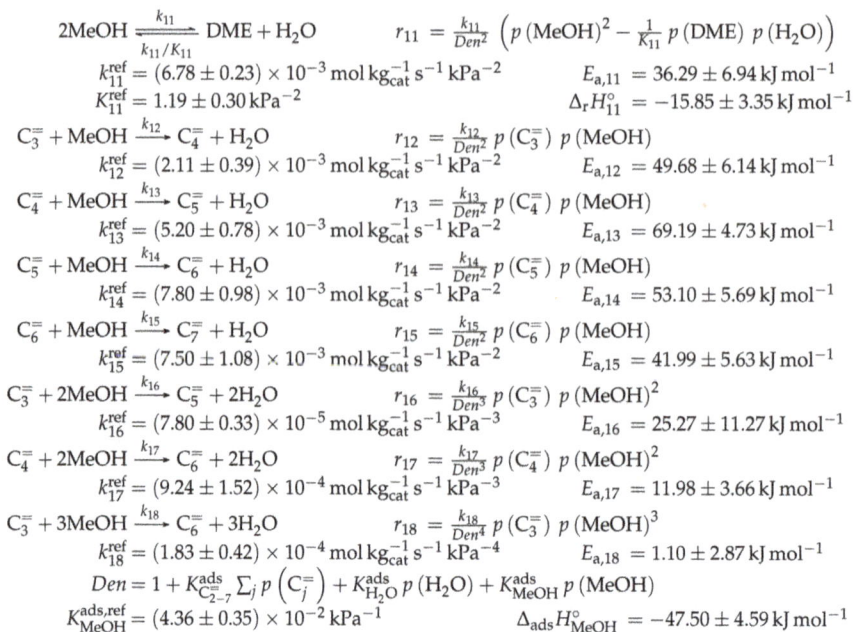

Scheme 38. Reaction network, rate equations and estimated parameters for the model by Huang et al. [240] with *j* ranging from 2 to 7.

Because the reaction rates are rather complex, only r_l is shown for the net rates of formation in Scheme 39.

$$R\left(\text{MeOH}\right) = -2\,r_{11} - r_{12} - r_{13} - r_{14} - r_{15} - 2\,r_{16} - 2\,r_{17} - 3\,r_{18}$$

$$R\left(\text{DME}\right) = r_{11}$$

$$R\left(\text{C}_2^=\right) = r_3 + r_6$$

$$R\left(\text{C}_3^=\right) = r_2 + r_3 + 2\,r_4 + 2\,r_7 + r_8 + r_{10} - 3\,r_1 - r_{12} - r_{16} - r_{18}$$

$$R\left(\text{C}_4^=\right) = r_1 + r_4 + r_5 + r_6 + r_8 + 3\,r_9 + r_{10} + r_{12} - 2\,r_2 - r_{13} - r_{17}$$

$$R\left(\text{C}_5^=\right) = r_1 + r_2 + r_8 + r_{13} + r_{16} - r_3 - 2\,r_4 - 2\,r_5 - r_{14}$$

$$R\left(\text{C}_6^=\right) = r_5 + r_{14} + r_{17} + r_{18} - r_6 - r_7 - 2\,r_8 - 2\,r_9 - r_{15}$$

$$R\left(\text{C}_7^=\right) = r_{15} - r_{10}$$

Scheme 39. Net rates of production of the different lumps for the model by Huang et al. [240].

5.1.4. Parameter Estimation

Details about the numerical routine can be found in Section 3.1.3. All parameters estimated there are kept constant during fitting the MTO model. For the latter, 20 unknown values exist: eight reference rate constants, eight activation energies, two reference equilibrium constants, one reaction enthalpy and one adsorption enthalpy. This includes the equilibrium constant of methanol dehydration because the experimental value deviated from the theoretical one calculated via thermodynamics. The final kinetic description, enriched with the side reactions [305], is used in subsequent studies to create a heterogeneous model of the recycle reactor [207,306] or of a monolith reactor [307].

5.2. Wen et al.: Ten-Lump Model Being Valid for ZSM-5 Powder and for ZSM-5 on Stainless Steel Fibers

5.2.1. Catalyst

In this work [303], a regular ZSM-5 zeolite as well as a catalyst consisting of ZSM-5 crystals grown on three-dimensional stainless steel microfibers were analyzed. The latter showed improved stability and propene yields in earlier studies [304,308]. This is attributed to higher resistances to the aromatic hydrocarbon pool which reduces ethene formation and to a narrow residence distribution being optimal for propene as intermediate. Furthermore, the small zeolite shell being only a few micrometers thick increases mass transfer as well as acid sites efficiency and thus activity compared to regular powder. Both samples were self-synthesized, but Western Metal Material provided the stainless steel fibers with a diameter of 20 μm and a voidage of 85%. Through the dip-coating method [304], a catalyst with 19%$_{wt}$ ZSM-5 and a Si/Al ratio of 147 was obtained. The powder exhibited a comparable Si/Al value of 155.

5.2.2. Setup and Conditions

Kinetic experiments were performed in a continuous fixed bed reactor made of quartz glass which had an inner diameter of 16 mm. An electrical furnace surrounding the reactor allowed for elevated temperatures. Methanol was fed in liquid state, evaporated and mixed with nitrogen as diluent. The fibered samples, provided as circular chips, were filled in layer by layer. Their diameter was 0.1 mm larger compared to the reactor to avoid bypass effects. In contrast, the application of quartz sand as diluent enabled comparable bed volumes for the powder sample. For product analysis, the authors used a GC having one column and an FID. As shown in Table 6, solely methanol was applied as feed. However, the model only works when olefins are present, otherwise, ethene and paraffins are produced exclusively. Therefore, the application range is similar to co-feeding conditions.

5.2.3. Reaction Network

For both samples, the same model is applied which consists of ten lumps: MeOH, DME, $C_2^=$, $C_3^=$, $C_4^=$, $C_5^=$, $C_6^=$, $C_{7+}^=$, C_1 and C_{2-6}. The reactions outlined in Scheme 40 can be divided into six parts: oxygenates interconversion (k_1 and k_2), methylation (k_3–k_6), olefin interconversion (k_7–k_{13} and k_{19}), oxygenates conversion to olefins (k_{14}) and paraffin formation out of olefins (k_{16} and k_{17}) as well as out of DME (k_{15} and k_{18}). The methylation is assumed to occur exclusively via DME which increases the carbon number of two similar olefins ($C_3^=$ to $C_6^=$) by one each. As mentioned in the previous paragraph, the model does not start at zero contact time, but at a minimum value where the end of the initiation phase is reached which means that the first olefins are produced already. Because of the low reactivity at the beginning, the oxygenates reached an equilibrated state at the first data points. For their model, the authors implemented both the forward reaction and the backward reaction of methanol dehydration as step of kinetic relevance without any equilibrium constants. The contribution of the aromatic hydrocarbon pool is restricted to the conversion of DME to ethene for simplicity. In general, methanol is not considered as reactant except for DME production. The olefin interconversion comprises the cracking of pentenes and hexenes including backward reactions whereby these are separately fitted again. Moreover, the formation of higher olefins as well as the dimerization of butenes is considered, but without reverse reaction. Finally, methane formation is limited to pentenes or DME, whereas the latter or propene can also react to lower paraffins. The origin of hydrogen necessary for methane formation is not resolved, also water is ignored in the rate equations of Scheme 40. Adsorption effects and mechanistic routes are not covered by this model. According to the authors, the reaction orders are adjusted to have highest agreement, but in fact, all are set to one. This means stoichiometry is neglected, as it also has arbitrary values for the net rates of production.

$$2\text{MeOH} \xrightarrow{k_1} \text{DME} + \text{H}_2\text{O}$$
$$E_{a,1}^{\text{MF}} = 13.46 \, \text{kJ} \, \text{mol}^{-1}$$

$$r_1 = k_1 \, y \, (\text{MeOH})$$
$$E_{a,1}^{\text{PO}} = -1.77 \, \text{kJ} \, \text{mol}^{-1}$$

$$k_1^{\text{MF,ref}} = 2.14 \, \text{mol}_\text{C} \, \text{h}^{-1} \, \text{g}_\text{cat}^{-1}$$
$$k_1^{\text{PO,ref}} = 0.36 \, \text{mol}_\text{C} \, \text{h}^{-1} \, \text{g}_\text{cat}^{-1}$$

$$\text{DME} + \text{H}_2\text{O} \xrightarrow{k_2} 2\text{MeOH}$$
$$E_{a,2}^{\text{MF}} = 13.46 \, \text{kJ} \, \text{mol}^{-1}$$

$$r_2 = k_2 \, y \, (\text{DME})$$
$$E_{a,2}^{\text{PO}} = -1.77 \, \text{kJ} \, \text{mol}^{-1}$$

$$k_2^{\text{MF,ref}} = 3.70 \, \text{mol}_\text{C} \, \text{h}^{-1} \, \text{g}_\text{cat}^{-1}$$
$$k_2^{\text{PO,ref}} = 0.64 \, \text{mol}_\text{C} \, \text{h}^{-1} \, \text{g}_\text{cat}^{-1}$$

$$2\text{C}_3^= + \text{DME} \xrightarrow{k_3} 2\text{C}_4^= + \text{H}_2\text{O}$$
$$E_{a,3}^{\text{MF}} = 34.62 \, \text{kJ} \, \text{mol}^{-1}$$

$$r_3 = k_3 \, y \, (\text{C}_3^=) \, y \, (\text{DME})$$
$$E_{a,3}^{\text{PO}} = 9.72 \, \text{kJ} \, \text{mol}^{-1}$$

$$k_3^{\text{MF,ref}} = 13.90 \, \text{mol}_\text{C} \, \text{h}^{-1} \, \text{g}_\text{cat}^{-1}$$
$$k_3^{\text{PO,ref}} = 1.73 \, \text{mol}_\text{C} \, \text{h}^{-1} \, \text{g}_\text{cat}^{-1}$$

$$2\text{C}_4^= + \text{DME} \xrightarrow{k_4} 2\text{C}_5^= + \text{H}_2\text{O}$$
$$E_{a,4}^{\text{MF}} = 44.03 \, \text{kJ} \, \text{mol}^{-1}$$

$$r_4 = k_4 \, y \, (\text{C}_4^=) \, y \, (\text{DME})$$
$$E_{a,4}^{\text{PO}} = 30.08 \, \text{kJ} \, \text{mol}^{-1}$$

$$k_4^{\text{MF,ref}} = 43.92 \, \text{mol}_\text{C} \, \text{h}^{-1} \, \text{g}_\text{cat}^{-1}$$
$$k_4^{\text{PO,ref}} = 7.32 \, \text{mol}_\text{C} \, \text{h}^{-1} \, \text{g}_\text{cat}^{-1}$$

$$2\text{C}_5^= + \text{DME} \xrightarrow{k_5} 2\text{C}_6^= + \text{H}_2\text{O}$$
$$E_{a,5}^{\text{MF}} = 16.49 \, \text{kJ} \, \text{mol}^{-1}$$

$$r_5 = k_5 \, y \, (\text{C}_5^=) \, y \, (\text{DME})$$
$$E_{a,5}^{\text{PO}} = 3.23 \, \text{kJ} \, \text{mol}^{-1}$$

$$k_5^{\text{MF,ref}} = 47.25 \, \text{mol}_\text{C} \, \text{h}^{-1} \, \text{g}_\text{cat}^{-1}$$
$$k_5^{\text{PO,ref}} = 16.96 \, \text{mol}_\text{C} \, \text{h}^{-1} \, \text{g}_\text{cat}^{-1}$$

$$2\text{C}_6^= + \text{DME} \xrightarrow{k_6} 2\text{C}_{7+}^= + \text{H}_2\text{O}$$
$$E_{a,6}^{\text{MF}} = 125.89 \, \text{kJ} \, \text{mol}^{-1}$$

$$r_6 = k_6 \, y \, (\text{C}_6^=) \, y \, (\text{DME})$$
$$E_{a,6}^{\text{PO}} = 151.27 \, \text{kJ} \, \text{mol}^{-1}$$

$$k_6^{\text{MF,ref}} = 25.21 \, \text{mol}_\text{C} \, \text{h}^{-1} \, \text{g}_\text{cat}^{-1}$$
$$k_6^{\text{PO,ref}} = 20.28 \, \text{mol}_\text{C} \, \text{h}^{-1} \, \text{g}_\text{cat}^{-1}$$

$$\text{C}_6^= \xrightarrow{k_7} 2\text{C}_3^=$$
$$E_{a,7}^{\text{MF}} = 109.08 \, \text{kJ} \, \text{mol}^{-1}$$

$$r_7 = k_7 \, y \, (\text{C}_6^=)$$
$$E_{a,7}^{\text{PO}} = 0.22 \, \text{kJ} \, \text{mol}^{-1}$$

$$k_7^{\text{MF,ref}} = 123.31 \, \text{mol}_\text{C} \, \text{h}^{-1} \, \text{g}_\text{cat}^{-1}$$
$$k_7^{\text{PO,ref}} = 82.27 \, \text{mol}_\text{C} \, \text{h}^{-1} \, \text{g}_\text{cat}^{-1}$$

$$2\text{C}_3^= \xrightarrow{k_8} \text{C}_6^=$$
$$E_{a,8}^{\text{MF}} = 0.63 \, \text{kJ} \, \text{mol}^{-1}$$

$$r_8 = k_8 \, y \, (\text{C}_3^=)$$
$$E_{a,8}^{\text{PO}} = -181.54 \, \text{kJ} \, \text{mol}^{-1}$$

$$k_8^{\text{MF,ref}} = 4.39 \, \text{mol}_\text{C} \, \text{h}^{-1} \, \text{g}_\text{cat}^{-1}$$
$$k_8^{\text{PO,ref}} = 1.25 \, \text{mol}_\text{C} \, \text{h}^{-1} \, \text{g}_\text{cat}^{-1}$$

$$\text{C}_6^= \xrightarrow{k_9} \text{C}_2^= + \text{C}_4^=$$
$$E_{a,9}^{\text{MF}} = 1.15 \, \text{kJ} \, \text{mol}^{-1}$$

$$r_9 = k_9 \, y \, (\text{C}_6^=)$$
$$E_{a,9}^{\text{PO}} = 120.32 \, \text{kJ} \, \text{mol}^{-1}$$

$$k_9^{\text{MF,ref}} = 0.86 \, \text{mol}_\text{C} \, \text{h}^{-1} \, \text{g}_\text{cat}^{-1}$$
$$k_9^{\text{PO,ref}} = 13.12 \, \text{mol}_\text{C} \, \text{h}^{-1} \, \text{g}_\text{cat}^{-1}$$

$$\text{C}_2^= + \text{C}_4^= \xrightarrow{k_{10}} \text{C}_6^=$$
$$E_{a,10}^{\text{MF}} = -103.56 \, \text{kJ} \, \text{mol}^{-1}$$

$$r_{10} = k_{10} \, y \, (\text{C}_2^=) \, y \, (\text{C}_4^=)$$
$$E_{a,10}^{\text{PO}} = -48.92 \, \text{kJ} \, \text{mol}^{-1}$$

$$k_{10}^{\text{MF,ref}} = 0.27 \, \text{mol}_\text{C} \, \text{h}^{-1} \, \text{g}_\text{cat}^{-1}$$
$$k_{10}^{\text{PO,ref}} = 1.57 \, \text{mol}_\text{C} \, \text{h}^{-1} \, \text{g}_\text{cat}^{-1}$$

$$\text{C}_5^= \xrightarrow{k_{11}} \text{C}_2^= + \text{C}_3^=$$
$$E_{a,11}^{\text{MF}} = -61.72 \, \text{kJ} \, \text{mol}^{-1}$$

$$r_{11} = k_{11} \, y \, (\text{C}_5^=)$$
$$E_{a,11}^{\text{PO}} = -143.09 \, \text{kJ} \, \text{mol}^{-1}$$

$$k_{11}^{\text{MF,ref}} = 2.64 \, \text{mol}_\text{C} \, \text{h}^{-1} \, \text{g}_\text{cat}^{-1}$$
$$k_{11}^{\text{PO,ref}} = 0.57 \, \text{mol}_\text{C} \, \text{h}^{-1} \, \text{g}_\text{cat}^{-1}$$

$$\text{C}_2^= + \text{C}_3^= \xrightarrow{k_{12}} \text{C}_5^=$$
$$E_{a,12}^{\text{MF}} = -123.64 \, \text{kJ} \, \text{mol}^{-1}$$

$$r_{12} = k_{12} \, y \, (\text{C}_2^=) \, y \, (\text{C}_3^=)$$
$$E_{a,12}^{\text{PO}} = -200.00 \, \text{kJ} \, \text{mol}^{-1}$$

$$k_{12}^{\text{MF,ref}} = 3.05 \, \text{mol}_\text{C} \, \text{h}^{-1} \, \text{g}_\text{cat}^{-1}$$
$$k_{12}^{\text{PO,ref}} = 0.50 \, \text{mol}_\text{C} \, \text{h}^{-1} \, \text{g}_\text{cat}^{-1}$$

$$2\text{C}_4^= \xrightarrow{k_{13}} \text{C}_3^= + \text{C}_5^=$$
$$E_{a,13}^{\text{MF}} = 235.91 \, \text{kJ} \, \text{mol}^{-1}$$

$$r_{13} = k_{13} \, y \, (\text{C}_4^=)$$
$$E_{a,13}^{\text{PO}} = 283.90 \, \text{kJ} \, \text{mol}^{-1}$$

$$k_{13}^{\text{MF,ref}} = 0.03 \, \text{mol}_\text{C} \, \text{h}^{-1} \, \text{g}_\text{cat}^{-1}$$
$$k_{13}^{\text{PO,ref}} = 0.01 \, \text{mol}_\text{C} \, \text{h}^{-1} \, \text{g}_\text{cat}^{-1}$$

$$\text{DME} \xrightarrow{k_{14}} \text{C}_2^= + \text{H}_2\text{O}$$
$$E_{a,14}^{\text{MF}} = 94.96 \, \text{kJ} \, \text{mol}^{-1}$$

$$r_{14} = k_{14} \, y \, (\text{DME})$$
$$E_{a,14}^{\text{PO}} = 133.90 \, \text{kJ} \, \text{mol}^{-1}$$

$$k_{14}^{\text{MF,ref}} = 0.33 \, \text{mol}_\text{C} \, \text{h}^{-1} \, \text{g}_\text{cat}^{-1}$$
$$k_{14}^{\text{PO,ref}} = 0.02 \, \text{mol}_\text{C} \, \text{h}^{-1} \, \text{g}_\text{cat}^{-1}$$

$$\text{DME} + 2\text{H}_2 \xrightarrow{k_{15}} 2\text{C}_1 + \text{H}_2\text{O}$$
$$E_{a,15}^{\text{MF}} = 141.02 \, \text{kJ} \, \text{mol}^{-1}$$

$$r_{15} = k_{15} \, y \, (\text{DME})$$
$$E_{a,15}^{\text{PO}} = 99.48 \, \text{kJ} \, \text{mol}^{-1}$$

$$k_{15}^{\text{MF,ref}} = 0.10 \, \text{mol}_\text{C} \, \text{h}^{-1} \, \text{g}_\text{cat}^{-1}$$
$$k_{15}^{\text{PO,ref}} = 0.03 \, \text{mol}_\text{C} \, \text{h}^{-1} \, \text{g}_\text{cat}^{-1}$$

$$\text{C}_5^= + \text{H}_2 \xrightarrow{k_{16}} \text{C}_4^= + \text{C}_1$$
$$E_{a,16}^{\text{MF}} = 63.90 \, \text{kJ} \, \text{mol}^{-1}$$

$$r_{16} = k_{16} \, y \, (\text{C}_5^=)$$
$$E_{a,16}^{\text{PO}} = 5.62 \, \text{kJ} \, \text{mol}^{-1}$$

$$k_{16}^{\text{MF,ref}} = 0.01 \, \text{mol}_\text{C} \, \text{h}^{-1} \, \text{g}_\text{cat}^{-1}$$
$$k_{16}^{\text{PO,ref}} = 0.01 \, \text{mol}_\text{C} \, \text{h}^{-1} \, \text{g}_\text{cat}^{-1}$$

$$\text{C}_3^= \xrightarrow{k_{17}} \text{C}_{2-6}$$
$$E_{a,17}^{\text{MF}} = -46.72 \, \text{kJ} \, \text{mol}^{-1}$$

$$r_{17} = k_{17} \, y \, (\text{C}_3^=)$$
$$E_{a,17}^{\text{PO}} = -43.38 \, \text{kJ} \, \text{mol}^{-1}$$

$$k_{17}^{\text{MF,ref}} = 0.01 \, \text{mol}_\text{C} \, \text{h}^{-1} \, \text{g}_\text{cat}^{-1}$$
$$k_{17}^{\text{PO,ref}} = 0.01 \, \text{mol}_\text{C} \, \text{h}^{-1} \, \text{g}_\text{cat}^{-1}$$

$$\text{DME} \xrightarrow{k_{18}} \text{C}_{2-6}$$
$$E_{a,18}^{\text{MF}} = 26.24 \, \text{kJ} \, \text{mol}^{-1}$$

$$r_{18} = k_{18} \, y \, (\text{DME})$$
$$E_{a,18}^{\text{PO}} = 16.58 \, \text{kJ} \, \text{mol}^{-1}$$

$$k_{18}^{\text{MF,ref}} = 1.13 \, \text{mol}_\text{C} \, \text{h}^{-1} \, \text{g}_\text{cat}^{-1}$$
$$k_{18}^{\text{PO,ref}} = 0.15 \, \text{mol}_\text{C} \, \text{h}^{-1} \, \text{g}_\text{cat}^{-1}$$

$$\text{C}_6^= \xrightarrow{k_{19}} \text{C}_{7+}^=$$
$$E_{a,19}^{\text{MF}} = 73.92 \, \text{kJ} \, \text{mol}^{-1}$$

$$r_{19} = k_{19} \, y \, (\text{C}_6^=)$$
$$E_{a,19}^{\text{PO}} = 127.89 \, \text{kJ} \, \text{mol}^{-1}$$

$$k_{19}^{\text{MF,ref}} = 0.11 \, \text{mol}_\text{C} \, \text{h}^{-1} \, \text{g}_\text{cat}^{-1}$$
$$k_{19}^{\text{PO,ref}} = 0.13 \, \text{mol}_\text{C} \, \text{h}^{-1} \, \text{g}_\text{cat}^{-1}$$

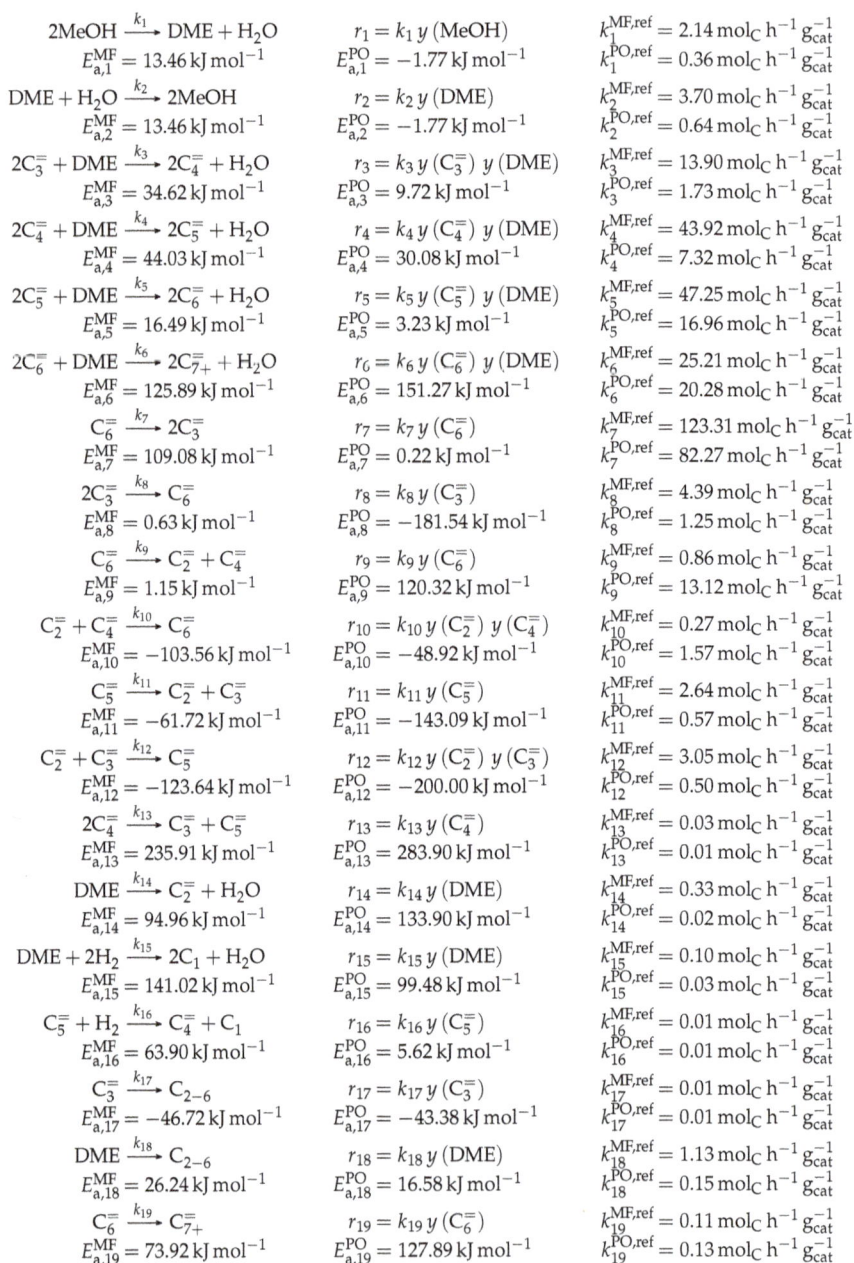

Scheme 40. Reaction network, rate equations and estimated parameters for the model by Wen et al. [303].

Scheme 41 contains the net rates of production of all species.

$$R\,(\text{MeOH}) = 2\,k_2\,y\,(\text{DME}) - 2\,k_1\,y\,(\text{MeOH})$$

$$R\,(\text{DME}) = k_1\,y\,(\text{MeOH}) - k_2\,y\,(\text{DME}) - k_3\,y\,(\text{C}_3^=)\,y\,(\text{DME}) - k_4\,y\,(\text{C}_4^=)\,y\,(\text{DME})$$
$$- k_5\,y\,(\text{C}_5^=)\,y\,(\text{DME}) - k_6\,y\,(\text{C}_6^=)\,y\,(\text{DME}) - k_{14}\,y\,(\text{DME}) - k_{15}\,y\,(\text{DME}) - k_{18}\,y\,(\text{DME})$$

$$R\,(\text{C}_2^=) = k_9\,y\,(\text{C}_6^=) + k_{11}\,y\,(\text{C}_5^=) + k_{14}\,y\,(\text{DME}) - k_{10}\,y\,(\text{C}_2^=)\,y\,(\text{C}_4^=) - k_{12}\,y\,(\text{C}_2^=)\,y\,(\text{C}_3^=)$$

$$R\,(\text{C}_3^=) = 2\,k_7\,y\,(\text{C}_6^=) + k_{11}\,y\,(\text{C}_5^=) + k_{13}\,y\,(\text{C}_4^=) - 2\,k_3\,y\,(\text{C}_3^=)\,y\,(\text{DME}) - 2\,k_8\,y\,(\text{C}_3^=)$$
$$- k_{12}\,y\,(\text{C}_2^=)\,y\,(\text{C}_3^=) - k_{17}\,y\,(\text{C}_3^=)$$

$$R\,(\text{C}_4^=) = 2\,k_3\,y\,(\text{C}_3^=)\,y\,(\text{DME}) + k_9\,y\,(\text{C}_6^=) + k_{16}\,y\,(\text{C}_5^=) - 2\,k_4\,y\,(\text{C}_4^=)\,y\,(\text{DME})$$
$$- k_{10}\,y\,(\text{C}_2^=)\,y\,(\text{C}_4^=) - 2\,k_{13}\,y\,(\text{C}_4^=)$$

$$R\,(\text{C}_5^=) = 3\,k_4\,y\,(\text{C}_4^=)\,y\,(\text{DME}) + k_{12}\,y\,(\text{C}_2^=)\,y\,(\text{C}_3^=) + k_{13}\,y\,(\text{C}_4^=) - 2\,k_5\,y\,(\text{C}_5^=)\,y\,(\text{DME})$$
$$- k_{11}\,y\,(\text{C}_5^=) - k_{16}\,y\,(\text{C}_5^=)$$

$$R\,(\text{C}_6^=) = 3\,k_5\,y\,(\text{C}_5^=)\,y\,(\text{DME}) + k_8\,y\,(\text{C}_3^=) + k_{10}\,y\,(\text{C}_2^=)\,y\,(\text{C}_4^=) - 2\,k_6\,y\,(\text{C}_6^=)\,y\,(\text{DME})$$
$$- k_7\,y\,(\text{C}_6^=) - k_9\,y\,(\text{C}_6^=) - k_{19}\,y\,(\text{C}_6^=)$$

$$R\,(\text{C}_{7+}^=) = 2\,k_6\,y\,(\text{C}_6^=)\,y\,(\text{DME}) + 2\,k_{19}\,y\,(\text{C}_6^=)$$

$$R\,(\text{C}_1) = 2\,k_{15}\,y\,(\text{DME}) + k_{16}\,y\,(\text{C}_5^=)$$

$$R\,(\text{C}_{2-6}) = k_{17}\,y\,(\text{C}_3^=) + k_{18}\,y\,(\text{DME})$$

Scheme 41. Net rates of production of the different lumps for the model by Wen et al. [303].

5.2.4. Parameter Estimation

Mole fractions are required for the rate equations in Scheme 40. For parameter estimation, *lsqnonlin* provided by MATLAB is used. It minimizes the objective function, i.e., the unweighted sum of squared residuals between modeled and measured mole fractions. The differential equations are integrated via the fourth–fifth-order Runge–Kutta method of *ode45*, also within MATLAB. The reparameterized Arrhenius approach (see Equation (21)) is applied using a reference temperature of 723 K, which is 10 K higher than the mean value of the investigated range. With this routine, 38 parameters are estimated: 19 reference rate constants and 19 activation energies.

5.3. Summary

Because the methanol co-feed implementation of Huang et al. [240] has similar methodology to the pure olefin interconversion case, advantages and disadvantages of the models are comparable. A problem might arise as the olefin interconversion equations are transferred to MTO without adaption of the denominator where the adsorption of methanol is missing. Nevertheless, the retained stoichiometry, the large reaction network and the HW type of mechanism yield a robust model. Further improvement could be achieved by having carbon number dependent adsorption values and by including DME adsorption. Furthermore, the methylation via DME is missing. On the other hand, several steps for double methylation are considered. The equilibrium constant of the oxygenates interaction is fitted to experimental data. In the approach by Wen et al. [303], both the forward and the backward reaction are estimated as rate constants which might impede thermodynamic consistency. This holds not only for oxygenates interaction, but also for olefin interconversion. This model restricts all methylation and olefin production reactions to DME as reactant. The aromatic hydrocarbon pool is indirectly considered via a step converting DME to ethene. Many different reactions including side product formation are depicted here; however, this causes also many fitting parameters. Extrapolation might be additionally difficult because of missing adsorption, mechanistic basics and stoichiometry. On the other hand, a reasonable agreement with experimental data on two different catalyst systems is achieved.

5.4. Other Studies

Guo et al. [212,309] performed measurements with a ZSM-5 zeolite (Si/Al of 200) at temperatures between 683 and 753 K. The feed consisted of methanol and different *n*-olefins and was diluted with water and nitrogen. The reaction network contains 14 lumps and requires 32 parameters. Besides olefin methylation, this model considers several olefin interconversion and side product formation steps. The rate equations are formulated as HW type of mechanism with the inhibiting water adsorption.

Another recent contribution by Ortega et al. [310] uses a recycle reactor and therefore olefin co-feed conditions, but temperatures are more within the MTG range as they are between 598 and 648 K.

6. Concluding Remarks and Outlook

The descriptions above show that the conversion of hydrocarbons over zeolites has complex reactivity which causes demanding reaction networks. Different ways of approaching these difficulties are shown above. Despite the many different possibilities, it is tried to sort some of the findings of this review and to give recommendations for future work. These are divided into general modeling advices and a reaction-specific part.

6.1. General

- Reparameterization should be performed. The choice of reference temperature is not of highest importance; nevertheless, an optimum value can improve the model performance.
- Forward and backward reactions should be expressed as such and not be fitted independently. The equilibrium constant of the reaction can be extracted from thermodynamics to have less unknown parameters. However, when lumps consisting of several species are involved, the equilibrium constant should be estimated because the lump might deviate from an equilibrium distribution. Thermodynamic consistency has to be retained.
- Expressing the rate equations via partial pressures is advantageous as the influence of pressure changes is directly included. In contrast, when carbon based values are chosen, this effect might be ignored.
- Inclusion of adsorption effects, especially via the HW type of mechanism, should lead to a comparably robust model. The agreement with experimental data can still be satisfying when adsorption is ignored, especially when high partial pressures are applied. However, one should be aware that such a model tends to extrapolation errors when different feed compositions are chosen.
- Negative activation energies might occur when these apparent values contain adsorption effects. However, also in empirical models, positive adsorption enthalpies should be avoided because these are physically not reasonable and contradictory to thermodynamics. In such a case, other phenomena seem to impair the underlying model.
- When no microkinetic model is applied, interpretation about preferred reaction pathways should be avoided. The estimated parameters describe the reactivity in an empirical way, but the values are influenced by too many factors to allow mechanistic analyses. Nevertheless, effect of conditions on product distributions can be elucidated; for example, negative activation energies show that this pathway is less preferred at higher temperatures.
- Although high agreement can be achieved in any way, the stoichiometry within one reaction step should be retained to have a reasonable characterization of the reactivity. Moreover, when the concept of elementary reactions is chosen, this should be applied consistently. It can cause problems when the same lump appears both as reactant and as product within one step.

6.2. Olefin Cracking and MTO

- For hydrocarbon conversion, a maximum carbon number of seven seems to be sufficient, although the level of detail can be increased by exceeding this value. Nevertheless, some higher intermediates can be included in the network which crack down immediately, thus having no fitting answer. Furthermore, this recommended value also depends on the feed: when pure hexenes are applied, the dimerization to $C_{12}^=$ has to be included.

- Lower olefins should not be summarized to one lump as their formation mechanisms and reactivity are different. The same holds for methanol and DME.
- Concerning ethene reactivity, reasonable results are obtained by assuming ethene both as reversibly and as irreversibly formed. However, the latter approach might be advantageous to reduce the number of estimated parameters.
- The complex interaction between zeolite and water is still not fully understood. Nevertheless, a useful approach is the inclusion of water as diluent and as competing adsorptive.
- Especially for MTO, the underlying chemistry is very complex through many different types of reactions. Consequently, it is difficult to describe the whole reactivity with one model. It is recommended to implement the types of reactions stepwise (e.g., first olefin interconversion, then methanol related reactions, and then side reactions) with individual experimental datasets. This reduces the number of unknown parameters in each fitting step and allows focusing on the respective type of reaction.
- Whenever MTO models for pure methanol feeds are created, one has to be aware that the unresolved initiation phase might influence the performance at low contact times which could impede the model. For such cases, it could be reasonable to simulate the product generation not from zero catalyst mass on. In contrast, this effect can be ignored for industrial MTP conditions where hydrocarbons are available from the beginning.
- For pure methanol feeds, an equilibrated state is reached comparably quickly because of the slow formation of the first C-C bond. However, when hydrocarbons are co-fed, this equilibrium among methanol, DME and water might not be reached.
- For MTO, it depends on the catalyst and the reaction conditions whether an implementation of the aromatic hydrocarbon pool is reasonable or not. If so, the underlying reactions have to be simplified to only some characteristic steps that are representative for the whole catalytic cycle.

In the end, it cannot be said which modeling methodology is the best; it always depends on the requirements it should fulfill. However, one always has to be aware of the range within which the model is valid. Simple kinetics might describe the investigated case in a satisfying and comparably fast way. Moreover, conclusions about the influence of reaction conditions on product distributions are possible. However, further application should be performed with caution because extrapolation out of the experimentally covered regime could cause unrealistic results and false trends. However, for microkinetics, one also always has to be aware that the theoretical description is still a model. Indeed, in the case of satisfying agreement, the probability is high that the chosen approach is a valid way to describe the surface reactions. On the other hand, no reaction mechanism can be proven by solely evaluating a microkinetic model. Thus, in these cases, overinterpretation should also be avoided.

This leads to the outlook for future studies. As mentioned above, even microkinetics are not sufficient to decode complexity. Consequently, kinetic studies should always be compared to experimental results and ab initio methods. Especially the latter is of high importance to exclude any transport or deactivation effects impairing intrinsic kinetic results and therefore distorting the model. Furthermore, it offers the possibility of introducing surface inhomogeneities into the model. Currently, most studies consider all active sites of a catalyst to be of identical reactivity. Here, a high overlapping of kinetic modeling, experimental insights and *ab initio* methods is desirable. Ideal surface kinetics can be used to simulate the performance on larger pellet shapes. Multicomponent transport phenomena have to be included then to accurately account for all ongoing effects in an industrial reactor. Simulation and optimization of a whole process is thus a multi-scale task of high complexity, as well as high potential for the future.

Funding: This research was funded by the Bavarian Ministry of Economic Affairs, Energy and Technology grant number [47-3665g/1075/1-NW-1501-0003].

Acknowledgments: Special thanks go to Maximilian Wende, Johanna Hemauer and Lei Li for their help in evaluating literature models. Furthermore, fruitful discussions about thermodynamics with Philipp J. Donaubauer and Daniel P. Schwinger are highly appreciated. The authors gratefully acknowledge the fruitful environment within the framework of MuniCat. S. Standl is thankful for the support from TUM Graduate School.

Conflicts of Interest: The authors declare no conflict of interest. The sponsors had no role in the design, execution, interpretation or writing of the study.

Nomenclature

A	Preexponential factor, variable
a	Catalytic activity, -
$\tilde{A}$	Single-event preexponential factor, variable
$C(i)$	Concentration, $\mathrm{mol\,m^{-3}}$
C_t	Concentration of total Brønsted acid sites, $\mathrm{mol\,kg_{cat}^{-1}}$
$C_{t,\mathrm{SBAS}}$	Concentration of strong Brønsted acid sites, $\mathrm{mol\,kg_{cat}^{-1}}$
$c_{p,i}$	Heat capacity, $\mathrm{J\,mol^{-1}\,K^{-1}}$
dof	Degree of freedom, -
d_p	Particle diameter, m
E_a	Activation energy, $\mathrm{J\,mol^{-1}}$
$F(i)$	Molar flow rate, $\mathrm{mol\,s^{-1}}$
f_i	Fugacity, Pa
G_i	Gibb's free energy, $\mathrm{J\,mol^{-1}}$
G_t	Total Gibb's free energy, J
H_i	Enthalpy, $\mathrm{J\,mol^{-1}}$
h	Planck constant, $\mathrm{J\,s}$
K	Equilibrium constant, variable
$\tilde{K}$	Single-event equilibrium constant, variable
k_B	Boltzmann constant, $\mathrm{J\,K^{-1}}$
k_l	Rate constant, variable
$\tilde{k}_l$	Single-event rate constant, variable
M_w	Molar mass, $\mathrm{kg\,mol^{-1}}$
m	Type of reactant intermediate, -
$N_{el,i}$	Number of atoms of element el, -
N_Exp	Number of experiments, -
N_Par	Number of parameters, -
N_Res	Number of fitting responses, -
n	Type of product intermediate, -
n_e	Number of single events, -
$n_{\mathrm{H_2O}}$	Reaction order of water adsorption, -
n_i	Number of moles, mol
n^d	Deactivation order, -
$n_{\mathrm{H_2O}}^\mathrm{d}$	Deactivation order of water adsorption, -
p	Pressure, Pa
R	Gas constant, $\mathrm{J\,mol^{-1}\,K^{-1}}$
R_1^+	Surface methyl group
$R(i)$	Net rate of production, $\mathrm{mol\,kg_{cat}^{-1}\,s^{-1}}$
r_l	Reaction rate, variable
SSQ	Sum of squared residuals, -
S_i	Entropy, $\mathrm{J\,mol^{-1}\,K^{-1}}$
T	Temperature, K
t	Time, s

W	Catalyst mass, kg_{cat}
$w\,(i)$	Mass fraction, -
X	Conversion, -
$y\,(i)$	Mole fraction, -
$\hat{y}\,(i)$	Modeled mole fraction, -

Greek letters

α	Parameter for carbon number dependence, -
α^l	Composed preexponential factor, variable
β	Parameter for carbon number dependence, -
β^l	Linearization parameter, -
γ	Parameter for carbon number dependence, -
γ^l	Parameter for carbon number dependence, -
Δ	Difference, variable
δ	Parameter for carbon number dependence, -
δ^l	Linearization parameter, -
θ	Coverage, -
κ	Reaction order, -
μ_i	Chemical potential, $\text{J}\,\text{mol}^{-1}$
ν	Stoichiometric coefficient, -
ϕ^l	Parameter for carbon number dependence, -
φ^l	Deactivation parameter, -
ω_i	Weighting factor, -

Subscripts

ads	Adsorption
C	Carbon
cat	Catalyst
el	Element
f	Formation
g	Gas phase
h	Running index for acid sites
i	Arbitrary species
j	Running index for arbitrary species
k	Running index for experiments
l	Reaction step
MW	Methanol and water
max	Maximum value
Ol	Olefin
Ox	Oxygenates
r	Reaction
t	Total
$\ddagger$	Transition state

Superscripts

ads	Adsorption
ar	Aromatization
chem	Chemisorption
co	Composite value
cr	Cracking

d	Deactivation
dim	Dimerization
in	Inlet value
MF	Metal fiber
PO	Powder
RDS	Rate-determining step
ref	Reference
TD	Thermodynamic
π	π complex
$\circ$	Standard condition ($p^\circ = 1 \times 10^5$ Pa)

Abbreviations

The following abbreviations are used in this manuscript:

Ads.	Adsorption
AEI	Framework code of SAPO-18
BAS	Brønsted acid sites
BET	Brunauer–Emmett–Teller
C_i	Paraffin with carbon number i
C_{ni}	n-Paraffin with carbon number i
$C_i^=$	Olefin with carbon number i
C_i^{al}	Aliphatic compound with carbon number i
C_i^{ar}	Aromatic compound with carbon number i
C_i^{HC}	Hydrocarbon with carbon number i
C_i^{SP}	Side products with carbon number i
C_x^+	Protonated intermediate
CHA	Framework code of chabazite
DFT	Density functional theory
DME	Dimethyl ether
DSC	Differential scanning calorimetry
ER	Eley–Rideal
FCC	Fluid catalytic cracking
FID	Flame ionization detector
GA	Genetic algorithm
GC	Gas chromatograph
GC-FTIR	Gas chromatograph with Fourier transform infrared
GC-MS	Gas chromatograph with mass spectrometer
HF	Hartree–Fock
HW	Hougen–Watson
Int	Intermediates
IR	Infrared
LAS	Lewis acid sites
L	Langmuir
LH	Langmuir–Hinshelwood
Mech.	Mechanistical scheme
MFI	Framework code of ZSM-5
MTG	Methanol-to-gasoline
MTH	Methanol-to-hydrocarbons
MTO	Methanol-to-olefins
MTT	Framework code of ZSM-22

OM	Oxonium methylide
Ox	Oxygenates (methanol and DME)
PCP	Protonated cyclopropane
PSSA	Pseudo-steady state approximation
SBU	Secondary building unit
SBAS	Strong Brønsted acid sites
Side prod.	Side products
TCD	Thermal conductivity detector
TG	Thermogravimetry
TOS	Time-on-stream
XRD	X-ray diffraction

References

1. Ren, T.; Patel, M.K.; Blok, K. Steam Cracking and Methane to Olefins: Energy Use, CO_2 Emissions and Production Costs. *Energy* **2008**, *33*, 817–833. [CrossRef]

2. Plotkin, J.S. The Propylene Gap: How Can It Be Filled? 2015. Available online: https://www.acs. org/content/acs/en/pressroom/cutting-edge-chemistry/the-propylene-gap-how-can-it-be-filled.html (accessed on 29 October 2018).

3. Blay, V.; Epelde, E.; Miravalles, R.; Perea, L.A. Converting Olefins to Propene: Ethene to Propene and Olefin Cracking. *Catal. Rev. Sci. Eng.* **2018**, *60*, 278–335. [CrossRef]

4. Ren, T.; Patel, M.K.; Blok, K. Olefins from Conventional and Heavy Feedstocks: Energy Use in Steam Cracking and Alternative Processes. *Energy* **2006**, *31*, 425–451. [CrossRef]

5. Neelis, M.; Patel, M.K.; Blok, K.; Haije, W.; Bach, P. Approximation of Theoretical Energy-Saving Potentials for the Petrochemical Industry using Energy Balances for 68 Key Processes. *Energy* **2007**, *32*, 1104–1123. [CrossRef]

6. Centi, G.; Iaquaniello, G.; Perathoner, S. Can We Afford to Waste Carbon Dioxide? Carbon Dioxide as a Valuable Source of Carbon for the Production of Light Olefins. *ChemSusChem* **2011**, *4*, 1265–1273. [CrossRef] [PubMed]

7. Torres Galvis, H.M.; de Jong, K.P. Catalysts for Production of Lower Olefins from Synthesis Gas: A Review. *ACS Catal.* **2013**, *3*, 2130–2149. [CrossRef]

8. Mokrani, T.; Scurrell, M. Gas Conversion to Liquid Fuels and Chemicals: The Methanol Route–Catalysis and Processes Development. *Catal. Rev. Sci. Eng.* **2009**, *51*, 1–145. [CrossRef]

9. Rahimi, N.; Karimzadeh, R. Catalytic Cracking of Hydrocarbons over Modified ZSM-5 Zeolites to Produce Light Olefins: A Review. *Appl. Catal. A* **2011**, *398*, 1–17. [CrossRef]

10. Nesterenko, N.; Aguilhon, J.; Bodart, P.; Minoux, D.; Dath, J.P. Methanol to Olefins: An Insight into Reaction Pathways and Products Formation. In *Zeolites and Zeolite-Like Materials*; Sels, B.F., Kustov, L.M., Eds.; Elsevier: Amsterdam, The Netherlands, 2016; pp. 189–263. [CrossRef]

11. Olsbye, U.; Svelle, S.; Bjørgen, M.; Beato, P.; Janssens, T.V.W.; Joensen, F.; Bordiga, S.; Lillerud, K.P. Conversion of Methanol to Hydrocarbons: How Zeolite Cavity and Pore Size Controls Product Selectivity. *Angew. Chem. Int. Ed.* **2012**, *51*, 5810–5831. [CrossRef]

12. Balcar, H.; Čejka, J. Mesoporous Molecular Sieves as Advanced Supports for Olefin Metathesis Catalysts. *Coord. Chem. Rev.* **2013**, *257*, 3107–3124. [CrossRef]

13. Gholampour, N.; Yusubov, M.; Verpoort, F. Investigation of the Preparation and Catalytic Activity of Supported Mo, W, and Re Oxides as Heterogeneous Catalysts in Olefin Metathesis. *Catal. Rev. Sci. Eng.* **2016**, *58*, 113–156. [CrossRef]

14. Sattler, J.J.H.B.; Ruiz-Martínez, J.; Santillan-Jimenez, E.; Weckhuysen, B.M. Catalytic Dehydrogenation of Light Alkanes on Metals and Metal Oxides. *Chem. Rev.* **2014**, *114*, 10613–10653. [CrossRef] [PubMed]

15. Nawaz, Z. Light Alkane Dehydrogenation to Light Olefin Technologies: A Comprehensive Review. *Rev. Chem. Eng.* **2015**, *31*, 413–436. [CrossRef]

16. Carrero, C.A.; Schlögl, R.; Wachs, I.E.; Schomäcker, R. Critical Literature Review of the Kinetics for the Oxidative Dehydrogenation of Propane over Well-Defined Supported Vanadium Oxide Catalysts. *ACS Catal.* **2014**, *4*, 3357–3380. [CrossRef]

17. Ghashghaee, M. Heterogeneous Catalysts for Gas-Phase Conversion of Ethylene to Higher Olefins. *Rev. Chem. Eng.* **2018**, *34*, 595–655. [CrossRef]

18. Dumesic, J.A.; Rudd, D.F.; Aparicio, L.M.; Rekoske, J.E.; Treviño, A.A. *The Microkinetics of Heterogeneous Catalysis*; ACS Professional Reference Book; American Chemical Society: Washington, DC, USA, 1993.

19. Kumar, P.; Thybaut, J.W.; Svelle, S.; Olsbye, U.; Marin, G.B. Single-Event Microkinetics for Methanol to Olefins on H-ZSM-5. *Ind. Eng. Chem. Res.* **2013**, *52*, 1491–1507. [CrossRef]

20. Vynckier, E.; Froment, G.F. Modeling of the Kinetics of Complex Processes based upon Elementary Steps. In *Kinetic and Thermodynamic Lumping of Multicomponent Mixtures*; Astarita, G., Sandler, S.I., Eds.; Elsevier: Amsterdam, The Netherlands, 1991; pp. 131–161.

21. Thybaut, J.W.; Marin, G.B. Single-Event MicroKinetics: Catalyst Design for Complex Reaction Networks. *J. Catal.* **2013**, *308*, 352–362. [CrossRef]

22. Khadzhiev, S.N.; Magomedova, M.V.; Peresypkina, E.G. Kinetic Models of Methanol and Dimethyl Ether Conversion to Olefins over Zeolite Catalysts (Review). *Pet. Chem.* **2015**, *55*, 503–521. [CrossRef]

23. Keil, F.J. Methanol-to-Hydrocarbons: Process Technology. *Microporous Mesoporous Mater.* **1999**, *29*, 49–66. [CrossRef]

24. Chen, N.Y.; Reagan, W.J. Evidence of Autocatalysis in Methanol to Hydrocarbon Reactions over Zeolite Catalysts. *J. Catal.* **1979**, *59*, 123–129. [CrossRef]

25. Chang, C.D. A Kinetic Model for Methanol Conversion to Hydrocarbons. *Chem. Eng. Sci.* **1980**, *35*, 619–622. [CrossRef]

26. Ono, Y.; Mori, T. Mechanism of Methanol Conversion into Hydrocarbons over ZSM-5 Zeolite. *J. Chem. Soc. Faraday Trans. 1* **1981**, *77*, 2209–2221. [CrossRef]

27. Mihail, R.; Straja, S.; Maria, G.; Musca, G.; Pop, G. Kinetic Model for Methanol Conversion to Olefins. *Ind. Eng. Chem. Process Des. Dev.* **1983**, *22*, 532–538. [CrossRef]

28. Mihail, R.; Straja, S.; Maria, G.; Musca, G.; Pop, G. A Kinetic Model for Methanol Conversion to Hydrocarbons. *Chem. Eng. Sci.* **1983**, *38*, 1581–1591. [CrossRef]

29. Schipper, P.H.; Krambeck, F.J. A Reactor Design Simulation with Reversible and Irreversible Catalyst Deactivation. *Chem. Eng. Sci.* **1986**, *41*, 1013–1019. [CrossRef]

30. Sedrán, U.A.; Mahay, A.; de Lasa, H.I. Modelling Methanol Conversion to Hydrocarbons: Revision and Testing of a Simple Kinetic Model. *Chem. Eng. J.* **1990**, *45*, 1161–1165. [CrossRef]

31. Sedrán, U.A.; Mahay, A.; de Lasa, H.I. Modelling Methanol Conversion to Hydrocarbons: Alternative Kinetic Models. *Chem. Eng. J.* **1990**, *45*, 33–42. [CrossRef]

32. Schönfelder, H.; Hinderer, J.; Werther, J.; Keil, F.J. Methanol to Olefins-Prediction of the Performance of a Circulating Fluidized-Bed Reactor on the Basis of Kinetic Experiments in a Fixed-Bed Reactor. *Chem. Eng. Sci.* **1994**, *49*, 5377–5390. [CrossRef]

33. Bos, R.; Tromp, P.J.; Akse, H.N. Conversion of Methanol to Lower Olefins. Kinetic Modeling, Reactor Simulation, and Selection. *Ind. Eng. Chem. Res.* **1995**, *34*, 3808–3816. [CrossRef]

34. Gayubo, A.G.; Benito, P.L.; Aguayo, A.T.; Aguirre, I.; Bilbao, J. Analysis of Kinetic Models of the Methanol-to-Gasoline (MTG) Process in an Integral Reactor. *Chem. Eng. J.* **1996**, *63*, 45–51. [CrossRef]

35. Van Speybroeck, V.; Hemelsoet, K.; Joos, L.; Waroquier, M.; Bell, R.G.; Catlow, R.A. Advances in Theory and Their Application within the Field of Zeolite Chemistry. *Chem. Soc. Rev.* **2015**, *44*, 7044–7111. [CrossRef]

36. Van Speybroeck, V.; de Wispelaere, K.; van der Mynsbrugge, J.; Vandichel, M.; Hemelsoet, K.; Waroquier, M. First Principle Chemical Kinetics in Zeolites: The Methanol-to-Olefin Process as a Case Study. *Chem. Soc Rev.* **2014**, *43*, 7326–7357. [CrossRef]

37. Hemelsoet, K.; van der Mynsbrugge, J.; de Wispelaere, K.; Waroquier, M.; van Speybroeck, V. Unraveling the Reaction Mechanisms Governing Methanol-to-Olefins Catalysis by Theory and Experiment. *ChemPhysChem* **2013**, *14*, 1526–1545. [CrossRef]

38. De Wispelaere, K.; Ensing, B.; Ghysels, A.; Meijer, E.J.; van Speybroeck, V. Complex Reaction Environments and Competing Reaction Mechanisms in Zeolite Catalysis: Insights from Advanced Molecular Dynamics. *Chem. Eur. J.* **2015**, *21*, 9385–9396. [CrossRef]

39. De Wispelaere, K.; Wondergem, C.S.; Ensing, B.; Hemelsoet, K.; Meijer, E.J.; Weckhuysen, B.M.; van Speybroeck, V.; Ruiz-Martínez, J. Insight into the Effect of Water on the Methanol-to-Olefins Conversion in H-SAPO-34 from Molecular Simulations and in Situ Microspectroscopy. *ACS Catal.* **2016**, *6*, 1991–2002. [CrossRef]

40. Moors, S.L.; de Wispelaere, K.; van der Mynsbrugge, J.; Waroquier, M.; van Speybroeck, V. Molecular Dynamics Kinetic Study on the Zeolite-Catalyzed Benzene Methylation in ZSM-5. *ACS Catal.* **2013**, *3*, 2556–2567. [CrossRef]

41. Van der Mynsbrugge, J.; de Ridder, J.; Hemelsoet, K.; Waroquier, M.; van Speybroeck, V. Enthalpy and Entropy Barriers Explain the Effects of Topology on the Kinetics of Zeolite-Catalyzed Reactions. *Chem. Eur. J.* **2013**, *19*, 11568–11576. [CrossRef]

42. Van Speybroeck, V.; van der Mynsbrugge, J.; Vandichel, M.; Hemelsoet, K.; Lesthaeghe, D.; Ghysels, A.; Marin, G.B.; Waroquier, M. First Principle Kinetic Studies of Zeolite-Catalyzed Methylation Reactions. *J. Am. Chem. Soc.* **2011**, *133*, 888–899. [CrossRef]

43. Van der Mynsbrugge, J.; Moors, S.L.; de Wispelaere, K.; van Speybroeck, V. Insight into the Formation and Reactivity of Framework-Bound Methoxide Species in H-ZSM-5 from Static and Dynamic Molecular Simulations. *ChemCatChem* **2014**, *6*, 1906–1918. [CrossRef]

44. Lesthaeghe, D.; van der Mynsbrugge, J.; Vandichel, M.; Waroquier, M.; van Speybroeck, V. Full Theoretical Cycle for both Ethene and Propene Formation during Methanol-to-Olefin Conversion in H-ZSM-5. *ChemCatChem* **2011**, *3*, 208–212. [CrossRef]

45. Yarulina, I.; de Wispelaere, K.; Bailleul, S.; Goetze, J.; Radersma, M.; Abou-Hamad, E.; Vollmer, I.; Goesten, M.; Mezari, B.; Hensen, E.J.; et al. Structure-Performance Descriptors and the Role of Lewis Acidity in the Methanol-to-Propylene Process. *Nat. Chem.* **2018**, *10*, 804–812. [CrossRef]

46. Van der Mynsbrugge, J.; Janda, A.; Lin, L.C.; van Speybroeck, V.; Head-Gordon, M.; Bell, A.T. Understanding Brønsted-Acid Catalyzed Monomolecular Reactions of Alkanes in Zeolite Pores by Combining Insights from Experiment and Theory. *ChemPhysChem* **2018**, *19*, 341–358. [CrossRef]

47. Van der Mynsbrugge, J.; Janda, A.; Mallikarjun Sharada, S.; Lin, L.C.; van Speybroeck, V.; Head-Gordon, M.; Bell, A.T. Theoretical Analysis of the Influence of Pore Geometry on Monomolecular Cracking and Dehydrogenation of *n*-Butane in Brønsted Acidic Zeolites. *ACS Catal.* **2017**, *7*, 2685–2697. [CrossRef]

48. Cnudde, P.; de Wispelaere, K.; Vanduyfhuys, L.; Demuynck, R.; van der Mynsbrugge, J.; Waroquier, M.; van Speybroeck, V. How Chain Length and Branching Influence the Alkene Cracking Reactivity on H-ZSM-5. *ACS Catal.* **2018**, *8*, 9579–9595. [CrossRef]

49. Cnudde, P.; de Wispelaere, K.; van der Mynsbrugge, J.; Waroquier, M.; van Speybroeck, V. Effect of Temperature and Branching on the Nature and Stability of Alkene Cracking Intermediates in H-ZSM-5. *J. Catal.* **2017**, *345*, 53–69. [CrossRef]

50. Hajek, J.; van der Mynsbrugge, J.; de Wispelaere, K.; Cnudde, P.; Vanduyfhuys, L.; Waroquier, M.; van Speybroeck, V. On the Stability and Nature of Adsorbed Pentene in Brønsted Acid Zeolite H-ZSM-5 at 323 K. *J. Catal.* **2016**, *340*, 227–235. [CrossRef]

51. Martens, G.G.; Marin, G.B.; Martens, J.A.; Jacobs, P.A.; Baron, G.V. A Fundamental Kinetic Model for Hydrocracking of C_8 to C_{12} Alkanes on Pt/US-Y Zeolites. *J. Catal.* **2000**, *195*, 253–267. [CrossRef]

52. Martens, G.G.; Marin, G.B. Kinetics for Hydrocracking Based on Structural Classes: Model Development and Applications. *AIChE J.* **2001**, *47*, 1607–1622. [CrossRef]

53. Martens, G.G.; Thybaut, J.W.; Marin, G.B. Single-Event Rate Parameters for the Hydrocracking of Cycloalkanes on Pt/US-Y Zeolites. *Ind. Eng. Chem. Res.* **2001**, *40*, 1832–1844. [CrossRef]

54. Thybaut, J.W.; Marin, G.B.; Baron, G.V.; Jacobs, P.A.; Martens, J.A. Alkene Protonation Enthalpy Determination from Fundamental Kinetic Modeling of Alkane Hydroconversion on Pt/H-(US)Y-Zeolite. *J. Catal.* **2001**, *202*, 324–339. [CrossRef]

55. Narasimhan, C.S.L.; Thybaut, J.W.; Marin, G.B.; Jacobs, P.A.; Martens, J.A.; Denayer, J.F.; Baron, G.V. Kinetic Modeling of Pore Mouth Catalysis in the Hydroconversion of *n*-Octane on Pt-H-ZSM-22. *J. Catal.* **2003**, *220*, 399–413. [CrossRef]

56. Thybaut, J.W.; Narasimhan, C.S.L.; Denayer, J.F.; Baron, G.V.; Jacobs, P.A.; Martens, J.A.; Marin, G.B. Acid-Metal Balance of a Hydrocracking Catalyst: Ideal versus Nonideal Behavior. *Ind. Eng. Chem. Res.* **2005**, *44*, 5159–5169. [CrossRef]

57. Narasimhan, C.S.L.; Thybaut, J.W.; Martens, J.A.; Jacobs, P.A.; Denayer, J.F.; Marin, G.B. A Unified Single-Event Microkinetic Model for Alkane Hydroconversion in Different Aggregation States on Pt/H-USY-Zeolites. *J. Phys. Chem. B* **2006**, *110*, 6750–6758. [CrossRef]

58. Quintana-Solórzano, R.; Thybaut, J.W.; Marin, G.B. Catalytic Cracking and Coking of (Cyclo)Alkane/1-Octene Mixtures on an Equilibrium Catalyst. *Appl. Catal. A* **2006**, *314*, 184–199. [CrossRef]

59. Thybaut, J.W.; Choudhury, I.R.; Denayer, J.F.; Baron, G.V.; Jacobs, P.A.; Martens, J.A.; Marin, G.B. Design of Optimum Zeolite Pore System for Central Hydrocracking of Long-Chain *n*-Alkanes based on a Single-Event Microkinetic Model. *Top. Catal.* **2009**, *52*, 1251–1260. [CrossRef]

60. Choudhury, I.R.; Hayasaka, K.; Thybaut, J.W.; Narasimhan, C.S.L.; Denayer, J.F.; Martens, J.A.; Marin, G.B. Pt/H-ZSM-22 Hydroisomerization Catalysts Optimization Guided by Single-Event Microkinetic Modeling. *J. Catal.* **2012**, *290*, 165–176. [CrossRef]

61. Vandegehuchte, B.D.; Choudhury, I.R.; Thybaut, J.W.; Martens, J.A.; Marin, G.B. Integrated Stefan-Maxwell, Mean Field, and Single-Event Microkinetic Methodology for Simultaneous Diffusion and Reaction inside Microporous Materials. *J. Phys. Chem. C* **2014**, *118*, 22053–22068. [CrossRef]

62. Vandegehuchte, B.D.; Thybaut, J.W.; Marin, G.B. Unraveling Diffusion and Other Shape Selectivity Effects in ZSM5 Using *n*-Hexane Hydroconversion Single-Event Microkinetics. *Ind. Eng. Chem. Res.* **2014**, *53*, 15333–15347. [CrossRef]

63. Toch, K.; Thybaut, J.W.; Marin, G.B. A Systematic Methodology for Kinetic Modeling of Chemical Reactions Applied to *n*-Hexane Hydroisomerization. *AIChE J.* **2015**, *61*, 880–892. [CrossRef]

64. Thybaut, J.W.; Marin, G.B. Multiscale Aspects in Hydrocracking: From Reaction Mechanism Over Catalysts to Kinetics and Industrial Application. In *Advances in Catalysis*; Song, C., Ed.; Academic Press: Cambridge, MA, USA, 2016; Volume 59, pp. 109–238. [CrossRef]

65. Surla, K.; Vleeming, H.; Guillaume, D.; Galtier, P. A Single Events Kinetic Model: *n*-Butane Isomerization. *Chem. Eng. Sci.* **2004**, *59*, 4773–4779. [CrossRef]

66. Valéry, E.; Guillaume, D.; Surla, K.; Galtier, P.; Verstraete, J.; Schweich, D. Kinetic Modeling of Acid Catalyzed Hydrocracking of Heavy Molecules: Application to Squalane. *Ind. Eng. Chem. Res.* **2007**, *46*, 4755–4763. [CrossRef]

67. Mitsios, M.; Guillaume, D.; Galtier, P.; Schweich, D. Single-Event Microkinetic Model for Long-Chain Paraffin Hydrocracking and Hydroisomerization on an Amorphous $Pt/SiO_2 \cdot Al_2O_3$ Catalyst. *Ind. Eng. Chem. Res.* **2009**, *48*, 3284–3292. [CrossRef]

68. Becker, P.J.; Serrand, N.; Celse, B.; Guillaume, D.; Dulot, H. A Single Events Microkinetic Model for Hydrocracking of Vacuum Gas Oil. *Comput. Chem. Eng.* **2017**, *98*, 70–79. [CrossRef]

69. Pellegrini, L.; Locatelli, S.; Rasella, S.; Bonomi, S.; Calemma, V. Modeling of Fischer–Tropsch Products Hydrocracking. *Chem. Eng. Sci.* **2004**, *59*, 4781–4787. [CrossRef]

70. Becker, P.J.; Celse, B.; Guillaume, D.; Dulot, H.; Becker, P.J. Hydrotreatment Modeling for a Variety of VGO Feedstocks: A Continuous Lumping Approach. *Fuel* **2015**, *139*, 133–143. [CrossRef]

71. Becker, P.J.; Celse, B.; Guillaume, D.; Costa, V.; Bertier, L.; Guillon, E.; Pirngruber, G. A Continuous Lumping Model for Hydrocracking on a Zeolite Catalysts: Model Development and Parameter Identification. *Fuel* **2016**, *164*, 73–82. [CrossRef]

72. Becker, P.J.; Serrand, N.; Celse, B.; Guillaume, D.; Dulot, H. Comparing Hydrocracking Models: Continuous Lumping vs. Single Events. *Fuel* **2016**, *165*, 306–315. [CrossRef]

73. Lopez Abelairas, M.; de Oliveira, L.P.; Verstraete, J.J. Application of Monte Carlo Techniques to LCO Gas Oil Hydrotreating: Molecular Reconstruction and Kinetic Modelling. *Catal. Today* **2016**, *271*, 188–198. [CrossRef]

74. Browning, B.; Afanasiev, P.; Pitault, I.; Couenne, F.; Tayakout-Fayolle, M. Detailed kinetic modelling of vacuum gas oil hydrocracking using bifunctional catalyst: A distribution approach. *Chem. Eng. J.* **2016**, *284*, 270–284. [CrossRef]

75. Shahrouzi, J.R.; Guillaume, D.; Rouchon, P.; Da Costa, P. Stochastic Simulation and Single Events Kinetic Modeling: Application to Olefin Oligomerization. *Ind. Eng. Chem. Res.* **2008**, *47*, 4308–4316. [CrossRef]

76. Walas, S.M. *Phase Equilibria in Chemical Engineering*; Butterworth: Boston, MA, USA, 1985.

77. Zhou, H.; Wang, Y.; Wei, F.; Wang, D.; Wang, Z. Kinetics of the Reactions of the Light Alkenes over SAPO-34. *Appl. Catal. A* **2008**, *348*, 135–141. [CrossRef]

78. Zhang, R.; Wang, Z.; Liu, H.; Liu, Z.; Liu, G.; Meng, X. Thermodynamic Equilibrium Distribution of Light Olefins in Catalytic Pyrolysis. *Appl. Catal. A* **2016**, *522*, 165–171. [CrossRef]

79. International Union of Pure and Applied Chemistry (Ed.) *Compendium of Chemical Terminology: Gold Book*, 2nd ed.; IUPAC: Research Triangle Park, NC, USA, 2014. [CrossRef]

80. Lide, D.R. (Ed.) *CRC Handbook of Chemistry and Physics*, 87th ed.; Taylor and Francis Group: Boca Raton, FL, USA, 2006.

81. Alberty, R.A. Extrapolation of Standard Chemical Thermodynamic Properties of Alkene Isomer Groups to Higher Carbon Numbers. *J. Phys. Chem.* **1983**, *87*, 4999–5002. [CrossRef]

82. Alberty, R.A. Chemical Thermodynamic Properties of Isomer Groups. *Ind. Eng. Chem. Fundam.* **1983**, *22*, 318–321. [CrossRef]

83. Alberty, R.A.; Oppenheim, I. A Continuous Thermodynamics Approach to Chemical Equilibrium within an Isomer Group. *J. Chem. Phys.* **1984**, *81*, 4603–4609. [CrossRef]

84. Alberty, R.A. Chemical Equilibrium In Complex Organic Systems. *J. Phys. Chem.* **1985**, *89*, 880–883. [CrossRef]

85. Alberty, R.A.; Gehrig, C.A. Standard Chemical Thermodynamic Properties of Alkane Isomer Groups. *J. Phys. Chem. Ref. Data* **1984**, *13*, 1173–1197. [CrossRef]

86. Alberty, R.A. Standard Chemical Thermodynamic Properties of Alkylbenzene Isomer Groups. *J. Phys. Chem. Ref. Data* **1985**, *14*, 177–192. [CrossRef]

87. Alberty, R.A.; Bloomstein, T.M. Standard Chemical Thermodynamic Properties of Alkylnaphthalene Isomer Groups. *J. Phys. Chem. Ref. Data* **1985**, *14*, 821–837. [CrossRef]

88. Alberty, R.A.; Gehrig, C.A. Standard Chemical Thermodynamic Properties of Alkene Isomer Groups. *J. Phys. Chem. Ref. Data* **1985**, *14*, 803–820. [CrossRef]

89. Alberty, R.A.; Ha, Y.S. Standard Chemical Thermodynamic Properties of Alkylcyclopentane Isomer Groups, Alkylcyclohexane Isomer Groups, and Combined Isomer Groups. *J. Phys. Chem. Ref. Data* **1985**, *14*, 1107–1132. [CrossRef]

90. Alberty, R.A.; Burmenko, E. Standard Chemical Thermodynamic Properties of Alkyne Isomer Groups. *J. Phys. Chem. Ref. Data* **1986**, *15*, 1339–1349. [CrossRef]

91. Alberty, R.A.; Chung, M.B.; Flood, T.M. Standard Chemical Thermodynamic Properties of Alkanol Isomer Groups. *J. Phys. Chem. Ref. Data* **1987**, *16*, 391–417. [CrossRef]

92. Alberty, R.A.; Reif, A.K. Standard Chemical Thermodynamic Properties of Polycyclic Aromatic Hydrocarbons and Their Isomer Groups I. Benzene Series. *J. Phys. Chem. Ref. Data* **1988**, *17*, 241–253. [CrossRef]

93. Alberty, R.A.; Reif, A.K. Erratum: Standard Chemical Thermodynamic Properties of Polycyclic Aromatic Hydrocarbons and Their Isomer Groups I. Benzene Series [J. Phys. Chem. Ref. Data 17, 241 (1988)]. *J. Phys. Chem. Ref. Data* **1989**, *18*, 551–553. [CrossRef]

94. Alberty, R.A.; Chung, M.B.; Reif, A.K. Standard Chemical Thermodynamic Properties of Polycyclic Aromatic Hydrocarbons and Their Isomer Groups. II. Pyrene Series, Naphthopyrene Series, and Coronene Series. *J. Phys. Chem. Ref. Data* **1989**, *18*, 77–109. [CrossRef]

95. Alberty, R.A.; Chung, M.B.; Reif, A.K. Standard Chemical Thermodynamic Properties of Polycyclic Aromatic Hydrocarbons and Their Isomer Groups. III. Naphthocoronene Series, Ovalene Series, and First Members of Some Higher Series. *J. Phys. Chem. Ref. Data* **1990**, *19*, 349–370. [CrossRef]

96. Benson, S.W.; Cruickshank, F.R.; Golden, D.M.; Haugen, G.R.; O'Neal, H.E.; Rodgers, A.S.; Shaw, R.; Walsh, R. Additivity Rules for the Estimation of Thermochemical Properties. *Chem. Rev.* **1969**, *69*, 279–324. [CrossRef]

97. Benson, S.W. *Thermochemical Kinetics: Methods for the Estimation of Thermochemical Data and Rate Parameters*, 2nd ed.; Wiley: New York, NY, USA, 1976.

98. Cohen, N.; Benson, S.W. Estimation of Heats of Formation of Organic Compounds by Additivity Methods. *Chem. Rev.* **1993**, *93*, 2419–2438. [CrossRef]

99. Cohen, N. Revised Group Additivity Values for Enthalpies of Formation (at 298 K) of Carbon-Hydrogen and Carbon-Hydrogen-Oxygen Compounds. *J. Phys. Chem. Ref. Data* **1996**, *25*, 1411–1481. [CrossRef]

100. Domalski, E.S.; Hearing, E.D. Estimation of the Thermodynamic Properties of Hydrocarbons at 298.15 K. *J. Phys. Chem. Ref. Data* **1988**, *17*, 1637–1678. [CrossRef]

101. Domalski, E.S.; Hearing, E.D. Estimation of the Thermodynamic Properties of C–H–N–O–S–Halogen Compounds at 298.15 K. *J. Phys. Chem. Ref. Data* **1993**, *22*, 805–1159. [CrossRef]

102. Sabbe, M.K.; Saeys, M.; Reyniers, M.F.; Marin, G.B.; van Speybroeck, V.; Waroquier, M. Group Additive Values for the Gas Phase Standard Enthalpy of Formation of Hydrocarbons and Hydrocarbon Radicals. *J. Phys. Chem. A* **2005**, *109*, 7466–7480. [CrossRef]

103. Sabbe, M.K.; de Vleeschouwer, F.; Reyniers, M.F.; Waroquier, M.; Marin, G.B. First Principles Based Group Additive Values for the Gas Phase Standard Entropy and Heat Capacity of Hydrocarbons and Hydrocarbon Radicals. *J. Phys. Chem. A* **2008**, *112*, 12235–12251. [CrossRef]

104. Poling, B.E.; Prausnitz, J.M.; O'Connell, J.P. *The Properties of Gases and Liquids*, 5th ed.; McGraw-Hill: New York, NY, USA, 2001.

105. Burgess, D.R. Thermochemical Data. In *NIST Chemistry WebBook, NIST Standard Reference Database Number 69*; Linstrom, P.J., Mallard, W.G., Eds.; National Institute of Standards and Technology: Gaithersburg, MD, USA, 2018.

106. Chao, J.; Hall, K.R.; Marsh, K.N.; Wilhoit, R.C. Thermodynamic Properties of Key Organic Oxygen Compounds in the Carbon Range C_1 to C_4. Part 2. Ideal Gas Properties. *J. Phys. Chem. Ref. Data* **1986**, *15*, 1369–1436. [CrossRef]

107. Pilcher, G.; Pell, S.; Coleman, D.J. Measurements of Heats of Combustion by Flame Calorimetry. Part 2.—Dimethyl Ether, Methyl Ethyl Ether, Methyl *n*-Propyl Ether, Methyl isoPropyl Ether. *Trans. Faraday Soc.* **1964**, *60*, 499–505. [CrossRef]

108. Kennedy, R.M.; Sagenkahn, M.; Aston, J.G. The Heat Capacity and Entropy, Heats of Fusion and Vaporization, and the Vapor Pressure of Dimethyl Ether. The Density of Gaseous Dimethyl Ether. *J. Am. Chem. Soc.* **1941**, *63*, 2267–2272. [CrossRef]

109. Aguayo, A.T.; Ereña, J.; Mier, D.; Arandes, J.M.; Olazar, M.; Bilbao, J. Kinetic Modeling of Dimethyl Ether Synthesis in a Single Step on a $CuO-ZnO-Al_2O_3/\gamma-Al_2O_3$ Catalyst. *Ind. Eng. Chem. Res.* **2007**, *46*, 5522–5530. [CrossRef]

110. Hayashi, H.; Moffat, J.B. The Properties of Heteropoly Acids and the Conversion of Methanol to Hydrocarbons. *J. Catal.* **1982**, *77*, 473–484. [CrossRef]

111. Diep, B.T.; Wainwright, M.S. Thermodynamic Equilibrium Constants for the Methanol-Dimethyl Ether-Water System. *J. Chem. Eng. Data* **1987**, *32*, 330–333. [CrossRef]

112. Given, P.H. Given—Korrelation zu Methanol-Dehydratisierung, 1943. *J. Chem. Soc.* **1943**, 589. [CrossRef]

113. Tavan, Y.; Hasanvandian, R. Two Practical Equations for Methanol Dehydration Reaction over HZSM-5 Catalyst—Part I: Second Rrder Rate Equation. *Fuel* **2015**, *142*, 208–214. [CrossRef]

114. Gayubo, A.G.; Aguayo, A.T.; Castilla, M.; Morán, A.L.; Bilbao, J. Role of Water in the Kinetic Modeling of Methanol Transformation into Hydrocarbons on HZSM-5 Zeolite. *Chem. Eng. Commun.* **2004**, *191*, 944–967. [CrossRef]

115. Schiffino, R.S.; Merrill, R.P. A Mechanistic Study of the Methanol Dehydration Reaction on γ–Alumina Catalyst. *J. Phys. Chem.* **1993**, *97*, 6425–6435. [CrossRef]

116. Khademi, M.H.; Farsi, M.; Rahimpour, M.R.; Jahanmiri, A. DME Synthesis and Cyclohexane Dehydrogenation Reaction in an Optimized Thermally Coupled Reactor. *Chem. Eng. Process.* **2011**, *50*, 113–123. [CrossRef]

117. Spivey, J.J. Review: Dehydration Catalysts for the Methanol/Dimethyl Ether Reaction. *Chem. Eng. Commun.* **1991**, *110*, 123–142. [CrossRef]

118. Steinfeld, J.I.; Francisco, J.S.; Hase, W.L. *Chemical Kinetics and Dynamics*; Prentice Hall: Englewood Cliffs, NJ, USA, 1989.

119. Davis, M.E.; Davis, R.J. *Fundamentals of Chemical Reaction Engineering*; Chemical Engineering Series; McGraw-Hill: New York, NY, USA, 2003.

120. Marin, G.B.; Yablonsky, G.S. *Kinetics of Chemical Reactions: Decoding Complexity*; Wiley-VCH: Weinheim, Germany, 2011.

121. Froment, G.F.; Bischoff, K.B.; de Wilde, J. *Chemical Reactor Analysis and Design*, 3rd ed.; Wiley: New York, NY, USA, 2011.

122. Dittmeyer, R.; Emig, G. Simultaneous Heat and Mass Transfer and Chemical Reaction. In *Handbook of Heterogeneous Catalysis*; Ertl, G., Knözinger, H., Schüth, F., Weitkamp, J., Eds.; Wiley-VCH: Weinheim, Germany, 2008; pp. 1727–1784.

123. Dumesic, J.A.; Huber, G.W.; Boudart, M. Principles of Heterogeneous Catalysis. In *Handbook of Heterogeneous Catalysis*; Ertl, G., Knözinger, H., Schüth, F., Weitkamp, J., Eds.; Wiley-VCH: Weinheim, Germany, 2008; pp. 1–15.

124. Lynggaard, H.; Andreasen, A.; Stegelmann, C.; Stoltze, P. Analysis of Simple Kinetic Models in Heterogeneous Catalysis. *Prog. Surf. Sci.* **2004**, *77*, 71–137. [CrossRef]

125. Neimark, A.V.; Sing, K.S.W.; Thommes, M. Surface Area and Porosity. In *Handbook of Heterogeneous Catalysis*; Ertl, G., Knözinger, H., Schüth, F., Weitkamp, J., Eds.; Wiley-VCH: Weinheim, Germany, 2008; pp. 721–738.

126. Schwaab, M.; Pinto, J.C. Optimum Reparameterization of Power Function Models. *Chem. Eng. Sci.* **2008**, *63*, 4631–4635. [CrossRef]
127. Kapteijn, F.; Berger, R.J.; Moulijn, J.A. Rate Procurement and Kinetic Modeling. In *Handbook of Heterogeneous Catalysis*; Ertl, G., Knözinger, H., Schüth, F., Weitkamp, J., Eds.; Wiley-VCH: Weinheim, Germany, 2008; pp. 1693–1714.
128. Buzzi-Ferraris, G.; Manenti, F. Kinetic Models Analysis. *Chem. Eng. Sci.* **2009**, *64*, 1061–1074. [CrossRef]
129. Eyring, H. The Activated Complex and the Absolute Rate of Chemical Reactions. *Chem. Rev.* **1935**, *17*, 65–77. [CrossRef]
130. Nørskov, J.K.; Studt, F.; Abild-Pedersen, F.; Bligaard, T. *Fundamental Concepts in Heterogeneous Catalysis*; John Wiley & Sons: Hoboken, NJ, USA, 2014.
131. Schwaab, M.; Pinto, J.C. Optimum Reference Temperature for Reparameterization of the Arrhenius equation. Part 1: Problems Involving One Kinetic Constant. *Chem. Eng. Sci.* **2007**, *62*, 2750–2764. [CrossRef]
132. Park, T.Y.; Froment, G.F. Kinetic Modeling of the Methanol to Olefins Process. 2. Experimental Results, Model Discrimination, and Parameter Estimation. *Ind. Eng. Chem. Res.* **2001**, *40*, 4187–4196. [CrossRef]
133. Schwaab, M.; Lemos, L.P.; Pinto, J.C. Optimum Reference Temperature for Reparameterization of the Arrhenius Equation. Part 2: Problems Involving Multiple Reparameterizations. *Chem. Eng. Sci.* **2008**, *63*, 2895–2906. [CrossRef]
134. Toch, K.; Thybaut, J.W.; Vandegehuchte, B.D.; Narasimhan, C.S.L.; Domokos, L.; Marin, G.B. A Single-Event Microkinetic Model for "Ethylbenzene Dealkylation/Xylene Isomerization" on Pt/H-ZSM-5 Zeolite Catalyst. *Appl. Catal. A* **2012**, *425–426*, 130–144. [CrossRef]
135. Wright, P.A.; Pearce, G.M. Structural Chemistry of Zeolites. In *Zeolites and Catalysis*; Čejka, J., Corma, A., Zones, S., Eds.; Wiley-VCH: Weinheim, Germany, 2010; Volume 1, pp. 171–207.
136. Schmidt, F.; Reichelt, L.; Pätzold, C. Catalysis of Methanol Conversion to Hydrocarbons. In *Methanol, Asinger's Vision Today*; Bertau, M., Offermanns, H., Plass, L., Schmidt, F., Wernicke, H.J., Eds.; Springer: Berlin, Germany, 2014; pp. 423–440.
137. Wright, P.A. *Microporous Framework Solids*; RSC Materials Monographs; The Royal Society of Chemistry: Cambridge, UK, 2008.
138. Ong, L.H. Nature and Stability of Aluminum Species in HZSM-5: Changes upon Hydrothermal Treatment and Effect of Binder. Ph.D. Dissertation, Technical University of Munich, Munich, Germany, 2009.
139. Menges, M.; Kraushaar-Czarnetzki, B. Kinetics of Methanol to Olefins over AlPO$_4$-Bound ZSM-5 Extrudates in a Two-Stage Unit with Dimethyl Ether Pre-Reactor. *Microporous Mesoporous Mater.* **2012**, *164*, 172–181. [CrossRef]
140. McCusker, L.B.; Baerlocher, C. Zeolite Structures. In *Introduction to Zeolite Science and Practice*; van Bekkum, H., Flanigen, E.M., Jacobs, P.A., Jansen, J.C., Eds.; Studies in Surface Science and Catalysis; Elsevier: Amsterdam, The Netherlands, 2001; pp. 37–67.
141. Baerlocher, C.; McCusker, L.B. Database of Zeolite Structures. Available online: http://www.iza-structure.org/databases/ (accessed on 18 October 2018).
142. Lobo, R.F. Introduction to the Structural Chemistry of Zeolites. In *Handbook of Zeolite Science and Technology*; Auerbach, S.M., Carrado, K.A., Dutta, P.K., Eds.; Marcel Dekker: New York, NY, USA, 2003; pp. 80–112.
143. Kokotailo, G.T.; Lawton, S.L.; Olson, D.H.; Meier, W.M. Structure of Synthetic Zeolite ZSM-5. *Nature* **1978**, *272*, 437–438. [CrossRef]
144. Chang, C.D. Hydrocarbons from Methanol. *Catal. Rev. Sci. Eng.* **1983**, *25*, 1–118. [CrossRef]
145. Rohrman, A.C.; LaPierre, R.B.; Schlenker, J.L.; Wood, J.D.; Valyocsik, E.W.; Rubin, M.K.; Higgins, J.B.; Rohrbaugh, W.J. The Framework Topology of ZSM-23: A High Silica Zeolite. *Zeolites* **1985**, *5*, 352–354. [CrossRef]
146. Kumar, P.; Thybaut, J.W.; Teketel, S.; Svelle, S.; Beato, P.; Olsbye, U.; Marin, G.B. Single-Event MicroKinetics (SEMK) for Methanol to Hydrocarbons (MTH) on H-ZSM-23. *Catal. Today* **2013**, *215*, 224–232. [CrossRef]
147. Teketel, S.; Skistad, W.; Benard, S.; Olsbye, U.; Lillerud, K.P.; Beato, P.; Svelle, S. Shape Selectivity in the Conversion of Methanol to Hydrocarbons: The Catalytic Performance of One-Dimensional 10-Ring Zeolites: ZSM-22, ZSM-23, ZSM-48, and EU-1. *ACS Catal.* **2012**, *2*, 26–37. [CrossRef]
148. Lok, B.M.; Messina, C.A.; Patton, L.; Gajek, R.T.; Cannan, T.R.; Flanigen, E.M. Silicoaluminophosphate molecular sieves: another new class of microporous crystalline inorganic solids. *J. Am. Chem. Soc.* **1984**, *106*, 6092–6093. [CrossRef]

149. Epelde, E.; Ibáñez, M.; Valecillos, J.; Aguayo, A.T.; Gayubo, A.G.; Bilbao, J.; Castaño, P. SAPO-18 and SAPO-34 Catalysts for Propylene Production from the Oligomerization-Cracking of Ethylene or 1-Butenes. *Appl. Catal. A* **2017**, *547*, 176–182. [CrossRef]

150. Stöcker, M. Methanol to Olefins (MTO) and Methanol to Gasoline (MTG). In *Zeolites and Catalysis*; Čejka, J., Corma, A., Zones, S., Eds.; Wiley-VCH: Weinheim, Germany, 2010; Volume 2, pp. 687–711.

151. Chen, J.; Thomas, J.M.; Wright, P.A.; Townsend, R.P. Silicoaluminophosphate Number Eighteen (SAPO-18): A New Microporous Solid Acid Catalyst. *Catal. Lett.* **1994**, *28*, 241–248. [CrossRef]

152. Chen, J.; Wright, P.A.; Thomas, J.M.; Natarajan, S.; Marchese, L.; Bradley, S.M.; Sankar, G.; Catlow, R.A.; Gai-Boyes, P.; Townsend, R.P.; et al. SAPO-18 Catalysts and Their Brønststed Acid Sites. *J. Phys. Chem.* **1994**, *98*, 10216–10224. [CrossRef]

153. Dessau, R.M.; LaPierre, R.B. On the Mechanism of Methanol Conversion to Hydrocarbons over HZSM-5. *J. Catal.* **1982**, *78*, 136–141. [CrossRef]

154. Weitkamp, J.; Jacobs, P.A.; Martens, J.A. Isomerization and Hydrocracking of C_9 through C_{16} *n*-Alkanes on Pt/HZSM-5 Zeolite. *Appl. Catal.* **1983**, *8*, 123–141. [CrossRef]

155. Garwood, W.E. Conversion of C_2–C_{10} to Higher Olefins over Synthetic Zeolite ZSM-5. In *Intrazeolite Chemistry*; ACS Symposium Series; Stucky, G.D., Dwyer, F.G., Eds.; American Chemical Society: Washington, DC, USA, 1983; Volume 218, pp. 383–396. [CrossRef]

156. Tabak, S.A.; Krambeck, F.J.; Garwood, W.E. Conversion of Propylene and Butylene over ZSM-5 Catalyst. *AIChE J.* **1986**, *32*, 1526–1531. [CrossRef]

157. Quann, R.J.; Green, L.A.; Tabak, S.A.; Krambeck, F.J. Chemistry of Olefin Oligomerization over ZSM-5 Catalyst. *Ind. Eng. Chem. Res.* **1988**, *27*, 565–570. [CrossRef]

158. Buchanan, J.S.; Santiesteban, J.G.; Haag, W.O. Mechanistic Considerations in Acid-Catalyzed Cracking of Olefins. *J. Catal.* **1996**, *158*, 279–287. [CrossRef]

159. Buchanan, J.S. Gasoline Selective ZSM-5 FCC Additives: Model Reactions of C_6–C_{10} Olefins over Steamed 55:1 and 450:1 ZSM-5. *Appl. Catal. A* **1998**, *171*, 57–64. [CrossRef]

160. Arudra, P.; Bhuiyan, T.I.; Akhtar, M.N.; Aitani, A.M.; Al-Khattaf, S.S.; Hattori, H. Silicalite-1 As Efficient Catalyst for Production of Propene from 1-Butene. *ACS Catal.* **2014**, *4*, 4205–4214. [CrossRef]

161. Von Aretin, T.; Standl, S.; Tonigold, M.; Hinrichsen, O. Optimization of the Product Spectrum for 1-Pentene Cracking on ZSM-5 Using Single-Event Mmethodology. Part 2: Recycle Reactor. *Chem. Eng. J.* **2017**, *309*, 873–885. [CrossRef]

162. Sundberg, J.; Standl, S.; von Aretin, T.; Tonigold, M.; Rehfeldt, S.; Hinrichsen, O.; Klein, H. Optimal Process for Catalytic Cracking of Higher Olefins on ZSM-5. *Chem. Eng. J.* **2018**, *348*, 84–94. [CrossRef]

163. Von Aretin, T.; Standl, S.; Tonigold, M.; Hinrichsen, O. Optimization of the Product Spectrum for 1-Pentene Cracking on ZSM-5 Using Single-Event Methodology. Part 1: Two-Zone reactor. *Chem. Eng. J.* **2017**, *309*, 886–897. [CrossRef]

164. Von Aretin, T.; Schallmoser, S.; Standl, S.; Tonigold, M.; Lercher, J.A.; Hinrichsen, O. Single-Event Kinetic Model for 1-Pentene Cracking on ZSM-5. *Ind. Eng. Chem. Res.* **2015**, *54*, 11792–11803. [CrossRef]

165. Borges, P.; Ramos Pinto, R.; Lemos, A.; Lemos, F.; Védrine, J.C.; Derouane, E.G.; Ramôa Ribeiro, F. Light Olefin Transformation over ZSM-5 Zeolites: A Kinetic Model for Olefin Consumption. *Appl. Catal. A* **2007**, *324*, 20–29. [CrossRef]

166. Huang, X.; Aihemaitijiang, D.; Xiao, W.D. Reaction Pathway and Kinetics of C_3–C_7 Olefin Transformation over High-silicon HZSM-5 Zeolite at 400–490 °C. *Chem. Eng. J.* **2015**, *280*, 222–232. [CrossRef]

167. Nguyen, C.M.; de Moor, B.A.; Reyniers, M.F.; Marin, G.B. Physisorption and Chemisorption of Linear Alkenes in Zeolites: A Combined QM-Pot(MP2//B3LYP:GULP)-Statistical Thermodynamics Study. *J. Phys. Chem. C* **2011**, *115*, 23831–23847. [CrossRef]

168. Kazansky, V.B.; Frash, M.V.; van Santen, R.A. Quantumchemical Study of the Isobutane Cracking on Zeolites. *Appl. Catal. A* **1996**, *146*, 225–247. [CrossRef]

169. Rigby, A.M.; Kramer, G.J.; van Santen, R.A. Mechanisms of Hydrocarbon Conversion in Zeolites: A Quantum Mechanical Study. *J. Catal.* **1997**, *170*, 1–10. [CrossRef]

170. Quintana-Solórzano, R.; Thybaut, J.W.; Marin, G.B.; Løndeng, R.; Holmen, A. Single-Event Microkinetics for Coke Formation in Catalytic Cracking. *Catal. Today* **2005**, *107–108*, 619–629. [CrossRef]

171. Sun, X.; Müller, S.; Liu, Y.; Shi, H.; Haller, G.L.; Sanchez-Sanchez, M.; van Veen, A.C.; Lercher, J.A. On Reaction Pathways in the Conversion of Methanol to Hydrocarbons on HZSM-5. *J. Catal.* **2014**, *317*, 185–197. [CrossRef]

172. Feng, W.; Vynckier, E.; Froment, G.F. Single-Event Kinetics of Catalytic Cracking. *Ind. Eng. Chem. Res.* **1993**, *32*, 2997–3005. [CrossRef]

173. Chang, C.D.; Silvestri, A.J. The Conversion of Methanol and Other O-Compounds to Hydrocarbons over Zeolite Catalysts. *J. Catal.* **1977**, *47*, 249–259. [CrossRef]

174. Stöcker, M. Methanol-to-Hydrocarbons: Catalytic Materials and their Behavior. *Microporous Mesoporous Mater.* **1999**, *29*, 3–48. [CrossRef]

175. Koempel, H.; Liebner, W. Lurgi's Methanol To Propylene (MTP®) Report on a Successful Commercialisation. In *Natural Gas Conversion VIII*; Studies in Surface Science and Catalysis; Noronha, F., Schmal, M., Sousa-Aguiar, E.F., Eds.; Elsevier: Amsterdam, The Netherlands, 2007; Volume 167, pp. 261–267.

176. Forester, T.R.; Howe, R.F. In Situ FTIR Studies of Methanol and Dimethyl Ether in ZSM-5. *J. Am. Chem. Soc.* **1987**, *109*, 5076–5082. [CrossRef]

177. Anderson, M.W.; Barrie, P.J.; Klinowski, J. ^{1}H Magic-Angle-Spinning NMR Studies of the Adsorption of Alcohols on Molecular Sieve Catalysts. *J. Phys. Chem.* **1991**, *95*, 235–239. [CrossRef]

178. Martin, K.A.; Zabransky, R.F. Conversion of Methanol to Dimethylether over ZSM-5 by DRIFT Spectroscopy. *Appl. Spectrosc.* **1991**, *45*, 68–72. [CrossRef]

179. Blaszkowski, S.R.; van Santen, R.A. The Mechanism of Dimethyl Ether Formation from Methanol Catalyzed by Zeolitic Protons. *J. Am. Chem. Soc.* **1996**, *118*, 5152–5153. [CrossRef]

180. Blaszkowski, S.R.; van Santen, R.A. Theoretical Study of the Mechanism of Surface Methoxy and Dimethyl Ether Formation from Methanol Catalyzed by Zeolitic Protons. *J. Phys. Chem. B* **1997**, *101*, 2292–2305. [CrossRef]

181. Maihom, T.; Boekfa, B.; Sirijaraensre, J.; Nanok, T.; Probst, M.; Limtrakul, J. Reaction Mechanisms of the Methylation of Ethene with Methanol and Dimethyl Ether over H-ZSM-5: An ONIOM Study. *J. Phys. Chem. C* **2009**, *113*, 6654–6662. [CrossRef]

182. Jones, A.J.; Iglesia, E. Kinetic, Spectroscopic, and Theoretical Assessment of Associative and Dissociative Methanol Dehydration Routes in Zeolites. *Angew. Chem. Int. Ed.* **2014**, *53*, 12177–12181. [CrossRef]

183. Liu, Y.; Müller, S.; Berger, D.; Jelic, J.; Reuter, K.; Tonigold, M.; Sanchez-Sanchez, M.; Lercher, J.A. Formation Mechanism of the First Carbon-Carbon Bond and the First Olefin in the Methanol Conversion into Hydrocarbons. *Angew. Chem. Int. Ed.* **2016**, *55*, 5723–5726. [CrossRef]

184. Dahl, I.M.; Kolboe, S. On the Reaction Mechanism for Propene Formation in the MTO Reaction over SAPO-34. *Catal. Lett.* **1993**, *20*, 329–336. [CrossRef]

185. Dahl, I.M.; Kolboe, S. On the Reaction Mechanism for Hydrocarbon Formation from Methanol over SAPO-34: I. Isotopic Labeling Studies of the Co-Reaction of Ethene and Methanol. *J. Catal.* **1994**, *149*, 458–464. [CrossRef]

186. Dahl, I.M.; Kolboe, S. On the Reaction Mechanism for Hydrocarbon Formation from Methanol over SAPO-34: 2. Isotopic Labeling Studies of the Co-Reaction of Propene and Methanol. *J. Catal.* **1996**, *161*, 304–309. [CrossRef]

187. Bjørgen, M.; Olsbye, U.; Petersen, D.; Kolboe, S. The Methanol-to-Hydrocarbons Reaction: Insight into the Reaction Mechanism from [^{12}C]Benzene and [^{13}C]Methanol Coreactions over Zeolite H-Beta. *J. Catal.* **2004**, *221*, 1–10. [CrossRef]

188. Wang, W.; Jiang, Y.; Hunger, M. Mechanistic Investigations of the Methanol-to-Olefin (MTO) Process on Acidic Zeolite Catalysts by in situ Solid-State NMR Spectroscopy. *Catal. Today* **2006**, *113*, 102–114. [CrossRef]

189. Olsbye, U.; Bjørgen, M.; Svelle, S.; Lillerud, K.P.; Kolboe, S. Mechanistic Insight into the Methanol-to-Hydrocarbons Reaction. *Catal. Today* **2005**, *106*, 108–111. [CrossRef]

190. Olsbye, U.; Svelle, S.; Lillerud, K.P.; Wei, Z.H.; Chen, Y.Y.; Li, J.F.; Wang, J.G.; Fan, W.B. The Formation and Degradation of Active Species during Methanol Conversion over Protonated Zeotype Catalysts. *Chem. Soc. Rev.* **2015**, *44*, 7155–7176. [CrossRef]

191. Tian, P.; Wei, Y.; Ye, M.; Liu, Z. Methanol to Olefins (MTO): From Fundamentals to Commercialization. *ACS Catal.* **2015**, *5*, 1922–1938. [CrossRef]

192. Mole, T.; Whiteside, J.A. Conversion of Methanol to Ethylene over ZSM-5 Zeolite in the Presence of Deuterated Water. *J. Catal.* **1982**, *75*, 284–290. [CrossRef]

193. Sassi, A.; Wildman, M.A.; Ahn, H.J.; Prasad, P.; Nicholas, J.B.; Haw, J.F. Methylbenzene Chemistry on Zeolite HBeta: Multiple Insights into Methanol-to-Olefin Catalysis. *J. Phys. Chem. B* **2002**, *106*, 2294–2303. [CrossRef]
194. Haw, J.F.; Song, W.; Marcus, D.M.; Nicholas, J.B. The Mechanism of Methanol to Hydrocarbon Catalysis. *Acc. Chem. Res.* **2003**, *36*, 317–326. [CrossRef]
195. Arstad, B.; Nicholas, J.B.; Haw, J.F. Theoretical Study of the Methylbenzene Side-Chain Hydrocarbon Pool Mechanism in Methanol to Olefin Catalysis. *J. Am. Chem. Soc.* **2004**, *126*, 2991–3001. [CrossRef]
196. Lesthaeghe, D.; Horré, A.; Waroquier, M.; Marin, G.B.; van Speybroeck, V. Theoretical Insights on Methylbenzene Side-Chain Growth in ZSM-5 Zeolites for Methanol-to-Olefin Conversion. *Chem. Eur. J.* **2009**, *15*, 10803–10808. [CrossRef]
197. Sullivan, R.F.; Egan, C.J.; Langlois, G.E.; Sieg, R.P. A New Reaction That Occurs in the Hydrocracking of Certain Aromatic Hydrocarbons. *J. Am. Chem. Soc.* **1961**, *83*, 1156–1160. [CrossRef]
198. Dessau, R.M. On the H-ZSM-5 Catalyzed Formation of Ethylene from Methanol or Higher Olefins. *J. Catal.* **1986**, *99*, 111–116. [CrossRef]
199. Svelle, S.; Joensen, F.; Nerlov, J.; Olsbye, U.; Lillerud, K.P.; Kolboe, S.; Bjørgen, M. Conversion of Methanol into Hydrocarbons over Zeolite H-ZSM-5: Ethene Formation is Mechanistically Separated from the Formation of Higher Alkenes. *J. Am. Chem. Soc.* **2006**, *128*, 14770–14771. [CrossRef]
200. Bjørgen, M.; Svelle, S.; Joensen, F.; Nerlov, J.; Kolboe, S.; Bonino, F.; Palumbo, L.; Bordiga, S.; Olsbye, U. Conversion of Methanol to Hydrocarbons over Zeolite H-ZSM-5: On the Origin of the Olefinic Species. *J. Catal.* **2007**, *249*, 195–207. [CrossRef]
201. Svelle, S.; Arstad, B.; Kolboe, S.; Swang, O. A Theoretical Investigation of the Methylation of Alkenes with Methanol over Acidic Zeolites. *J. Phys. Chem. B* **2003**, *107*, 9281–9289. [CrossRef]
202. Svelle, S.; Kolboe, S.; Swang, O.; Olsbye, U. Methylation of Alkenes and Methylbenzenes by Dimethyl Ether or Methanol on Acidic Zeolites. *J. Phys. Chem. B* **2005**, *109*, 12874–12878. [CrossRef]
203. Martínez-Espín, J.S.; Mortén, M.; Janssens, T.V.W.; Svelle, S.; Beato, P.; Olsbye, U. New Insights into Catalyst Deactivation and Product Distribution of Zeolites in the Methanol-to-Hydrocarbons (MTH) Reaction with Methanol and Dimethyl Ether Feeds. *Catal. Sci. Technol.* **2017**, *7*, 2700–2716. [CrossRef]
204. Hutchings, G.J.; Hunter, R. Hydrocarbon Formation from Methanol and Dimethyl Ether: A Review of the Experimental Observations Concerning the Mechanism of Formation of the Primary Products. *Catal. Today* **1990**, *6*, 279–306. [CrossRef]
205. Müller, S.; Liu, Y.; Kirchberger, F.M.; Tonigold, M.; Sanchez-Sanchez, M.; Lercher, J.A. Hydrogen Transfer Pathways during Zeolite Catalyzed Methanol Conversion to Hydrocarbons. *J. Am. Chem. Soc.* **2016**, *138*, 15994–16003. [CrossRef]
206. Martínez-Espín, J.S.; de Wispelaere, K.; Janssens, T.V.W.; Svelle, S.; Lillerud, K.P.; Beato, P.; van Speybroeck, V.; Olsbye, U. Hydrogen Transfer versus Methylation: On the Genesis of Aromatics Formation in the Methanol-To-Hydrocarbons Reaction over H-ZSM-5. *ACS Catal.* **2017**, *7*, 5773–5780. [CrossRef]
207. Huang, X.; Li, H.; Li, H.; Xiao, W.D. Modeling and Analysis of the Lurgi-Type Methanol to Propylene Process: Optimization of Olefin Recycle. *AIChE J.* **2017**, *63*, 306–313. [CrossRef]
208. Svelle, S.; Rønning, P.O.; Kolboe, S. Kinetic Studies of Zeolite-Catalyzed Methylation Reactions 1. Coreaction of [^{12}C]Ethene and [^{13}C]Methanol. *J. Catal.* **2004**, *224*, 115–123. [CrossRef]
209. Svelle, S.; Rønning, P.O.; Olsbye, U.; Kolboe, S. Kinetic Studies of Zeolite-Catalyzed Methylation Reactions. Part 2. Co-Reaction of [^{12}C]Propene or [^{12}C]*n*-Butene and [^{13}C]Methanol. *J. Catal.* **2005**, *234*, 385–400. [CrossRef]
210. Wu, W.; Guo, W.; Xiao, W.D.; Luo, M. Dominant Reaction Pathway for Methanol Conversion to Propene over High Silicon H-ZSM-5. *Chem. Eng. Sci.* **2011**, *66*, 4722–4732. [CrossRef]
211. Ilias, S.; Bhan, A. Tuning the Selectivity of Methanol-to-Hydrocarbons Conversion on H-ZSM-5 by Co-Processing Olefin or Aromatic Compounds. *J. Catal.* **2012**, *290*, 186–192. [CrossRef]
212. Guo, W.; Xiao, W.D.; Luo, M. Comparison among Monolithic and Randomly Packed Reactors for the Methanol-to-Propylene Process. *Chem. Eng. J.* **2012**, *207–208*, 734–745. [CrossRef]
213. Sun, X.; Müller, S.; Shi, H.; Haller, G.L.; Sanchez-Sanchez, M.; van Veen, A.C.; Lercher, J.A. On the Impact of Co-Feeding Aromatics and Olefins for the Methanol-to-Olefins Reaction on HZSM-5. *J. Catal.* **2014**, *314*, 21–31. [CrossRef]
214. Chang, C.D. Methanol Conversion to Light Olefins. *Catal. Rev. Sci. Eng.* **1984**, *26*, 323–345. [CrossRef]

215. Froment, G.F.; Dehertog, W.J.H.; Marchi, A.J. Zeolite Catalysis in the Conversion of Methanol into Olefins. In *Catalysis*; Spivey, J.J., Ed.; The Royal Society of Chemistry: Cambridge, UK, 1992; Volume 9, pp. 1–64.

216. Svelle, S.; Visur, M.; Olsbye, U.; Saepurahman.; Bjørgen, M. Mechanistic Aspects of the Zeolite Catalyzed Methylation of Alkenes and Aromatics with Methanol: A Review. *Top. Catal.* **2011**, *54*, 897–906. [CrossRef]

217. Ilias, S.; Bhan, A. Mechanism of the Catalytic Conversion of Methanol to Hydrocarbons. *ACS Catal.* **2013**, *3*, 18–31. [CrossRef]

218. Svelle, S.; Bjørgen, M. Mechanistic Proposal for the Zeolite Catalyzed Methylation of Aromatic Compounds. *J. Phys. Chem. A* **2010**, *114*, 12548–12554. [CrossRef]

219. Hill, I.M.; Ng, Y.S.; Bhan, A. Kinetics of Butene Isomer Methylation with Dimethyl Ether over Zeolite Catalysts. *ACS Catal.* **2012**, *2*, 1742–1748. [CrossRef]

220. Hill, I.M.; Al Hashimi, S.; Bhan, A. Kinetics and Mechanism of Olefin Methylation Reactions on Zeolites. *J. Catal.* **2012**, *285*, 115–123. [CrossRef]

221. Hill, I.; Malek, A.; Bhan, A. Kinetics and Mechanism of Benzene, Toluene, and Xylene Methylation over H-MFI. *ACS Catal.* **2013**, *3*, 1992–2001. [CrossRef]

222. Brogaard, R.Y.; Wang, C.M.; Studt, F. Methanol-Alkene Reactions in Zeotype Acid Catalysts: Insights from a Descriptor-Based Approach and Microkinetic Modeling. *ACS Catal.* **2014**, *4*, 4504–4509. [CrossRef]

223. Brogaard, R.Y.; Henry, R.; Schuurman, Y.; Medford, A.J.; Moses, P.G.; Beato, P.; Svelle, S.; Nørskov, J.K.; Olsbye, U. Methanol-to-Hydrocarbons Conversion: The Alkene Methylation Pathway. *J. Catal.* **2014**, *314*, 159–169. [CrossRef]

224. Martínez-Espín, J.S.; de Wispelaere, K.; Westgård Erichsen, M.; Svelle, S.; Janssens, T.V.W.; van Speybroeck, V.; Beato, P.; Olsbye, U. Benzene Co-Reaction with Methanol and Dimethyl Ether over Zeolite and Zeotype Catalysts: Evidence of Parallel Reaction Paths to Toluene and Diphenylmethane. *J. Catal.* **2017**, *349*, 136–148. [CrossRef]

225. Li, J.; Wei, Y.; Liu, G.; Qi, Y.; Tian, P.; Li, B.; He, Y.; Liu, Z. Comparative Study of MTO Conversion over SAPO-34, H-ZSM-5 and H-ZSM-22: Correlating Catalytic Performance and Reaction Mechanism to Zeolite Topology. *Catal. Today* **2011**, *171*, 221–228. [CrossRef]

226. Almutairi, S.M.T.; Mezari, B.; Pidko, E.A.; Magusin, P.C.M.M.; Hensen, E.J.M. Influence of Steaming on the Acidity and the Methanol Conversion Reaction of HZSM-5 Zeolite. *J. Catal.* **2013**, *307*, 194–203. [CrossRef]

227. Khare, R.; Bhan, A. Mechanistic Studies of Methanol-to-Hydrocarbons Conversion on Diffusion-Free MFI Samples. *J. Catal.* **2015**, *329*, 218–228. [CrossRef]

228. Westgård Erichsen, M.; de Wispelaere, K.; Hemelsoet, K.; Moors, S.L.; Deconinck, T.; Waroquier, M.; Svelle, S.; van Speybroeck, V.; Olsbye, U. How Zeolitic Acid Strength and Composition Alter the Reactivity of Alkenes and Aromatics towards Methanol. *J. Catal.* **2015**, *328*, 186–196. [CrossRef]

229. Wu, W.; Guo, W.; Xiao, W.D.; Luo, M. Methanol Conversion to Olefins (MTO) over H-ZSM-5: Evidence of Product Distribution Governed by Methanol Conversion. *Fuel Process. Technol.* **2013**, *108*, 19–24. [CrossRef]

230. Kaarsholm, M.; Joensen, F.; Nerlov, J.; Cenni, R.; Chaouki, J.; Patience, G.S. Phosphorous Modified ZSM-5: Deactivation and Product Distribution for MTO. *Chem. Eng. Sci.* **2007**, *62*, 5527–5532. [CrossRef]

231. Janssens, T.V.W. A New Approach to the Modeling of Deactivation in the Conversion of Methanol on Zeolite Catalysts. *J. Catal.* **2009**, *264*, 130–137. [CrossRef]

232. Schulz, H. "Coking" of Zeolites during Methanol Conversion: Basic Reactions of the MTO-, MTP- and MTG Processes. *Catal. Today* **2010**, *154*, 183–194. [CrossRef]

233. Bleken, F.L.; Barbera, K.; Bonino, F.; Olsbye, U.; Lillerud, K.P.; Bordiga, S.; Beato, P.; Janssens, T.V.W.; Svelle, S. Catalyst Deactivation by Coke Formation in Microporous and Desilicated Zeolite H-ZSM-5 During the Conversion of Methanol to Hydrocarbons. *J. Catal.* **2013**, *307*, 62–73. [CrossRef]

234. Müller, S.; Liu, Y.; Vishnuvarthan, M.; Sun, X.; van Veen, A.C.; Haller, G.L.; Sanchez-Sanchez, M.; Lercher, J.A. Coke Formation and Deactivation Pathways on H-ZSM-5 in the Conversion of Methanol to Olefins. *J. Catal.* **2015**, *325*, 48–59. [CrossRef]

235. Epelde, E.; Aguayo, A.T.; Olazar, M.; Bilbao, J.; Gayubo, A.G. Kinetic Model for the Transformation of 1-Butene on a K-Modified HZSM-5 Catalyst. *Ind. Eng. Chem. Res.* **2014**, *53*, 10599–10607. [CrossRef]

236. Ying, L.; Zhu, J.; Cheng, Y.; Wang, L.; Li, X. Kinetic Modeling of C_2–C_7 Olefins Interconversion over ZSM-5 Catalyst. *J. Ind. Eng. Chem.* **2016**, *33*, 80–90. [CrossRef]

237. Oliveira, P.; Borges, P.; Ramos Pinto, R.; Lemos, A.; Lemos, F.; Védrine, J.C.; Ramôa Ribeiro, F. Light Olefin Transformation over ZSM-5 Zeolites with Different Acid Strengths—A Kinetic Model. *Appl. Catal. A* **2010**, *384*, 177–185. [CrossRef]

238. Epelde, E.; Gayubo, A.G.; Olazar, M.; Bilbao, J.; Aguayo, A.T. Modified HZSM-5 Zeolites for Intensifying Propylene Production in the Transformation of 1-Butene. *Chem. Eng. J.* **2014**, *251*, 80–91. [CrossRef]

239. Epelde, E.; Gayubo, A.G.; Olazar, M.; Bilbao, J.; Aguayo, A.T. Intensifying Propylene Production by 1-Butene Transformation on a K Modified HZSM-5 Zeolite-Catalyst. *Ind. Eng. Chem. Res.* **2014**, *53*, 4614–4622. [CrossRef]

240. Huang, X.; Aihemaitijiang, D.; Xiao, W.D. Co-Reaction of Methanol and Olefins on the High Silicon HZSM-5 Catalyst: A Kinetic Study. *Chem. Eng. J.* **2016**, *286*, 150–164. [CrossRef]

241. Borges, P.; Ramos Pinto, R.; Lemos, A.; Lemos, F.; Védrine, J.C.; Derouane, E.G.; Ramôa Ribeiro, F. Activity-acidity Relationship for Alkane Cracking over Zeolites: *n*-Hexane Cracking over HZSM-5. *J. Mol. Catal. A Chem.* **2005**, *229*, 127–135. [CrossRef]

242. Schallmoser, S. Insight into Catalytic Cracking of Hydrocarbons over MFI type Zeolites. Ph.D. Dissertation, Technical University of Munich, Munich, Germany, 2014.

243. Nguyen, C.M.; de Moor, B.A.; Reyniers, M.F.; Marin, G.B. Isobutene Protonation in H-FAU, H-MOR, H-ZSM-5, and H-ZSM-22. *J. Phys. Chem. C* **2012**, *116*, 18236–18249. [CrossRef]

244. Standl, S.; Tonigold, M.; Hinrichsen, O. Single-Event Kinetic Modeling of Olefin Cracking on ZSM-5: Proof of Feed Independence. *Ind. Eng. Chem. Res.* **2017**, *56*, 13096–13108. [CrossRef]

245. Baltanas, M.A.; van Raemdonck, K.K.; Froment, G.F.; Mohedas, S.R. Fundamental Kinetic Modeling of Hydroisomerization and Hydrocracking on Noble-Metal-Loaded Faujasites. 1. Rate Parameters for Hydroisomerization. *Ind. Eng. Chem. Res.* **1989**, *28*, 899–910. [CrossRef]

246. Mhadeshwar, A.B.; Wang, H.; Vlachos, D.G. Thermodynamic Consistency in Microkinetic Development of Surface Reaction Mechanisms. *J. Phys. Chem. B* **2003**, *107*, 12721–12733. [CrossRef]

247. Chen, C.J.; Rangarajan, S.; Hill, I.M.; Bhan, A. Kinetics and Thermochemistry of C_4–C_6 Olefin Cracking on H-ZSM-5. *ACS Catal.* **2014**, *4*, 2319–2327. [CrossRef]

248. Mazar, M.N.; Al-Hashimi, S.; Cococcioni, M.; Bhan, A. β-Scission of Olefins on Acidic Zeolites: A Periodic PBE-D Study in H-ZSM-5. *J. Phys. Chem. C* **2013**, *117*, 23609–23620. [CrossRef]

249. Li, J.; Li, T.; Ma, H.; Sun, Q.; Li, C.; Ying, W.; Fang, D. Kinetics of Coupling Cracking of Butene and Pentene on Modified HZSM-5 Catalyst. *Chem. Eng. J.* **2018**, *346*, 397–405. [CrossRef]

250. Meng, X.; Xu, C.; Li, L.; Gao, J. Kinetic Study of Catalytic Pyrolysis of C_4 Hydrocarbons on a Modified ZSM-5 Zeolite Catalyst. *Energy Fuels* **2010**, *24*, 6233–6238. [CrossRef]

251. Jiang, B.; Feng, X.; Yan, L.; Jiang, Y.; Liao, Z.; Wang, J.; Yang, Y. Methanol to Propylene Process in a Moving Bed Reactor with Byproducts Recycling: Kinetic Study and Reactor Simulation. *Ind. Eng. Chem. Res.* **2014**, *53*, 4623–4632. [CrossRef]

252. Aguayo, A.T.; Mier, D.; Gayubo, A.G.; Gamero, M.; Bilbao, J. Kinetics of Methanol Transformation into Hydrocarbons on a HZSM-5 Zeolite Catalyst at High Temperature (400–550 °C). *Ind. Eng. Chem. Res.* **2010**, *49*, 12371–12378. [CrossRef]

253. Pérez-Uriarte, P.; Ateka, A.; Aguayo, A.T.; Gayubo, A.G.; Bilbao, J. Kinetic Model for the Reaction of DME to Olefins over a HZSM-5 Zeolite Catalyst. *Chem. Eng. J.* **2016**, *302*, 801–810. [CrossRef]

254. Park, T.Y.; Froment, G.F. Kinetic Modeling of the Methanol to Olefins Process. 1. Model Formulation. *Ind. Eng. Chem. Res.* **2001**, *40*, 4172–4186. [CrossRef]

255. Gayubo, A.G.; Aguayo, A.T.; Sánchez del Campo, A.E.; Tarrío, A.M.; Bilbao, J. Kinetic Modeling of Methanol Transformation into Olefins on a SAPO-34 Catalyst. *Ind. Eng. Chem. Res.* **2000**, *39*, 292–300. [CrossRef]

256. Ying, L.; Yuan, X.; Ye, M.; Cheng, Y.; Li, X.; Liu, Z. A Seven Lumped Kinetic Model for Industrial Catalyst in DMTO Process. *Chem. Eng. Res. Des.* **2015**, *100*, 179–191. [CrossRef]

257. Chen, D.; Grønvold, A.; Moljord, K.; Holmen, A. Methanol Conversion to Light Olefins over SAPO-34: Reaction Network and Deactivation Kinetics. *Ind. Eng. Chem. Res.* **2007**, *46*, 4116–4123. [CrossRef]

258. Alwahabi, S.M.; Froment, G.F. Single Event Kinetic Modeling of the Methanol-to-Olefins Process on SAPO-34. *Ind. Eng. Chem. Res.* **2004**, *43*, 5098–5111. [CrossRef]

259. Gayubo, A.G.; Aguayo, A.T.; Alonso, A.; Atutxa, A.; Bilbao, J. Reaction Scheme and Kinetic Modelling for the MTO Process over a SAPO-18 Catalyst. *Catal. Today* **2005**, *106*, 112–117. [CrossRef]

260. Gayubo, A.G.; Aguayo, A.T.; Alonso, A.; Bilbao, J. Kinetic Modeling of the Methanol-to-Olefins Process on a Silicoaluminophosphate (SAPO-18) Catalyst by Considering Deactivation and the Formation of Individual Olefins. *Ind. Eng. Chem. Res.* **2007**, *46*, 1981–1989. [CrossRef]

261. Sánchez del Campo, A.E.; Gayubo, A.G.; Aguayo, A.T.; Tarrío, A.; Bilbao, J. Acidity, Surface Species, and Mechanism of Methanol Transformation into Olefins on a SAPO-34. *Ind. Eng. Chem. Res.* **1998**, *37*, 2336–2340. [CrossRef]

262. Menges, M. Untersuchungen zum MTO-Prozess an AlPO$_4$-gebundenen ZSM-5-Extrudaten und Beschreibung der Reaktionskinetik. Ph.D. Dissertation, Karlsruher Institut für Technologie, Karlsruhe, Germany, 2012. [CrossRef]

263. Chen, D.; Grønvold, A.; Rebo, H.P.; Moljord, K.; Holmen, A. Catalyst Deactivation Studied by Conventional and Oscillating Microbalance Reactors. *Appl. Catal. A* **1996**, *137*, L1–L8. [CrossRef]

264. Freiding, J.; Patcas, F.C.; Kraushaar-Czarnetzki, B. Extrusion of Zeolites: Properties of Catalysts with a Novel Aluminium Phosphate Sintermatrix. *Appl. Catal. A* **2007**, *328*, 210–218. [CrossRef]

265. Freiding, J. Extrusion von technischen ZSM-5-Kontakten und ihre Anwendung im MTO-Prozess. Ph.D. Dissertation, Karlsruher Institut für Technologie, Karlsruhe, Germany, 2009. [CrossRef]

266. Freiding, J.; Kraushaar-Czarnetzki, B. Novel Extruded Fixed-Bed MTO Catalysts with High Olefin Selectivity and High Resistance against Coke Deactivation. *Appl. Catal. A* **2011**, *391*, 254–260. [CrossRef]

267. Gayubo, A.G.; Aguayo, A.T.; Morán, A.L.; Olazar, M.; Bilbao, J. Role of Water in the Kinetic Modeling of Catalyst Deactivation in the MTG Process. *AIChE J.* **2002**, *48*, 1561–1571. [CrossRef]

268. Gayubo, A.G.; Aguayo, A.T.; Olazar, M.; Vivanco, R.; Bilbao, J. Kinetics of the Irreversible Deactivation of the HZSM-5 Catalyst in the MTO Process. *Chem. Eng. Sci.* **2003**, *58*, 5239–5249. [CrossRef]

269. Gayubo, A.G.; Arandes, J.M.; Aguayo, A.T.; Olazar, M.; Bilbao, J. Calculation of the Kinetics of Deactivation by Coke in an Integral Reactor for a Triangular Scheme Reaction. *Chem. Eng. Sci.* **1993**, *48*, 1077–1087. [CrossRef]

270. Aguayo, A.T.; Gayubo, A.G.; Ortega, J.M.; Olazar, M.; Bilbao, J. Catalyst Deactivation by Coking in the MTG Process in Fixed and Fluidized Bed Reactors. *Catal. Today* **1997**, *37*, 239–248. [CrossRef]

271. Benito, P.L.; Gayubo, A.G.; Aguayo, A.T.; Castilla, M.; Bilbao, J. Concentration-Dependent Kinetic Model for Catalyst Deactivation in the MTG Process. *Ind. Eng. Chem. Res.* **1996**, *35*, 81–89. [CrossRef]

272. Mier, D.; Aguayo, A.T.; Gayubo, A.G.; Olazar, M.; Bilbao, J. Catalyst Discrimination for Olefin Production by Coupled Methanol/*n*-Butane Cracking. *Appl. Catal. A* **2010**, *383*, 202–210. [CrossRef]

273. Mier, D.; Aguayo, A.T.; Gamero, M.; Gayubo, A.G.; Bilbao, J. Kinetic Modeling of *n*-Butane Cracking on HZSM-5 Zeolite Catalyst. *Ind. Eng. Chem. Res.* **2010**, *49*, 8415–8423. [CrossRef]

274. Pérez-Uriarte, P.; Ateka, A.; Gamero, M.; Aguayo, A.T.; Bilbao, J. Effect of the Operating Conditions in the Transformation of DME to Olefins over a HZSM-5 Zeolite Catalyst. *Ind. Eng. Chem. Res.* **2016**, *55*, 6569–6578. [CrossRef]

275. Pérez-Uriarte, P.; Gamero, M.; Ateka, A.; Díaz, M.; Aguayo, A.T.; Bilbao, J. Effect of the Acidity of HZSM-5 Zeolite and the Binder in the DME Transformation to Olefins. *Ind. Eng. Chem. Res.* **2016**, *55*, 1513–1521. [CrossRef]

276. Pérez-Uriarte, P.; Ateka, A.; Gayubo, A.G.; Cordero-Lanzac, T.; Aguayo, A.T.; Bilbao, J. Deactivation Kinetics for the Conversion of Dimethyl Ether to Olefins over a HZSM-5 Zeolite Catalyst. *Chem. Eng. J.* **2017**, *311*, 367–377. [CrossRef]

277. Lox, E.; Coenen, F.; Vermeulen, R.; Froment, G.F. A Versatile Bench-Scale Unit for Kinetic Studies of Catalytic Reactions. *Ind. Eng. Chem. Res.* **1988**, *27*, 576–580. [CrossRef]

278. Hutchings, G.J.; Gottschalk, F.; Hall, M.V.M.; Hunter, R. Hydrocarbon Formation from Methylating Agents over the Zeolite Catalyst ZSM-5. *J. Chem. Soc. Faraday Trans. 1* **1987**, *83*, 571–583. [CrossRef]

279. Park, T.Y.; Froment, G.F. A Hybrid Genetic Algorithm for the Estimation of Parameters in Detailed Kinetic Models. *Comput. Chem. Eng.* **1998**, *22*, S103–S110. [CrossRef]

280. Boudart, M.; Mears, D.E.; Vannice, M.A. Kinetics of Heterogeneous Catalytic Reactions. *Ind. Chim. Belge* **1967**, *32*, 281–284.

281. Evans, M.G.; Polanyi, M. Inertia and Driving Force of Chemical Reactions. *Trans. Faraday Soc.* **1938**, *34*, 11–24. [CrossRef]

282. Park, T.Y.; Froment, G.F. Analysis of Fundamental Reaction Rates in the Methanol-to-Olefins Process on ZSM-5 as a Basis for Reactor Design and Operation. *Ind. Eng. Chem. Res.* **2004**, *43*, 682–689. [CrossRef]

283. Denayer, J.F.; Souverijns, W.; Jacobs, P.A.; Martens, J.A.; Baron, G.V. High-Temperature Low-Pressure Adsorption of Branched C_5–C_8 Alkanes on Zeolite Beta, ZSM-5, ZSM-22, Zeolite Y, and Mordenite. *J. Phys. Chem. B* **1998**, *102*, 4588–4597. [CrossRef]

284. Denayer, J.F.; Baron, G.V.; Martens, J.A.; Jacobs, P.A. Chromatographic Study of Adsorption of *n*-Alkanes on Zeolites at High Temperatures. *J. Phys. Chem. B* **1998**, *102*, 3077–3081. [CrossRef]

285. Chen, D.; Rebo, H.P.; Grønvold, A.; Moljord, K.; Holmen, A. Methanol Conversion to Light Olefins over SAPO-34: Kinetic Modeling of Coke Formation. *Microporous Mesoporous Mater.* **2000**, *35–36*, 121–135. [CrossRef]

286. Chen, D.; Bjorgum, E.; Christensen, K.O.; Holmen, A.; Lodeng, R. Characterization of Catalysts under Working Conditions with an Oscillating Microbalance Reactor. In *Advances in Catalysis*; Gates, B.C., Knözinger, H., Eds.; Academic Press: Cambridge, MA, USA, 2007; Volume 51, pp. 351–382.

287. Alwahabi, S.M. Conversion of Methanol to Light Olefins on SAPO-34: Kinetic Modeling and Reactor Design. Ph.D. Dissertation, Texas A&M University, College Station, TX, USA, 2003.

288. Gayubo, A.G.; Vivanco, R.; Alonso, A.; Valle, B.; Aguayo, A.T. Kinetic Behavior of the SAPO-18 Catalyst in the Transformation of Methanol into Olefins. *Ind. Eng. Chem. Res.* **2005**, *44*, 6605–6614. [CrossRef]

289. Bleken, F.; Bjørgen, M.; Palumbo, L.; Bordiga, S.; Svelle, S.; Lillerud, K.P.; Olsbye, U. The Effect of Acid Strength on the Conversion of Methanol to Olefins Over Acidic Microporous Catalysts with the CHA Topology. *Top. Catal.* **2009**, *52*, 218–228. [CrossRef]

290. Teketel, S.; Olsbye, U.; Lillerud, K.P.; Beato, P.; Svelle, S. Selectivity Control through Fundamental Mechanistic Insight in the Conversion of Methanol to Hydrocarbons over Zeolites. *Microporous Mesoporous Mater.* **2010**, *136*, 33–41. [CrossRef]

291. Narasimhan, C.S.L.; Thybaut, J.W.; Marin, G.B.; Martens, J.A.; Denayer, J.F.; Baron, G.V. Pore Mouth Physisorption of Alkanes on ZSM-22: Estimation of Physisorption Enthalpies and Entropies by Additivity Method. *J. Catal.* **2003**, *218*, 135–147. [CrossRef]

292. Denayer, J.F.; Ocakoglu, R.A.; Huybrechts, W.; Martens, J.A.; Thybaut, J.W.; Marin, G.B.; Baron, G.V. Pore Mouth versus Intracrystalline Adsorption of Isoalkanes on ZSM-22 and ZSM-23 Zeolites under Vapour and Liquid Phase Conditions. *Chem. Commun.* **2003**, 1880–1881. [CrossRef]

293. Kaarsholm, M.; Rafii, B.; Joensen, F.; Cenni, R.; Chaouki, J.; Patience, G.S. Kinetic Modeling of Methanol-to-Olefin Reaction over ZSM-5 in Fluid Bed. *Ind. Eng. Chem. Res.* **2010**, *49*, 29–38. [CrossRef]

294. Yuan, X.; Li, H.; Ye, M.; Liu, Z. Kinetic Modeling of Methanol to Olefins Process over SAPO-34 Catalyst Based on the Dual-Cycle Reaction Mechanism. *AIChE J.* **2018** [CrossRef]

295. Yuan, X.; Li, H.; Ye, M.; Liu, Z. Comparative Study of MTO Kinetics over SAPO-34 Catalyst in Fixed and Fluidized Bed Reactors. *Chem. Eng. J.* **2017**, *329*, 35–44. [CrossRef]

296. Lu, B.; Zhang, J.; Luo, H.; Wang, W.; Li, H.; Ye, M.; Liu, Z.; Li, J. Numerical Simulation of Scale-Up Effects of Methanol-to-Olefins Fluidized Bed Reactors. *Chem. Eng. Sci.* **2017**, *171*, 244–255. [CrossRef]

297. Zhang, J.; Lu, B.; Chen, F.; Li, H.; Ye, M.; Wang, W. Simulation of a Large Methanol-to-Olefins Fluidized Bed Reactor with Consideration of Coke Distribution. *Chem. Eng. Sci.* **2018**, *189*, 212–220. [CrossRef]

298. Strizhak, P.; Zhokh, A.; Trypolskyi, A. Methanol Conversion to Olefins on H-ZSM-5/Al_2O_3 Catalysts: Kinetic Modeling. *React. Kinet. Mech. Catal.* **2018**, *123*, 247–268. [CrossRef]

299. Sedighi, M.; Bahrami, H.; Towfighi, J. Kinetic Modeling Formulation of the Methanol to Olefin Process: Parameter estimation. *J. Ind. Eng. Chem.* **2014**, *20*, 3108–3114. [CrossRef]

300. Fatourehchi, N.; Sohrabi, M.; Royaee, S.J.; Mirarefin, S.M. Preparation of SAPO-34 catalyst and presentation of a kinetic model for methanol to olefin process (MTO). *Chem. Eng. Res. Des.* **2011**, *89*, 811–816. [CrossRef]

301. Taheri Najafabadi, A.; Fatemi, S.; Sohrabi, M.; Salmasi, M. Kinetic Modeling and Optimization of the Operating Condition of MTO Process on SAPO-34 Catalyst. *J. Ind. Eng. Chem.* **2012**, *18*, 29–37. [CrossRef]

302. Azarhoosh, M.J.; Halladj, R.; Askari, S. Presenting a New Kinetic Model for Methanol to Light Olefins Reactions over a Hierarchical SAPO-34 Catalyst Using the Langmuir–Hinshelwood–Hougen–Watson Mechanism. *J. Phys. Condens. Matter* **2017**, *29*, 425202. [CrossRef]

303. Wen, M.; Ding, J.; Wang, C.; Li, Y.; Zhao, G.; Liu, Y.; Lu, Y. High-Performance SS-Fiber@HZSM-5 Core–Shell Catalyst for Methanol-to-Propylene: A Kinetic and Modeling Study. *Microporous Mesoporous Mater.* **2016**, *221*, 187–196. [CrossRef]

304. Wen, M.; Wang, X.; Han, L.; Ding, J.; Sun, Y.; Liu, Y.; Lu, Y. Monolithic Metal-Fiber@HZSM-5 Core–Shell Catalysts for Methanol-to-Propylene. *Microporous Mesoporous Mater.* **2015**, *206*, 8–16. [CrossRef]

305. Huang, X.; Li, H.; Xiao, W.D.; Chen, D. Insight into the Side Reactions in Methanol-to-Olefin Process over HZSM-5: A Kinetic Study. *Chem. Eng. J.* **2016**, *299*, 263–275. [CrossRef]
306. Huang, X.; Li, H.; Li, H.; Xiao, W.D. A Computationally Efficient Multi-Scale Simulation of a Multi-Stage Fixed-Bed Reactor for Methanol to Propylene Reactions. *Fuel Process. Technol.* **2016**, *150*, 104–116. [CrossRef]
307. Huang, X.; Li, X.G.; Li, H.; Xiao, W.D. High-Performance HZSM-5/Cordierite Monolithic Catalyst for Methanol to Propylene Reaction: A Combined Experimental and Modelling Study. *Fuel Process. Technol.* **2017**, *159*, 168–177. [CrossRef]
308. Wang, X.; Wen, M.; Wang, C.; Ding, J.; Sun, Y.; Liu, Y.; Lu, Y. Microstructured Fiber@HZSM-5 Core-Shell Catalysts with Dramatic Selectivity and Stability Improvement for the Methanol-to-Propylene Process. *Chem. Commun.* **2014**, *50*, 6343–6345. [CrossRef]
309. Guo, W.; Wu, W.; Luo, M.; Xiao, W.D. Modeling of Diffusion and Reaction in Monolithic Catalysts for the Methanol-to-Propylene Process. *Fuel Process. Technol.* **2013**, *108*, 133–138. [CrossRef]
310. Ortega, C.; Hessel, V.; Kolb, G. Dimethyl Ether to Hydrocarbons over ZSM-5: Kinetic Study in an External Recycle Reactor. *Chem. Eng. J.* **2018**, *354*, 21–34. [CrossRef]

catalysts

MDPI

Article

Modeling of a Pilot-Scale Fixed-Bed Reactor for Dehydration of 2,3-Butanediol to 1,3-Butadiene and Methyl Ethyl Ketone

Daesung Song

Global Technology, SK Innovation, 325 Exporo, Yuseong-gu, Daejeon 305-712, Korea; daesungs@gmail.com; Tel.: +82-10-9312-6098

Received: 13 January 2018; Accepted: 7 February 2018; Published: 9 February 2018

Abstract: A 1D heterogeneous reactor model accounting for interfacial and intra-particle gradients was developed to simulate the dehydration of 2,3-Butanediol (2,3-BDO) to 1,3-Butadiene (1,3-BD) and Methyl Ethyl Ketone (MEK) over an amorphous calcium phosphate (a-CP) catalyst in a pilot-scale fixed-bed reactor. The developed model was validated with experimental data in terms of a fluid temperature profile along with the length of the catalyst bed, 2,3-BDO conversion, and selectivity for the major products, 1,3-BD and MEK, at the outlet of the reactor. The fluid temperature profile obtained from the model along the length of the catalyst bed coincides satisfactorily with the experimental observations. The difference between the experimental data and the 1D heterogeneous reactor model prediction for 2,3-BDO conversion and selectivity of 1,3-BD and MEK were 0.1%, 9 wt %, and 2 wt %, respectively. In addition, valuable insights related to the feeding system of a commercial-scale plant were made through troubleshooting of the pilot tests. Notably, if the feed including only 2,3-BDO and furnaces that increase the temperature of the feed to the reaction temperature were used in a commercial plant, the feeding system could not be operational because of the presence of heavy chemicals considered oligomers of 2,3-BDO.

Keywords: 2,3-Butanediol dehydration; 1,3-Butadiene; Methyl Ethyl Ketone; amorphous calcium phosphate; reactor modeling; pilot-scale fixed-bed reactor

---

1. Introduction

1,3-BD and MEK are widely used in various industrial fields. However, these compounds are mainly prepared from petroleum, which is a finite resource and a major cause of regional disparities and environmental pollution. 2,3-BDO has been considered as a potential intermediate for the production of hydrocarbons including 1,3-BD and MEK because 2,3-BDO can be produced through bio-fermentation using various biomasses, synthetic gases (syngas) from coal gasification, and industrial gas waste as feedstock [1–3].

Research on the dehydration of 2,3-BDO to 1,3-BD and MEK using various catalysts has been conducted by several research group since the 1940s. The catalysts are bentonite clay [4], metal and earth oxides [5–9], zeolites [10–13], a perfluorinated resin with sulfonic acid groups [10], heteropolyacids [11,14], calcium phosphates [15–20], Cs/SiO2 [21], sodium phosphates [22] and so on. Research on old chemistry and new catalytic advances in the on-purpose synthesis of butadiene has been reviewed by Makshina et al. [23]. Duan et al. [24] prospected future of the production of 1,3-butadiene from butanediols. However, most research was conducted to identify dehydration catalysts or reaction conditions that produce good performance. Recently, reaction kinetics and a deactivation model of the dehydration of 2,3-BDO to 1,3-BD and MEK over a-CP catalyst were proposed [18,19]. In addition, a process design for the recovery of 1,3-BD and MEK from BDO-dehydration products, which were obtained from lab-scale experiments, was proposed as a conceptual design for the industrial scale [25].

Development of reactor model and pilot-scale tests of a reactor are essential for commercialization of the 2,3-BDO dehydration process. However, to our knowledge, research on modeling and pilot-scale tests of a reactor for dehydration of 2,3-BDO to 1,3-BD and MEK has not been done. The purpose of this work is, therefore, to develop a suitable reactor model for the dehydration of 2,3-BDO to 1,3-BD and MEK over a-CP catalyst and to validate the reactor model against experimental data obtained using a pilot-scale fixed-bed reactor.

The pilot-scale fixed-bed reactor was simulated by a one-dimensional (1D) heterogeneous reactor model. The simulation results were compared with the experimental data in terms of fluid temperature profile along with length of catalyst bed, 2,3-BDO conversion, and selectivity for the major products at the outlet of the reactor. In addition, valuable insights related to the feeding system of a commercial-scale plant were made when troubleshooting the pilot tests. The reactor model, experimental data, and investigation are anticipated to be very useful when the 2,3-BDO dehydration process is commercialized.

2. Results and Discussion

The 1D heterogeneous reactor model accounting for interfacial and intra-particle gradients was validated with the test results of the pilot-scale fixed-bed reactor. The simulation results were compared with the experimental data of test 2 in terms of the fluid temperature profile along with the length of the catalyst bed, conversion of 2,3-BDO, and selectivity for the major products at the outlet of the reactor. Average values of operating conditions, temperature in the catalyst bed, and product compositions were used for the comparison.

In Figure 1, the solid line represents the fluid temperature obtained from the reactor model along the length of the catalyst bed, while the scattered points show the measured temperature. As is evident in the figure, the fluid temperature profile coincides satisfactorily with the experimental observations. The fluid temperature decreases rapidly from the inlet of the catalyst bed to the point at 0.05 m because of the endothermic nature of the reactions, the higher temperature, and the greater concentration of reactant, 2,3-BDO, near the inlet of the reactor. The fluid temperature decreases slowly from 0.05 to 0.29 m owing to the reduced temperature and concentration of the reactant. After 0.29 m, the temperature increases because nearly 100% of the reactant is consumed, so there are no reactions taking place and heat transfers from the outside of the reactor to the catalyst bed.

Figure 2 also represents good performance of the model for the prediction of 2,3-BDO conversion and selectivity of the target products, 1,3-BD and MEK, at the outlet of the catalyst bed in spite of the discrepancies between model prediction and experimental result when it comes to the selectivities of the low concentration components 3B2OL and 2-methylpropanal (2MPL). The experimental 1,3-BD selectivity is higher than the simulated one because the experimental selectivity of 3B2OL, which is the intermediate product of 1,3-BD, and 2MPL, which is the other product produced from 2,3-BDO, is less than the simulated one. This result means that the route leading to the formation of 3B2OL and then to the formation 1,3-BD from 3B2OL would be more active than the simulation result expects. In addition to that, impurities are not considered in the reaction products since the total amount of minor butene isomers and heavy compounds made by polymerization of 1,3-BD [19] are less than 0.7 wt % over all experiments. This assumption would lead to the higher experimental selectivity of 1,3-BD and the lower experimental selectivity of 3B2OL and 2MPL.

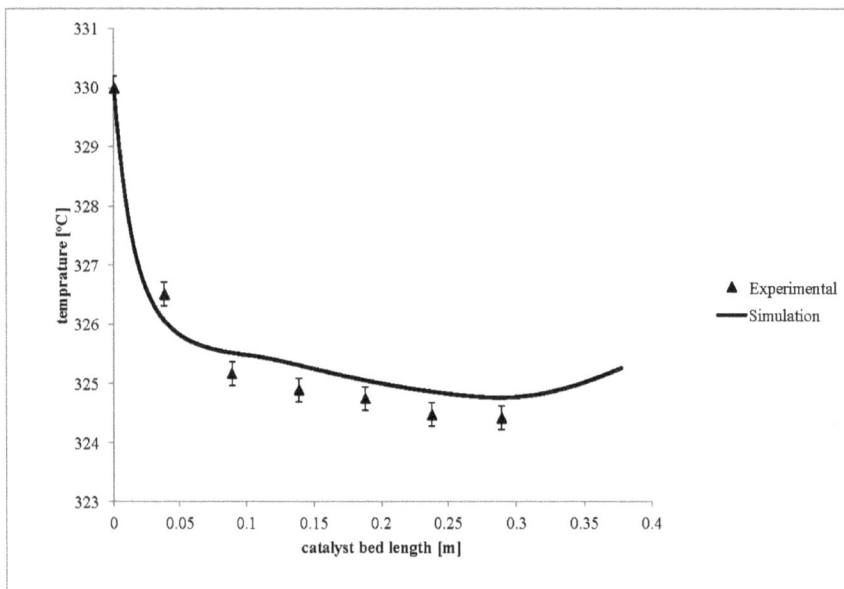

Figure 1. Comparison of the fluid temperature profiles of simulated and experimental data.

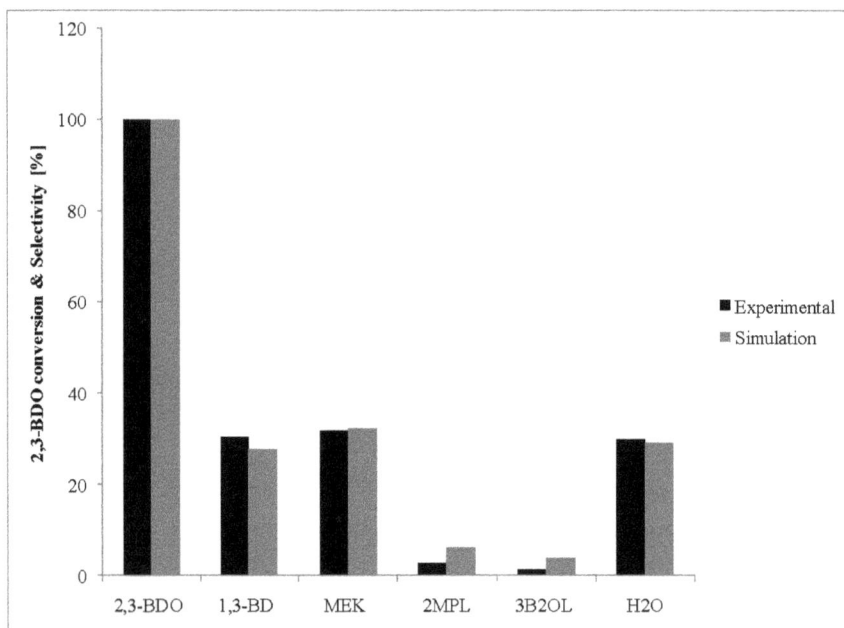

Figure 2. Comparison of 2,3-BDO conversion and selectivity of the major products between the simulated and experimental data.

3. Experimental Studies

3.1. Experimental Setup

Figure 3 illustrates a schematic drawing of the reaction system used for the dehydration of 2,3-BDO. 2,3-BDO including 1.45% water (98.65%, Sigma-Aldrich, St. Louis, MO, USA) was fed to the reactor by an HPLC pump (P1) to a ceramic fiber heater (CFH1) and N_2 as a carrier gas was fed through a line heater (LE1) to raise it to the reaction temperature by a mass flow meter (MFM). 2,3-BDO reached the reaction temperature by passing through three ceramic fiber heaters (CFH1-3). The heated mixture of 2,3-BDO and N_2 was fed to one of two reactors (R1&R2) through a line mixer and the temperature of the reactors was maintained by electric heaters around R1 and R2. The product of the reactor was cooled by a cooler (HE1). The condensed mixture was spilt into gas and liquid phases through sight glasses (SG1&SG2). A liquid sample was collected from SG1 and SG2. The gas stream from the sight glasses went to the flow transmitter (FT) and a gas sample was collected through the sample point with a gas bag.

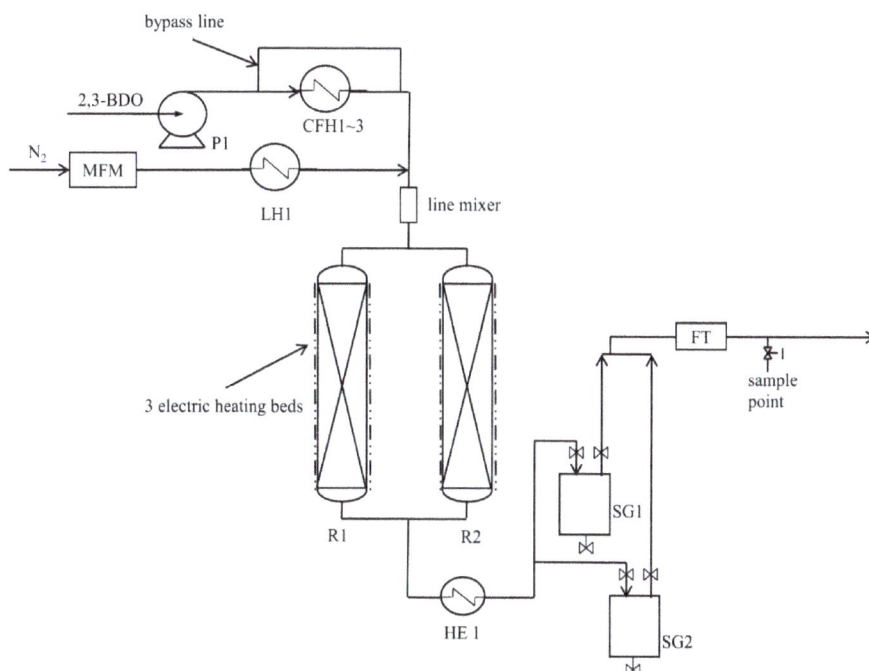

Figure 3. Schematic of the reaction system for the dehydration of 2,3-BDO.

Figure 4 shows a scheme of the fixed-bed reactor. Amorphous calcium phosphate catalysts, prepared by a co-precipitation method as in previous work [18,19], were crushed, sieved through a 16–40 mesh filter, and loaded into a space between 552 mm from the top of the reactor and 256 mm from the bottom. Other parts of the reactor were filled with 3.2 mm spherical silica beads. An additional thermocouple tube was installed at the center of the reactor so that temperature-detecting sensors (T1–T7) could be inserted into the reactor. T1 was located at 414 mm from the top of the reactor. The inlet temperature of the catalyst bed was checked by T1. T2–T7 were located at 38, 88, 138, 188, 238, and 288 mm from the top of the catalyst bed. Temperature profiles inside the catalyst bed were obtained through T2–T7. The detailed reactor specifications are summarized in Table 1.

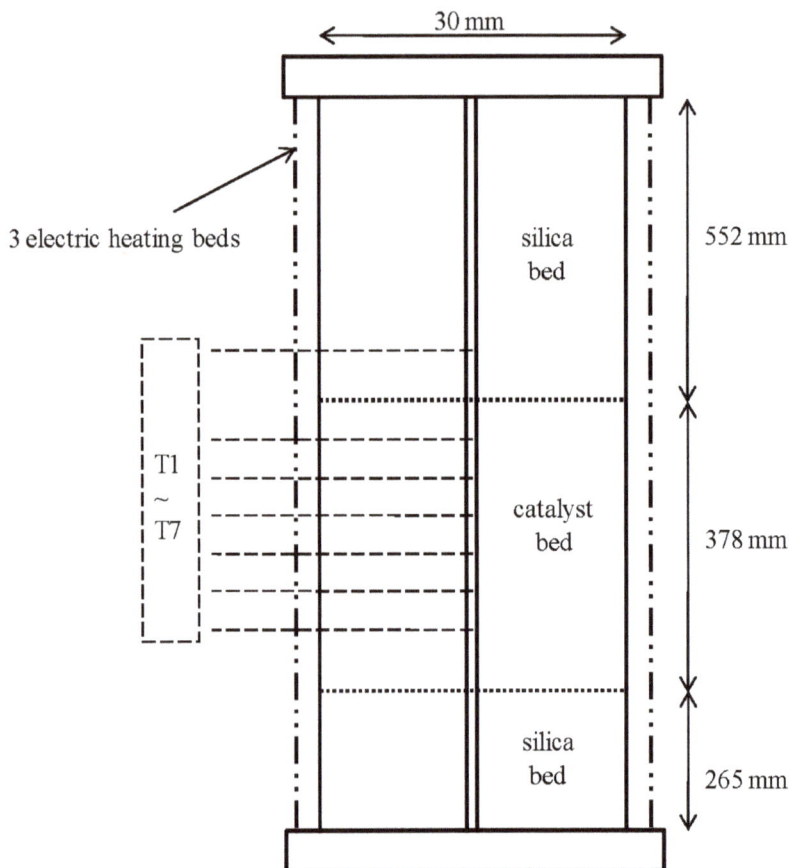

Figure 4. Scheme of the fixed-bed reactor.

Table 1. Reactor specifications.

	Property	Value	Unit
Catalyst	Type	Amorphous Calcium Phosphate (Ca/P = 1.3)	
	average diameter	2.855	mm
	weight	80	g
	density	460.5	kg/m^3
	heat capacity [26]	995	$J/(kg\ K)$
	conductivity	0.251	$W/(m\ K)$
	porosity	0.121	-
	tortuosity [27,28]	1.73	-
tube	inner diameter	30	mm
	tube length	1195	mm
tube wall	thickness	3.937	mm
	thermal conductivity	16	$W/(m\ K)$
	heat capacity	2000	$J/(kg\ K)$
catalyst bed	length	378	mm
	density	299.4	kg/m^3
	porosity	0.35	-

3.2. Operation and Troubleshooting

Initially, 800 g/h N_2 was fed into the reaction system for 6 h to purge the system and establish the reaction temperature. Then, the flow rate of N_2 and 2,3-BDO and the temperature of T1 used as the inlet temperature of the catalyst bed were set to the operating conditions shown in Table 2 by MFC, P1, CHF1-3, and LH1 as shown in Figure 3 for 2 h. The wall temperature of the reactor tube was maintained by 3 electric heating beds during the tests. The product stream of the reactor was cooled by HE1. The temperature of HE1 was maintained as 25 to 30 °C to avoid freezing unreacted 2,3-BDO, which freezes at around 20 °C at 1 bar.

During the operation using R1 after the initial work mentioned above for test 1, which was without an N_2 feed, the pressure of P1 was increased sharply because the line passing through CHF1-3 was blocked with heavy chemicals thought to be oligomers of 2,3-BDO. The reasons for formation of the oligomers were most likely a long residence time of 2,3-BDO in CHF1-3 and local hot spots on the line surface generated by CHF1-3. This means that if the feed including only 2,3-BDO and the furnaces to increase the temperature of the feed to the reaction temperature would be used in a commercial plant, the feeding system could not be operational because of the oligomers of 2,3-BDO. Design of a stable 2,3-BDO feeding system would be essential to commercializing the 2,3-BDO dehydration process. To solve the problem in the pilot-scale reaction system, a bypass line as shown in Figure 3 was installed to bypass CHF1-3 and the temperature of T1 was maintained at a constant temperature by electric heat beds around the reactor without using CHF1-3.

Test 2 was performed continuously in the other reactor, R2. After the initial work discussed above, the reactor was operated for 6 h under the operating conditions of test 2 to reach a steady state, which was assessed by the temperature profile of the catalyst bed, and then was operated for 16 h under the same conditions to obtain gas and liquid samples every 4 h. Ideally, tests for different inlet conditions are necessary to validate the model. However, a-CP catalyst in the lab-scale tests under the operating conditions of test 2 was deactivated sharply in 24 h [19]. To remove the deactivation effects, tests for different inlet conditions were not implemented. The data from test 2 were used for a preliminary validation.

Table 2. Reactor operating conditions.

Operating Conditions	Test 1	Test 2	Unit
inlet temperature of the catalyst bed	330	330	°C
pressure	1	1	bar
N_2 flow rate	0	393	g/h
2,3-BDO flow rate	80	39	g/h
temperature of 3 electric heating beds	330	330	°C

3.3. Analysis Methods

The same analysis methods as were used in previous research [19] were used here. The compositions of the gas and liquid samples were analyzed in a gas chromatograph (GC, Agilent 7890A, Santa Clara, CA, USA) with a DB-1 column (non-polar phase, 60 m × 0.250 mm × 1 µm) and a Flame Ionization Detector (FID) for analysis of the hydrocarbon content. The compositions of major components were normalized to remove the effects of impurities. The conversion of 2,3-BDO and the selectivity for each product were computed as follows:

$$X_{2,3-BDO} = \frac{F_{BDO,in} - F_{BDO,out}}{F_{BDO,in}} \times 100, \tag{1}$$

$$S_n = \frac{F_{n,out}}{F_{total} - F_{BDO,out} - F_{N_2}} \times 100, \tag{2}$$

where $X_{2,3\text{-}BDO}$ is the conversion of 2,3-BDO, n is a component of the product, S is the mass and F is the mass flow rate.

The composition of water was calculated by the reaction stoichiometry of Equations (3)–(6) in Section 4.1 based on the compositions of 3-Buten-2-ol(3B2OL), 1,3-BD, MEK, and 2-Methylpropanal (2MPL) in the gas and liquid samples.

4. Development of the Reactor Model

4.1. Reaction Kinetics

The reaction kinetics of the dehydration of 2,3-BDO to 1,3-BD and MEK using a-CP as a catalyst, as in previous research [18], were used for the reactor model. The major pathways of 2,3-BDO dehydration are described by the following reactions:

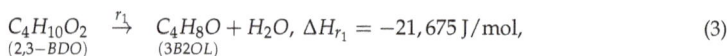

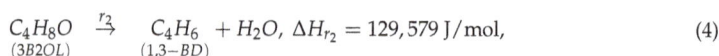

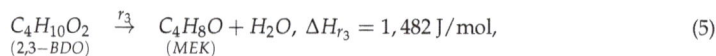

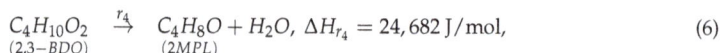

$$\underset{(2,3-BDO)}{C_4H_{10}O_2} \xrightarrow{r_1} \underset{(3B2OL)}{C_4H_8O} + H_2O, \ \Delta H_{r_1} = -21,675 \text{ J/mol}, \tag{3}$$

$$\underset{(3B2OL)}{C_4H_8O} \xrightarrow{r_2} \underset{(1,3-BD)}{C_4H_6} + H_2O, \ \Delta H_{r_2} = 129,579 \text{ J/mol}, \tag{4}$$

$$\underset{(2,3-BDO)}{C_4H_{10}O_2} \xrightarrow{r_3} \underset{(MEK)}{C_4H_8O} + H_2O, \ \Delta H_{r_3} = 1,482 \text{ J/mol}, \tag{5}$$

$$\underset{(2,3-BDO)}{C_4H_{10}O_2} \xrightarrow{r_4} \underset{(2MPL)}{C_4H_8O} + H_2O, \ \Delta H_{r_4} = 24,682 \text{ J/mol}, \tag{6}$$

The reaction rates based on the power law are:

$$r_i = k_i C_{react,i}{}^{n_i}, \tag{7}$$

where

$$C_j = \frac{P_j}{RT}, \tag{8}$$

$$k_i = k_{T_{ref},i} \exp\left(\frac{-E_i}{R}\left(\frac{1}{T} - \frac{1}{T_{ref}}\right)\right), \tag{9}$$

where i is the number of reaction, r is the reaction rate, *react* is the reactant, C is the mole concentration, n is the reaction order, j is the number of species, P is the pressure, R is the ideal gas law constant, T is the temperature in bulk gas phase, k is the reaction rate constant, $k_{T_{ref}}$ is the transformed adsorption pre-exponential factor, E is the activation energy and T_{ref} is the reference temperature. The kinetic parameter values are shown in Table 3.

Table 3. Kinetic parameters [18].

Model Parameter	Unit	Value
E_1	J/mol	2.33×10^5
E_2	J/mol	2.82×10^5
E_3	J/mol	1.93×10^5
E_4	J/mol	1.66×10^5
$k_{Tref,1}$	$\text{mol}^{(1-n1)} \text{ m}^{3(n1-1)} \text{ s}^{-1}$	7.45×10^{-4}
$k_{Tref,2}$	$\text{mol}^{(1-n2)} \text{ m}^{3(n2-1)} \text{ s}^{-1}$	4.41×10^{-4}
$k_{Tref,3}$	$\text{mol}^{(1-n3)} \text{ m}^{3(n3-1)} \text{ s}^{-1}$	6.44×10^{-4}
$k_{Tref,4}$	$\text{mol}^{(1-n4)} \text{ m}^{3(n4-1)} \text{ s}^{-1}$	1.27×10^{-4}
$n1, n3, n4$	-	0.0187
$n2$	-	0.146

4.2. Reactor Model

The 1D heterogeneous reactor model accounting for interfacial and intra-particle gradients was conducted for reactor modeling. A plug flow was assumed to apply, axial dispersion and thermal conductivity were ignored, and it was assumed that there was no channeling along the reactor tube. Under the above assumptions, the conservation equations are as follows.

For the fluid phase:

$$0 = -\frac{\partial(u_s C_j)}{\partial z} + k_f a_v (C_{ssj} - C_j), \tag{10}$$

$$0 = -\frac{\partial(u_s \rho_f Cp_f T)}{\partial z} + h_f a_v (T_{ss} - T) - \frac{4}{d_t} U(T - T_w), \tag{11}$$

where z is the axial reactor coordinate, u_s is the superficial fluid velocity, k_f is the mass transfer coefficient between catalyst surface and fluid, a_v is the external particle surface area per unit reactor volume, C_{ss} is the mole concentration at the surface of catalysts, ρ_f is the fluid density, Cp_f is the fluid heat capacity, h_f is the heat transfer coefficient between catalyst surface and fluid, T_{ss} is the temperature at the surface of catalysts, d_t is the diameter of a reactor, U is the overall heat transfer coefficient and T_w is the temperature of electric heaters around the reactors.

For a cross section of the bed including the solid and fluid phases:

$$k_f a_v (C_{ssj} - C_j) = \rho_B \sum_{i=1}^{rxn} v_{ji} \eta_i r_{si,s}, \tag{12}$$

$$h_f a_v (T_{ss} - T) = \rho_B \sum_{i=1}^{rxn} (-\Delta H_{R_i}) \eta_i r_{si,s}, \tag{13}$$

where ρ_B is the bulk density of catalyst bed, rxn is the number of reactions, v_{ji} is the stoichiometric coefficient of species j in reaction i, η is the effectiveness factor, $r_{si,s}$ is the reaction rate at the surface of the catalyst and ΔH_R is the heat of reaction.

For the solid phase:

$$0 = D_e \left(\frac{\partial^2 C_{sj}}{\partial r_p^2} + \frac{2}{r_p} \frac{\partial C_{sj}}{\partial r_p} \right) + \rho_s \sum_{i=1}^{rxn} v_{ji} r_{si}, \tag{14}$$

$$0 = \lambda_p \left(\frac{\partial^2 T_s}{\partial r_p^2} + \frac{2}{r_p} \frac{\partial T_s}{\partial r_p} \right) + \rho_s \sum_{i=1}^{rxn} (-\Delta H_{R_i}) r_{si}, \tag{15}$$

where D_e is the effective diffusivity within a catalyst, C_s is the mole concentration in catalysts, r_p is the catalyst radius, ρ_s is the catalyst density, r_s is the reaction rate in catalysts, λ_p is the catalyst heat conductivity and T_s is the temperature in catalysts.

Boundary conditions:

$$F_j = F_{j,in}, \ T = T_{in} \ at \ z = 0$$
$$\frac{dC_j}{dr_p} = 0, \ \frac{dT}{dr_p} = 0 \ at \ r_p = 0, \ \forall z \in (0, L_t]$$
$$C_{sj} = C_{ss,j}, \ T_s = T_{ss} \ at \ r_p = d_p/2, \ \forall z \in (0, L_t]$$

where F is the mass flow rate, F_{in} is the mass flow rate at the inlet of the catalyst bed, T_{in} is the temperature at the inlet of the catalyst bed, C_{sj} is the mole concentration at the surface of the catalyst, T_s is the temperature at the surface of the catalyst and L_t is the length of catalyst bed.

The pressure drop in the reactor tube was calculated by the classical Ergun equation [29]. Ergun correlations combine the equation for the friction factor in highly turbulent flow in a channel with an equation for laminar flow in an empty conduit. The fluid-to-particle interfacial heat and mass transfer resistance are considered by using Hougen correlation [30], which is based on Colburn j-factor analogy. The correlation relates the j-factor to Reynolds number for packed beds of spheres.

The tube inside heat transfer coefficient is calculated from the effective bed heat conductivity and bed-wall heat transfer coefficient [31]. These two coefficients have both a static and a dynamic contribution, where the static contribution relates to heat transfer in the hypothetic situation of zero flow, and the dynamic contribution accounts for hydrodynamics effects [32]. The relevant correlations and equations for the 1D heterogeneous model are given in Table 4. The methods used to calculate the physicochemical properties of the reactor model are provided in Table 5. The algebraic equations and ordinary differential equations with the boundary conditions of the reactor model were formulated in gPROMS and solved by the numerical DAE solvers named DAEBDF provided by gPROMS [33].

Table 4. Correlations and equations used for the 1D heterogeneous model.

Parameter	Formula
mass and heat transfer coefficient between catalyst surface and fluid [30]	$Sh_p = Re_p Sc^{1/3} max(1.66 Re_p^{-0.51}, 0.983 Re_p^{-0.41})$
	$Nu_p = Re_p Pr^{1/3} max(1.66 Re_p^{-0.51}, 0.983 Re_p^{-0.41})$
overall heat transfer coefficient [34]	$\frac{1}{U} = \frac{1}{h_i} + \frac{x_w}{\lambda_m} \frac{d_t}{d_L}$
tube inside heat transfer coefficient [31,32]	$\frac{1}{h_i} = \frac{1}{\alpha_{ws}^e} + \frac{(d_t/2)}{3\lambda_{rs}^e} \frac{Bi+3}{Bi+4}$
	$\alpha_{ws}^e = \alpha_{ws,0}^e + \alpha_{ws,d}^e$
	$\lambda_{rs}^e = \lambda_{rs,0}^e + \lambda_{rs,d}^e$
effective diffusivity within a catalyst [35]	$D_e = D\_mean(\varepsilon_p/\tau)$
effectiveness factor [35]	$\eta_i = \int_0^{r_p} r_{si} dr/(r_p r_{si,s})$

Table 5. Methods used to calculate physicochemical properties.

Property	Method
fluid density	Peng-Robinson [36]
fluid viscosity	Lucas [36]
fluid heat capacity	ideal gas [36]
fluid conductivity	Steil-Thodos [37]
binary diffusion coefficient, components i and j	Fuller-Schettler-Gidding (FSG) [37]
fluid compressibility factor	Peng-Robinson [36]

5. Conclusions

A 1D heterogeneous reactor model considering interfacial and intra-particle gradients was used to simulate 2,3-BDO dehydration in a pilot-scale fixed-bed reactor. The model was validated with experimental data obtained from the pilot plant in terms of the fluid temperature profile along with the length of the catalyst bed, 2,3-BDO conversion, and selectivity for the major products at the outlet of the reactor. The temperature profile along the length of the catalyst bed coincides satisfactorily with the experimental observations, and the developed model shows good performance for the prediction of 2,3-BDO conversion and selectivity of the target products, 1,3-BD and MEK, at the outlet of the catalyst bed, even though the selectivity of 3B2OL and 2MPL are different. The differences between the experimental data and the 1D heterogeneous reactor model prediction for 2,3-BDO conversion and the selectivity of 1,3-BD and MEK were 0.1%, 9 wt %, and 2 wt %, respectively. On the other hand, the reactor model was validated using preliminary validation data and but needs to be validated with more experimental data for future study.

Valuable insights related to the feeding system of a commercial-scale plant were found through troubleshooting of the pilot tests. If the feed including only 2,3-BDO and furnaces to increase the temperature of the feed to the reaction temperature were used in a commercial plant, the feeding system could not be operated owing to the presence of heavy chemicals that are oligomers of 2,3-BDO.

The design of a stable 2,3-BDO feeding system would be a very important part of the commercialization of the 2,3-BDO dehydration process.

Conflicts of Interest: The author declares no conflict of interest.

Nomenclature

a_v	external particle surface area per unit reactor volume, m^2/m^3
Bi	Biot number, m
C	mole concentration, mol/m^3
Cp_f	fluid heat capacity, J/kg K
C_{sj}	mole concentration in catalysts, mol/m^3
$C_{sj,s}$	mole concentration at the surface of the catalyst, mol/m^3
$\bar{d}_L$	logarithmic mean diameter, m
d_p	diameter of a catalyst, m
d_t	diameter of a reactor, m
D_e	effective diffusivity within a catalyst, m^2/s
D_mean	mean diffusivity coefficient, m^2/s
E	activation energy, J/mol
F	mass flow rate, g/s
F_{in}	mass flow rate at the inlet of the catalyst bed, g/s
h_f	heat transfer coefficient between catalyst surface and fluid, W/m^2 K
h_i	tube inside heat transfer coefficient, W/m^2 K
k	reaction rate constant, $mol^{(1-n)}\ m^{3(n-1)}\ s^{-1}$
k_f	mass transfer coefficient between catalyst surface and fluid, m/s
$k_{T_{ref}}$	transformed adsorption pre-exponential factor, m^3/mol
L_t	length of the catalyst bed, m
n	reaction order
Nu_p	Nusselt number for fluid-solid heat transfer
P	pressure, Pa
Pr	Prandtl number for the fluid
r	reaction rate, mol/kg-cat s
r_p	catalyst radius, m
r_{si}	reaction rate in catalysts, mol/kg-cat s
$r_{si,s}$	reaction rate at the surface of the catalyst, mol/kg-cat s
R	ideal gas law constant, J/mol K
Re_p	Reynolds number for packed bed
S	mass selectivity, %
Sc	Schmidt number
Sh_p	Sherwood number for packed bed
T	temperature, K
T_{in}	temperature at the inlet of the catalyst bed, K
T_s	temperature in catalysts, K
T_{ss}	temperature at the surface of catalysts, K
T_{ref}	reference temperature, K
T_w	temperature of electric heaters around the reactors, K
u_s	superficial fluid velocity, m/s
U	overall heat transfer coefficient, W/m^2 K
X	conversion, %
x_w	tube wall thickness, m
z	axial reactor coordinate, m

Greek Letters

α_{ws}^{e}	effective bed-wall heat transfer coefficient, $W/m^2\,K$
$\alpha_{ws,0}^{e}$	static term of the effective bed-wall heat transfer coefficient, $W/m^2\,K$
$\alpha_{ws,d}^{e}$	static term of the effective bed-wall heat transfer coefficient, $W/m^2\,K$
ΔH_R	heat of reaction, J/mol
ε_p	catalyst porosity
η	effectiveness factor
λ_m	wall thermal conductivity, $W/m\,K$
λ_p	catalyst heat conductivity, $W/m\,K$
λ_{rs}^{e}	effective bed heat conductivity, $W/m\,K$
$\lambda_{rs,0}^{e}$	static term of effective bed heat, $W/m\,K$
$\lambda_{rs,d}^{e}$	dynamic term of effective bed heat conductivity, $W/m\,K$
ν_{ji}	stoichiometric coefficient of species j in reaction i
ρ_B	bulk density of catalyst bed, kg/m^3
ρ_f	fluid density, kg/m^3
ρ_s	catalyst density, kg/m^3
τ	catalyst tortuosity

Subscripts

i	reaction i
j	species j
react	reactant
rxn	reaction

References

1. Daniell, J.; Köpke, M.; Simpson, S.D. Commercial biomass syngas fermentation. *Energies* **2012**, *5*, 5372–5417. [CrossRef]

2. Kopke, M.; Mihalcea, C.; Liew, F.; Tizard, J.H.; Ali, M.S.; Conolly, J.J.; Al-Sinawi, B.; Simpson, S.D. 2,3-Butanediol Production by Acetogenic Bacteria, an Alternative Route to Chemical Synthesis, Using Industrial Waste Gas. *Appl. Environ. Microbiol.* **2011**, *77*, 5467–5475. [CrossRef] [PubMed]

3. Zheng, Q.; Wales, M.D.; Heidlage, M.G.; Rezac, M.; Wang, H.; Bossmann, S.H.; Hohn, K.L. Conversion of 2,3-butanediol to butenes over bifunctional catalysts in a single reactor. *J. Catal.* **2015**, *330*, 222–237. [CrossRef]

4. Bourns, A.N.; Nicholls, R.V.V. The Catalytic Action of Aluminium Silicates: I. The Dehydration of Butanediol-2,3 and Butanone-2 over Activated Morden Bentonite. *Can. J. Res.* **1947**, *25b*, 80–89. [CrossRef]

5. Winfield, M.E. The catalytic dehydration of 2,3-butanediol to butadiene. *Aust. J. Sci. Res. Ser. A* **1945**, *3*, 290–305.

6. Winfield, M.E. The Catalytic Dehydration of 2,3-Butanediol to Butadiene. II. Adsorption Equilibriaitle. *Aust. J. Sci. Res. Ser. A* **1950**, *3*, 290–305.

7. Kannan, S.V.; Pillai, C.N. Dehydration of meso- and dl-hydrobenzoins and 2,3-butanediols over alumina. *Indian J. Chem.* **1969**, *7*, 1164–1166.

8. Duan, H.; Sun, D.; Yamada, Y.; Sato, S. Dehydration of 2,3-butanediol into 3-buten-2-ol catalyzed by ZrO_2. *Catal. Commun.* **2014**, *48*, 1–4. [CrossRef]

9. Duan, H.; Yamada, Y.; Sato, S. Applied Catalysis A: General Efficient production of 1,3-butadiene in the catalytic dehydration of 2,3-butanediol. *Appl. Catal. A Gen.* **2015**, *491*, 163–169. [CrossRef]

10. Bucsi, I.; Molnár, Á.; Bartók, M.; Olah, G.A. Transformation of 1,3-, 1,4- and 1,5-diols over perfluorinated resinsulfonic acids (Nafion-H). *Tetrahedron* **1995**, *51*, 3319–3326. [CrossRef]

11. Molnár, Á.; Bucsi, I.; Bartók, M. Pinacol rearrangement on zeolites. *Stud. Surf. Sci. Catal.* **1988**, *41*, 203–210. [CrossRef]

12. Lee, J.; Grutzner, J.B.; Walters, W.E.; Delgass, W.N. The conversion of 2,3-butanediol to methyl ethyl ketone over zeolites. *Stud. Surf. Sci. Catal.* **2000**, *130*, 2603–2608. [CrossRef]

13. Zhang, W.; Yu, D.; Ji, X.; Huang, H. Efficient dehydration of bio-based 2,3-butanediol to butanone over boric acid modified HZSM-5 zeolites. *Green Chem.* **2012**, *14*, 3441–3450. [CrossRef]

14. Tsrsk, B.; Bucsi, I.; Beregsz, T.; Kapocsi, I.; Molnfir, A. Transformation of diols in the presence of heteropoly acids under homogeneous and heterogeneous conditions. *J. Mol. Catal. A Chem.* **1996**, *107*, 305–311.

15. Hahn, H.-D.; Dämbkes, G.; Rupprich, N.; Bahl, H.; Frey, G.D. Butanols. *Ullmanns Encycl. Ind. Chem.* **2013**, 1–13. [CrossRef]

16. Kim, S.J.; Seo, S.O.; Park, Y.C.; Jin, Y.S.; Seo, J.H. Production of 2,3-butanediol from xylose by engineered *Saccharomyces cerevisiae*. *J. Biotechnol.* **2014**, *192*, 376–382. [CrossRef] [PubMed]

17. Nikitina, M.A.; Sushkevich, V.L.; Ivanova, I.I. Dehydration of 2,3-butanediol over zeolite catalysts. *Pet. Chem.* **2016**, *56*, 230–236. [CrossRef]

18. Song, D. Kinetic Model Development for Dehydration of 2,3-Butanediol to 1,3-Butadiene and Methyl Ethyl Ketone over an Amorphous Calcium Phosphate Catalyst. *Ind. Eng. Chem. Res.* **2016**, *55*, 11664–11671. [CrossRef]

19. Song, D. Development of a deactivation model for the dehydration of 2,3-butanediol to 1,3-butadiene and methyl ethyl ketone over an amorphous calcium phosphate catalyst. *Ind. Eng. Chem. Res.* **2017**, *56*, 11013–11020. [CrossRef]

20. Tsukamoto, D.; Sakami, S.; Ito, M.; Yamada, K.; Ito, M.; Yonehara, T. Production of Bio-based 1,3-Butadiene by Highly Selective Dehydration of 2,3-Butanediol over SiO_2-supported Cesium Dihydrogen Phosphate Catalyst. *Chem. Lett.* **2016**, *45*, 831–833. [CrossRef]

21. Kim, T.Y.; Baek, J.; Song, C.K.; Yun, Y.S.; Park, D.S.; Kim, W.; Han, J.W.; Yi, J. Gas-phase dehydration of vicinal diols to epoxides: Dehydrative epoxidation over a Cs/SiO_2 catalyst. *J. Catal.* **2015**, *323*, 85–99. [CrossRef]

22. Kim, W.; Shin, W.; Lee, K.J.; Song, H.; Kim, H.S.; Seung, D.; Filimonov, I.N. Applied Catalysis A: General 2,3-Butanediol dehydration catalyzed by silica-supported sodium phosphates. *Appl. Catal. A Gen.* **2016**, *511*, 156–167. [CrossRef]

23. Makshina, E.V.; Dusselier, M.; Janssens, W.; Degrève, J.; Jacobs, P.A.; Sels, B.F. Review of old chemistry and new catalytic advances in the on-purpose synthesis of butadiene. *Chem. Soc. Rev.* **2014**, *43*, 7917–7953. [CrossRef] [PubMed]

24. Duan, H.; Yamada, Y.; Sato, S. Future prospect of the production of 1,3-butadiene from butanediols. *Chem. Lett.* **2016**, *45*, 1036–1047. [CrossRef]

25. Song, D.; Yoon, Y.-G.; Lee, C.-J. Conceptual design for the recovery of 1,3-Butadiene and methyl ethyl ketone via a 2,3-Butanediol-dehydration process. *Chem. Eng. Res. Des.* **2017**, *123*, 268–276. [CrossRef]

26. Kelley, K.K. *[Part] 13. High-Temperature Heat-Content, Heat-Capacity, and Entropy Data for the Elements and Inorganic Compounds*; Bureau of Mines: Washington, DC, USA, 1960.

27. Bhatia, S.K. Directional autocorrelation and the diffusional tortuosity of capillary porous media. *J. Catal.* **1985**, *93*, 192–196. [CrossRef]

28. Dykhuizen, R.C.; Casey, W.H. An analysis of solute diffusion in rocks. *Geochim. Cosmochim. Acta* **1989**, *53*, 2797–2805. [CrossRef]

29. Ergun, S. Fluid Flow through Packed Columns. *Chem. Eng. Prog.* **1952**, *48*, 89–94.

30. Hougen, O. Engineering Aspects of Solid Catalysts. *Ind. Eng. Chem.* **1961**, *53*, 509–528. [CrossRef]

31. Dixon, A.G. An improved equation for the overall heat transfer coefficient in packed beds. *Chem. Eng. Process. Process Intensif.* **1996**, *35*, 323–331. [CrossRef]

32. Specchia, V.; Baldi, G.; Sicardi, S. Heat Transfer in Packed Bed Reactors with One Phase Flow. *Chem. Eng. Commun.* **1980**, *4*, 361–380. [CrossRef]

33. Process Systems Enterprise Ltd. *gPROMS Advanced User Guide*; Process Systems Enterprise Ltd.: London, UK, 2004.

34. McCabe, W.L.; Smith, J.; Harriott, P. *Unit Operations of Chemical Engineering*, 6th ed.; McGraw Hill: New York, NY, USA, 2001.

35. Riggs, J.M. *Introduction to Numerical Methods for Chemical Engineer*, 2nd ed.; Texas Tech University Press: Lubbock, TX, USA, 1994.

36. Poling, B.E.; Prausnitz, J.M.; O's Connell, J.P. *The Properties of Gases & Liquids*; McGraw Hill: New York, NY, USA, 2001.
37. Perry, R.H.; Green, D.W. *Perry's Chemical Engineering Hand-Book*; McGraw Hill: New York, NY, USA, 1997.

catalysts

MDPI

Article

CaRMeN: An Improved Computer-Aided Method for Developing Catalytic Reaction Mechanisms

Hendrik Gossler [1], Lubow Maier [2], Sofia Angeli [1], Steffen Tischer [1] and Olaf Deutschmann [1,2,*]

[1] Institute for Chemical Technology and Polymer Chemistry, Karlsruhe Institute of Technology (KIT), 76131 Karlsruhe, Germany; hendrik.gossler@kit.edu (H.G.); sofia.angeli@kit.edu (S.A.); steffen.tischer@kit.edu (S.T.)

[2] Institute of Catalysis Research and Technology, Karlsruhe Institute of Technology (KIT), 76344 Eggenstein-Leopoldshafen, Germany; lubow.maier@kit.edu

* Correspondence: deutschmann@kit.edu; Tel.: +49-721-608-43064

Received: 25 January 2019; Accepted: 25 February 2019; Published: 1 March 2019

Abstract: The software tool CaRMeN (Catalytic Reaction Mechanism Network) was exemplarily used to analyze several surface reaction mechanisms for the combustion of H_2, CO, and CH_4 over Rh. This tool provides a way to archive and combine experimental and modeling information as well as computer simulations from a wide variety of sources. The tool facilitates rapid analysis of experiments, chemical models, and computer codes for reactor simulations, helping to support the development of chemical kinetic models and the analysis of experimental data. In a comparative study, experimental data from different reactor configurations (channel, annular, and stagnation flow reactors) were modeled and numerically simulated using four different catalytic reaction mechanisms from the literature. It is shown that the software greatly enhanced productivity.

Keywords: catalytic combustion; automation; digitalization; mechanism analysis; rhodium; methane

---

1. Introduction

Computer-aided design using chemical kinetics software has become essential in reaction engineering, as it provides valuable guidance in scale-up. In particular, the prediction of the reactor performance based on kinetics is a crucial issue. Using chemical kinetics, reactive flows can be simulated numerically on a technical scale, helping to reduce elaborate and expensive experiments. Furthermore, kinetics can also lead to a profound understanding of the underlying elementary processes.

In chemical engineering, a macroscopic kinetic approach was used for many years. In the macroscopic regime, the rate of catalytic reaction is modeled by fitting empirical equations such as power laws to experimental data to describe its dependence on concentration and pressure and to determine rate constants that depend exponentially on temperature. The downside of this approach is its very limited extrapolation to conditions that were not covered by the fit to experimental data. A more robust approach is to use microkinetics, where the processes are described by a sequence of elementary reaction steps of the catalytic cycle as they occur on a microscopic scale. These steps include adsorption, surface diffusion, reactions between adsorbed species, and desorption. A major advantage of the microkinetic approach versus the macrokinetic approach are its improved prediction capabilities beyond the experimental data that were used in its development.

However, the development of microkinetic models is a difficult and very time-consuming task. A hierarchical approach is commonly followed to develop reaction kinetics. Starting from a single fuel, the complexity of the reaction scheme is augmented by increasing the number of reactive components. For example, H_2 oxidation, CO oxidation, preferential oxidation of CO in H_2 and

O_2 mixtures, water–gas shift (WGS), and reverse water–gas shift (RWGS) reactions as well as total and partial oxidation reactions of CH_4 are added consecutively. To optimize the reaction kinetics, reactions are examined for varying fuel/oxygen ratios over a wide range of temperatures and compared to experimental data. Catalytic ignition studies are also conducted to understand the adsorption and desorption kinetics of the reactive species.

Hence, a major part of the development process towards a validated, reliable mechanism is comprised of iteratively comparing simulation results with experimental data. In each refinement step, changes are made to the reaction mechanism until the simulation results match measurements from experiments. Due to its repetitive nature, this approach is time-consuming and error prone. CaRMeN (Catalytic Reaction Mechanism Network) is a recently developed software tool [1] that addresses these problems by automating experiment vs. model comparisons in a graphical user interface. This is achieved by providing a platform to archive and evaluate structured experimental data, kinetic models, and simulation software. These data can be conveniently compared with the results of any simulation code under the matching experimental conditions.

There are several related projects described in the literature, such as PrIMe (Process Informatics Model), RESPECTH (short for reaction kinetics, spectroscopy, thermochemistry), or CloudFlame [2–4]. However, these projects have a very specific focus on combustion research that does not include the use of catalysts.

The high diversity of data in the field of catalysis makes such automation software far from trivial. Many reactor types and measurement techniques have been developed in the last years that are each capable of elucidating certain facets of the catalytic system. Ideally, these techniques are used in tandem when developing a detailed kinetic scheme. Accordingly, the simulation software used to model the reactions in these reactors is equally diverse. Furthermore, the verification of the surface kinetic mechanisms can itself be a challenging task, given the complications of performing experiments under a purely kinetically controlled regime. This requirement can be difficult to fulfill, especially for very fast processes, such as oxidation reactions. Therefore, special attention for evaluation datasets has to be paid to choosing accurate, reproducible experiments, which were carried out under appropriate operating conditions, so as to minimize the influence of transport phenomena. In addition, there are various commonly used metrics to assess the performance of a catalyst that can be extracted from a single experiment. These metrics can therefore be seen as different "views" of the data. An example is a "light-off" experiment, in which the temperature is varied and species concentrations are determined at the end of the reactor. This data can either be viewed directly, or the data can be transformed to conversions as a function of temperature, from which, in turn, $T_{1\%}$ to $T_{100\%}$ values can be derived.

In this contribution, we illustrate how the existing hierarchical approach to develop reaction mechanisms is significantly improved using the recently released software tool CaRMeN [1]. This refined methodology improves the quality and applicability of the model considerably because the model can be validated against a larger experimental database than before. Furthermore, reaction mechanisms can be developed and evaluated more quickly.

2. Combustion over Rh-Based Catalysts

Systems of catalytic oxidation CO in hydrogen-rich mixtures and fuel-lean methane oxidation have key applications for automotive and factory exhaust gas after treatment. Even though the catalytic combustion systems have been studied intensively, both experimentally and theoretically, accurate and reliable kinetic models for these systems are not readily available. In this article, this system is used as an example to show case various features of the CaRMeN software. The software is, however, neither limited to rhodium based catalysts nor to catalytic combustion. For example, catalysts based on Pt, Pd or Ni, which also play a very important role in industry [5–10], have also been studied with the software. Furthermore, CaRMeN can also be used to develop gas-phase mechanisms.

2.1. Overview of Mechanisms in the Literature

A selection of microkinetic models for CO–H_2–O_2 mixtures and methane combustion over Rh-based catalysts, developed by several groups, are summarized in Table 1. The following four detailed surface reaction mechanisms were used in this work as examples to demonstrate the capabilities of the software tool and are therefore described in more detail below:

- Maier–Deutschmann (2001) [11]

The detailed surface reaction mechanism for methane oxidation, produced by Deutschmann and co-workers, assumes dissociative adsorption of oxygen and two different methane activation paths. The first pyrolytic path involves the stepwise abstraction of hydrogen from CH_x* (x = 0–4) species on the free Rh sites down to surface carbon C*. The second path considers oxygen-assisted methane activation through pre-adsorbed O*. The reaction mechanism consists of six gas-phase species including the reactants and products (H_2, CO, H_2O, and CO_2), 11 surface intermediates and a total of 38 elementary-step reactions. The mechanism was then further improved by Schwiedernoch et al. [12] including coverage-dependent heats of chemisorption for CO and oxygen and validated against own light-off experiments and transient measurements by Williams et al. [13].

- Karakaya–Deutschmann (2016) [14]

The model was developed on the basis of former kinetic scheme of Deutschmann et al. [11] using the same dual methane activation route, including additional CO–H_2 coupling reactions via carboxyl COOH* related pathways, which are important in the water gas shift (WGS) reaction. The 44-step mechanism contains elementary reaction for H_2 oxidation, CO oxidation, preferential oxidation, and WGS. Both methane models [11,14] were developed and extensively validated for fuel-rich partial oxidation and reforming of CH_4 with water and CO_2. One of the objectives of this work was to test the mechanisms against the experimental datasets at fuel-lean combustion conditions.

- Deshmukh–Vlachos (2007) [15]

The model presents a reduced mechanism for fuel-lean methane/air catalytic combustion on a Rh catalyst. It was developed from a detailed microkinetic model of Mhadeshwar and Vlachos [16]. The original 104-elementary-step mechanism [16] was reduced (to 15 reversible reactions) using reaction path, sensitivity, and partial equilibrium analysis to deduce the most abundant reaction intermediate and the rate-determining step. The mechanism was evaluated against the methane catalytic combustion experiments in microreactor on Rh/Al_2O_3 catalyst [15].

- Rankovic–Da Costa (2011) [17]

The mechanism contains elementary reaction for H_2 oxidation, CO oxidation, CO–H_2 coupling, and NOx chemistry on Rh. It was derived from the previous modeling works of Deutschmann and co-workers. Most of the kinetic data were taken from Schwiedernoch et al. [12] and Boll [18] with the exception of two parameters: the pre-exponential factor of oxygen adsorption was changed from 1×10^{-2} to 6×10^{-3} and the coverage-dependence of the CO desorption step was increased from $15\,\Theta_{CO}$ to $55\,\Theta_{CO}$ so that the model reproduces the CO conversion curves obtained in experimental measurements in packed bed flow reactor with Rh/Al_2O_3 [19] and Rh/SiO_2 [20].

Table 1. Selection of surface reaction mechanisms for CO and H_2 oxidation, preferential oxidation of CO, and CH_4 combustion over Rh-based catalysts (R = number of reactions).

	Mechanism	R	Features	Used in
1	Zum Mallen–Schmidt 1993 [21] H_2/O_2; H_2/H_2O	12	high temperature H_2-oxidation with partially noncompetitive adsorption of O_2	[21]
2	Hickman–Schmidt (1993) [22] CO-H_2/O_2; CH_4/O_2	19	high temperature CO, H_2-oxidation, CO-H_2 coupling/ pyrolysis CH_4 mechanism including a desorption of OH radical	[22]
3	Maier–Deutschmann (2001) [11] CO-H_2/O_2; $CH_4/O_2/H_2O$	38	CO, H_2-oxidation, CO-H_2 coupling, WGS, CH_4 oxidation and reforming [11]. Including coverage-dependent desorption energies for CO and O_2: [12]	[11,12,23–25]
4	Karakaya–Deutschmann (2016) [14] CO-H_2/O_2; $CH_4/O_2/H_2O/CO_2$	48	CO, H_2-oxidation, CO-H_2 coupling, WGS, CH_4 oxidation and reforming / CO-H_2 coupling via COOH	[14,25]
5	Mhadeshwar–Vlachos (2005) [16] CO-H_2/O_2	44	CO, H_2-oxidation, CO-H_2 coupling/WGS via COOH and HCOO; activation energies are coverage-dependent and temperature-dependent	[15,16]
6	Deshmukh–Vlachos (2007) [15] CH_4/O_2	15	CH_4 oxidation/reduced Mhadeshwar et al. [16] model for fuel-lean methane catalytic combustion (no CO and H_2 in products)	[15,25]
7	Maestri–Vlachos (2008) [26] H_2/O_2	18	H_2-oxidation/H_2 sub-mechanism from Mhadeshwar et al. [16], activation energies are coverage- and temperature-dependent	[26]
8	Rankovic–Da Costa (2011) [17] CO-H_2/O_2	28	CO, H_2-oxidation, CO-H_2 coupling, WGS/mechanism includes N_2 and NO_x chemistry	[17]

2.2. Overview of Experimental Setups

Data from different experimental setups were used in the present work: A stagnation flow reactor, an annular duct reactor, and an optically accessible single channel-flow reactor.

Stagnation flow reactors offer a simple configuration for the investigation of reactions on catalytic surfaces. The setup allows microprobe sampling of the gas-phase boundary layer adjacent to the catalyst surface. A reactive gas mixture (e.g., O_2 and H_2) enters the reactor and impinges upon the heated catalyst surface (e.g., Rh/Al_2O_3 coated disk) [27]. The catalyst surface is approximately 5 cm in diameter and the separation distance between the porous-frit gas inlet and the catalyst surface is approximately 3.9 cm. This configuration also enables efficient modeling of the surface chemistry, coupled with convective and diffusive transport within the boundary layer. In the past, manifold measurements in stagnation–flow reactor over rhodium were used in model verification for H_2 oxidation [21,27], CO oxidation [28,29], WGS [30], methane partial oxidation [14] and reforming [14,31].

Appel et al. [23] introduced the methodology of in situ spatially-resolved Raman measurements of species concentrations in gas phase over the catalyst boundary layer as a direct way to assess the catalytic reactivity at realistic operating conditions. Here, experiments were performed in a rectangular, optically accessible reactor, which comprises two horizontal, non-porous ceramic plates coated with Rh/Al_2O_3 and vertical quartz windows. This technique represents a powerful method to gain detailed insight into the reactor during operation in a non-intrusive manner. Sui et al. [24,25] studied the hetero-/homogeneous combustion of $CH_4/O_2/N_2$ mixtures over rhodium in this set up.The experiments included in situ spatially-resolved Raman measurements of gas phase species concentrations for evaluating the catalytic processes, and planar laser induced fluorescence (LIF) of the OH radical for assessing homogeneous combustion.

2.3. Overview of Simulation Tools

CaRMeN can be configured to run any simulation software. The simulations are run on a central server and the results are cached to avoid multiple runs of the same input parameters. In this work, two codes based on DETCHEM were used for the numerical simulations. These are briefly described in the following.

DETCHEM Stagnation calculates a catalytically active stagnation point flow reactor. Within a boundary layer above the surface, the general fluid flow problem can be reduced to a one-dimensional model. Thus, temperature, axial velocity and gas composition only depend on the axial position, i.e., the distance from the surface. Concentrations at the gas-surface interface and coverages are independent of the position on the plate [28,32,33].

DETCHEM Channel simulates the steady state chemically reacting gas flow through a cylindrical channel. If radial velocity gradients inside a tube cannot be neglected, it is necessary to resolve another dimension of the flow field. In typical channel flow configurations, axial transport is dominated by convection instead of diffusion. Then, the general Navier–Stokes equations can be parabolized, resulting in a partial differential equation system for conservation of mass, species, momentum and energy, which can be solved efficiently by a method-of-lines integration [34]. If cylindrical symmetry is exploited, channel flows and annular duct flows can be described. Furthermore, the case of parallel plates in a micro reactor channel can be covered by the same model, when the radius of the inner duct is chosen much larger than the distance of the two walls.

3. Illustrative Examples

In this section, the capabilities of CaRMeN are illustrated using combustion data for the CH_4, CO, and H_2 over rhodium catalysts. All results presented in this section were generated using the CaRMeN software. The numerical simulations were carried out using four different mechanisms available in the literature and are described in the previous section. The experiments were conducted in an annular reactor (Tavazzi et al. [35]), in a rectangular shaped and optically accessible channel reactor (Sui et al. [25]), and two different stagnation flow reactors (Karakaya et al. [14] and Pery et al. [29]).

A screen shot of the graphical user interface is shown in Figure 1. Experimental data are listed on the left sidebar, while the right side bar contains mechanisms. The selections on the left control which experiments to display. These items can be combined with models on the right by making selections. Four exemplary cases was selected, and each case was combined with the Karakaya–Deutschmann [14] and the Rankovic–Da Costa [17] mechanisms. In this example, only mechanisms are shown on the right-hand side. However, any other model, e.g., various diffusion models can be added to the sidebar to evaluate diffusion models.

3.1. Detailed Comparison—CO Combustion

A detailed comparison between the simulation output and the corresponding experimental data is very indicative to evaluate the quality of the model. The following example in Figure 2 shows two spatially resolved concentration profiles from a stagnation flow reactor. Both experiments were carried out under the same conditions, except for the surface temperatures.

The inlet gas feed had a temperature of 313 K and contained a CO/O_2 ratio of 2 diluted in argon, corresponding to a stoichiometric mixture for total oxidation of CO. The surface temperatures were 673 K (Figure 2, left) and 873 K (Figure 2, right). The lines in the figures represent model predictions by the Karakaya–Deutschmann (dash-dotted) and the Rankovic–DaCosta (dashed) mechanisms that were obtained using DETCHEM Stagnation. Good agreement between model and experiment is obtained with the Karakaya–Deutschmann mechanism, while the other fails to predict the experimental data in the case of 673 K. This is, however, not surprising, as the Rankovic–DaCosta mechanism was developed for use in high temperature combustion applications. This mechanism matches the experimental points with a surface temperature of 873 K and otherwise same conditions perfectly as shown in the right figure.

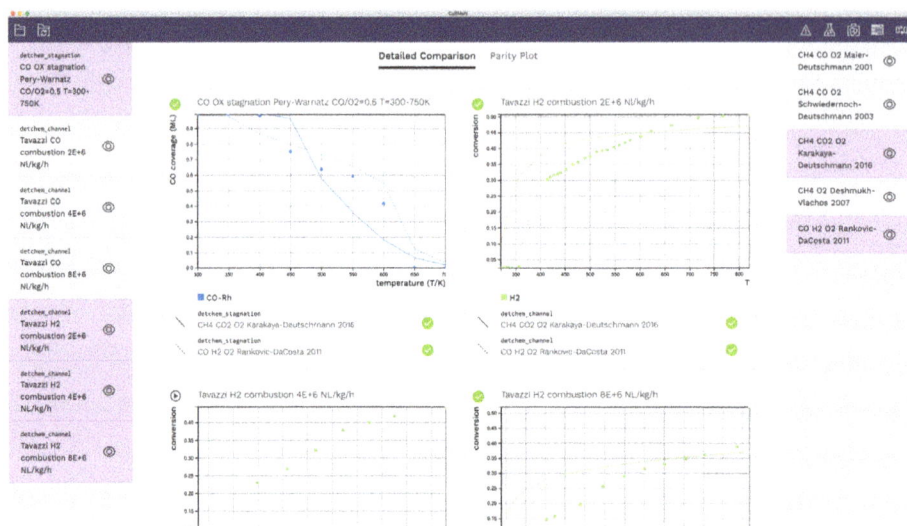

Figure 1. Screen shot of the detailed comparison view. Experimental data are shown on the left sidebar, and reaction mechanisms on the right. The selections on the left control which data to display. These items can then be combined with the mechanisms on the right. The dots in the graphs represent experimental data points while lines correspond to simulation data.

From a practical perspective, these comparisons are fairly straightforward, because the experimental data can be mapped directly to the simulation data without post processing. Hence, only a single simulation must be carried out for each mechanism and each experiment. Figure 2 therefore shows the results of four simulations. The situation is slightly more complex in the next example, where each data point is the result of one simulation. Here, CaRMeN provides a significant benefit to the work flow.

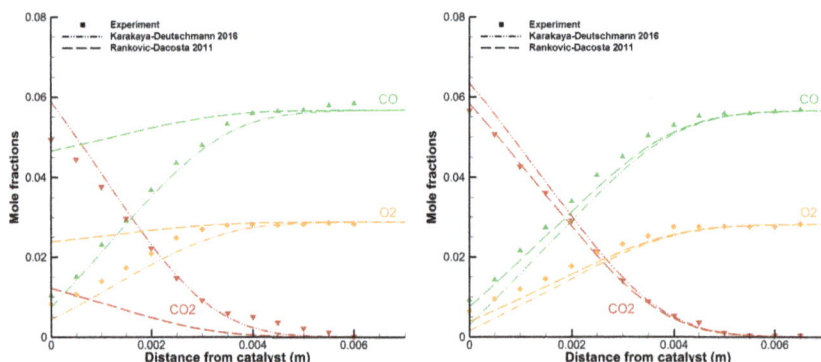

Figure 2. Data from a stagnation flow experiment showing mole fractions of CO (green triangles pointing up), O_2 (yellow diamonds) and CO_2 (red triangles pointing down) as functions of distance from the catalytic surface ((**left**) $T_{surface} = 673$ K; and (**right**) $T_{surface} = 873$ K).

Figure 3 shows nine measured data points of surface coverages normalized to unity as a function of surface temperature. Surface coverages of adsorbates can reveal important information for the development of elementary step kinetics on catalytic surfaces. Pery et al. [29] carried out CO surface coverage experiments in a stagnation flow reactor. In these experiments, the surface temperature of the catalytic plate was varied in the range of 300 K to 700 K. At intervals of 50 K, the surface coverage of CO on the rhodium surface was characterized by sum-frequency generation in-situ spectroscopy. The lines in the figure denote the same mechanisms as in Figure 2, and again using DETCHEM Stagnation. Both mechanisms showed reasonable agreement with the experiment, although the overall shape, i.e., the step-like appearance, was reproduced better by the Rankovic–DaCosta mechanism.

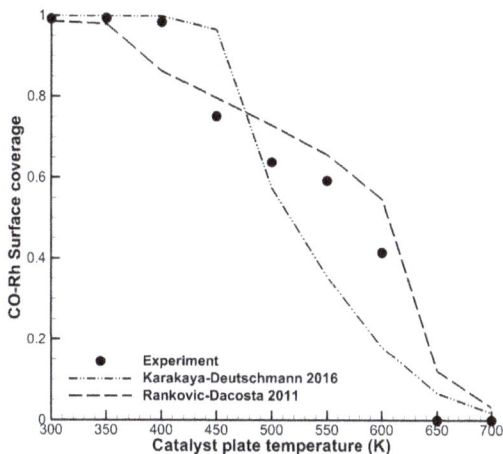

Figure 3. Surface coverages of CO on rhodium as a function of catalyst plate temperature. The experiment (dots) was carried out by Pery et al. [29].

The catalyst plate temperature enters the simulation as an input parameter, hence a separate simulation must be carried out for each of the nine data points. A post processing step then extracts the CO-coverage (*y* value) from each simulation and maps it to the appropriate surface temperature (*x* value). Because two mechanisms are shown in the figure, a total of 18 simulations were required for the corresponding model data. CaRMeN can execute each simulation in parallel because each simulation is independent of the others.

3.2. Light-Off Curves—H_2 Combustion

Light-off curves are a very common experiment type used to assess the performance of a catalyst. In these experiments, the temperature is varied and species concentrations are determined at the end of the reactor ("end-of-pipe" measurement), effectively resulting in species concentrations as functions of temperature. It is also common to plot the conversions instead of the concentrations directly. Similar to the last example in Figure 3, the data themselves (the temperature) enter the simulation as an input parameter. Consequently, one simulation was again required for each measured point.

Tavazzi et al. [35] carried out H_2 combustion experiments under rich conditions in an annular duct reactor. The temperature of the enclosing oven was varied, and the conversion of H_2 at the end of the reactor was determined. Figure 4 shows a total of about 45 experimental data points (dots) at three different space velocities compared to model predictions of the Karakaya–Deutschmann (dash-dotted) and Rankovic–DaCosta (dashed) mechanisms. Conversion tended towards a maximum of 50 % due to lack of oxygen in the rich conditions. The Rankovic–DaCosta mechanism reproduced the light-off temperature reasonably well, whereas the Karakaya–Deutschmann mechanism overpredicted H_2 conversion. At approximately 500 K, both models gave the same results for all flow configurations. Furthermore, both mechanisms were less accurate at higher space velocities. This behavior may be due to diffusion effects within the washcoat, which were not captured accurately by the used diffusion model of the simulation code (DETCHEM Channel). However, it may also permit the conclusion that it is important to take into account various flow velocities and accurate diffusion models during mechanism development.

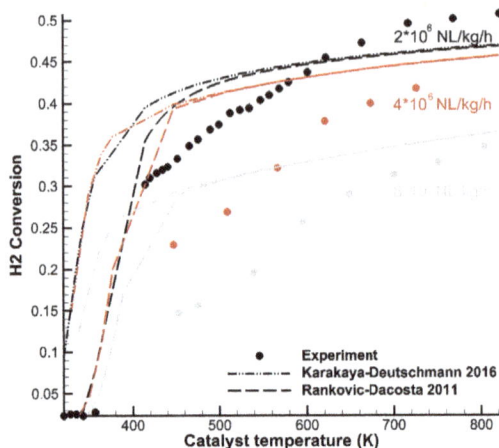

Figure 4. Hydrogen conversion over temperature at different GHSV reported by Tavazzi et al. [35]. The lines correspond to numerical data using Karakaya–Deutschmann (2016) (dash-dotted) and Rankovic–DaCosta (2011) (dashed).

To reproduce the lines in Figure 4, approximately 90 simulations were required. In the case of the used simulation code DETCHEM Channel, an additional post processing step was required to calculate the H_2 conversions from the format that was output by the code to be able to map the value to the experimental data. At this point, it is clear that tools to automate such comparisons are essential for a productive work flow. CaRMeN generates the input files, manages the intermediate results (90 individual simulations), carries out the post processing and handles the visualization. Adding another mechanism or testing the influence of a changed rate parameter then becomes trivial.

3.3. Parity Plot—CH₄ Combustion

The previously shown detailed comparisons are useful to gain a comprehensive understanding of how the model reproduces experimental data. However, as more experimental data are used, it becomes difficult to evaluate all the individual plots. In these cases, parity plots can be more useful, as they allow quickly judging the quality of a model against large sets of experimental data.

Figure 5 shows a parity plot with points from the experiments by Sui et al. [25] plotted against model data using three different mechanisms: Deshmukh–Vlachos (2007), Maier–Deutschmann (2003) and Karakaya–Deutschmann (2016). Here, x is the experimental value, and y is the corresponding value from the simulation. The experiments were conducted in a wide pressure range (2 bar to 12 bar), under total oxidation conditions with C/O_2 ratios from 0.15 to 0.2, and dilution in N_2. The 2D experimental data points extracted were transversely averaged and simulated using DETCHEM Channel considering the average temperature profile of the two coated plates. Overall, Karakaya–Deutschmann (2016) (red diamonds) showed best agreement with the experimental data, while the other two mechanisms generally overpredicted the CH_4 conversion. However, the experimental data used for this comparison span a large pressure range up to 12 bar, which are conditions none of the mechanisms were tested under during their development. Furthermore, the simulation code used for the comparison (DETCHEM Channel) does not account for axial diffusion. Therefore, some of the discrepancies between model and experiment can likely be attributed to the simplified flow model, and not to the chemical model.

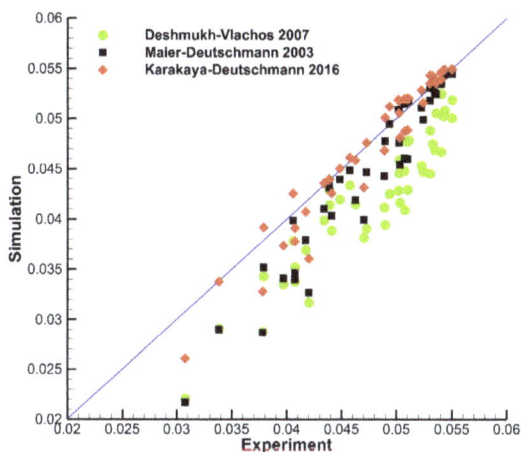

Figure 5. Parity plot showing points from the experiments by Sui et al. [25] plotted against model data using mechanisms Deshmukh–Vlachos (2007) (green circles), Maier–Deutschmann (2003) (black squares), and Karakaya–Deutschmann (2016) (red diamonds). The experiments were conducted in a wide pressure range (2 bar to 12 bar), with C/O ratios from 0.15 to 0.2, and dilution in N_2.

4. Conclusions

Elementary-step based reaction mechanisms are a powerful tool in predictive simulations of catalytic reactors. In the work presented, the methodology that was developed and implemented in the computer code CaRMeN [1] was exemplarily applied to analyze and evaluate catalytic reaction mechanisms developed for the combustion of CH_4, H_2 and CO over rhodium-coated surfaces. Appropriate sets of experimental data, reaction mechanisms and reactor simulators (flow solvers) for this specific chemical system were able to be compared in an automated fashion. CaRMeN generates the input files for the simulation code, manages the intermediate results (90 individual simulations), carries out the post processing and handles the visualization.

Without surprise, the mechanisms worked well to reproduce the experimental data, which were originally used in their development process. Mostly they were also able to predict experiments by other groups and reactor configurations conducted at similar operating conditions; sometimes, however, they failed completely, which implies that the microkinetic model was not truly intrinsic, i.e. physical or reactor-specific features were also represented in the kinetics. The more recent models, in particular those that obey thermodynamic consistency, usually performed much better and were often applicable to a wider range of conditions.

The tool CaRMeN can also be used to easily compare different physical models, for example the flow field, diffusion, heat transfer and physical parameters, such as catalyst loadings and structure, as well as inlet and boundary conditions. Furthermore, special features of a certain catalyst structure/support and also systematic errors in certain experimental set-ups can be identified more easily by comparing huge datasets from different sources and reactor types. The biggest hurdle for archiving much more data, however, is the incompleteness of literature information for reproduction of the experiments and models described. CaRMeN can be downloaded for free from www.detchem.com.

Author Contributions: Conceptualization: H.G. and O.D.; methodology: H.G. and O.D.; literature survey: L.M. and S.A.; original draft preparation: H.G., L.M., S.A., and S.T; review and editing: H.G., L.M., and O.D.; supervision: O.D.; funding acquisition: O.D.

Acknowledgments: The authors would like to thank Robert J. Kee, Huayang Zhu, Canan Karakaya, and Greg Jackson, Colorado School of Mines, for fruitful discussions.

Funding: This work was supported by Steinbeis GmbH & Co. KG.

Conflicts of Interest: The authors declare no conflict of interest.

References

1. Gossler, H.; Maier, L.; Angeli, S.; Tischer, S.; Deutschmann, O. CaRMeN: A Tool for Analysing and Deriving Kinetics in the Real World. *Phys. Chem. Chem. Phys.* **2018**, *20*, 10857–10876. doi:10.1039/C7CP07777G.
2. PrIMe. Available online: http://primekinetics.org (accessed on 9 December 2018).
3. RESPECTH. Available online: http://respecth.chem.elte.hu/respecth/ (accessed on 9 December 2018).
4. Goteng, G.; Speight, M.; Nettyam, N.; Farooq, A.; Franklach, M.; Sarathy, S. A Hybrid Cloud System for Combustion Kinetics Simulation. In Proceedings of the 23rd International Symposium on Gas Kinetics and Related Phenomena, Szeged, Hungary, 20–24 July 2014.
5. Lunsford, J.H. Catalytic Conversion of Methane to More Useful Chemicals and Fuels: A Challenge for the 21st Century. *Catal. Today* **2000**, *63*, 165–174. doi:10.1016/S0920-5861(00)00456-9.
6. Abbasi, R.; Wu, L.; Wanke, S.E.; Hayes, R.E. Kinetics of Methane Combustion over Pt and Pt–Pd Catalysts. *Chem. Eng. Res. Des.* **2012**, *90*, 1930–1942. doi:10.1016/j.cherd.2012.03.003.
7. Chen, D.; Lødeng, R.; Svendsen, H.; Holmen, A. Hierarchical Multiscale Modeling of Methane Steam Reforming Reactions. *Ind. Eng. Chem. Res.* **2011**, *50*, 2600–2612. doi:10.1021/ie1006504.
8. Osman, A.I.; Abu-Dahrieh, J.K.; Laffir, F.; Curtin, T.; Thompson, J.M.; Rooney, D.W. A Bimetallic Catalyst on a Dual Component Support for Low Temperature Total Methane Oxidation. *Appl. Catal. B Environ.* **2016**, *187*, 408–418. doi:10.1016/j.apcatb.2016.01.017.
9. Osman, A.I.; Meudal, J.; Laffir, F.; Thompson, J.; Rooney, D. Enhanced Catalytic Activity of Ni on η-Al2O3 and ZSM-5 on Addition of Ceria Zirconia for the Partial Oxidation of Methane. *Appl. Catal. B Environ.* **2017**, *212*, 68–79. doi:10.1016/j.apcatb.2016.12.058.
10. Maier, L.; Schädel, B.; Delgado, K.H.; Tischer, S.; Deutschmann, O. Steam Reforming of Methane Over Nickel: Development of a Multi-Step Surface Reaction Mechanism. *Top. Catal.* **2011**, *54*, 845. doi:10.1007/s11244-011-9702-1.
11. Deutschmann, O.; Schwiedernoch, R.; Maier, L.I.; Chatterjee, D. Natural Gas Conversion in Monolithic Catalysts: Interaction of Chemical Reactions and Transport Phenomena. In *Studies in Surface Science and Catalysis*; Iglesia, E., Spivey, J.J., Fleisch, T.H., Eds.; Elsevier: Amsterdam, The Netherlands, 2001; Volume 136, Natural Gas Conversion VI, pp. 251–258. doi:10.1016/S0167-2991(01)80312-8.
12. Schwiedernoch, R.; Tischer, S.; Correa, C.; Deutschmann, O. Experimental and Numerical Study on the Transient Behavior of Partial Oxidation of Methane in a Catalytic Monolith. *Chem. Eng. Sci.* **2003**, *58*, 633–642. doi:10.1016/S0009-2509(02)00589-4.
13. Williams, K.A.; Leclerc, C.A.; Schmidt, L.D. Rapid Lightoff of Syngas Production from Methane: A Transient Product Analysis. *AIChE J.* **2005**, *51*, 247–260. doi:10.1002/aic.10294.
14. Karakaya, C.; Maier, L.; Deutschmann, O. Surface Reaction Kinetics of the Oxidation and Reforming of CH4 over Rh/Al2O3 Catalysts. *Int. J. Chem. Kinet.* **2016**, *48*, 144–160. doi:10.1002/kin.20980.
15. Deshmukh, S.R.; Vlachos, D.G. A Reduced Mechanism for Methane and One-Step Rate Expressions for Fuel-Lean Catalytic Combustion of Small Alkanes on Noble Metals. *Combust. Flame* **2007**, *149*, 366–383. doi:10.1016/j.combustflame.2007.02.006.
16. Mhadeshwar, A.B.; Vlachos, D.G. Hierarchical Multiscale Mechanism Development for Methane Partial Oxidation and Reforming and for Thermal Decomposition of Oxygenates on Rh. *J. Phys. Chem. B* **2005**, *109*, 16819–16835. doi:10.1021/jp052479t.
17. Rankovic, N.; Nicolle, A.; Berthout, D.; Da Costa, P. Kinetic Modeling Study of the Oxidation of Carbon Monoxide–Hydrogen Mixtures over Pt/Al2O3 and Rh/Al2O3 Catalysts. *J. Phys. Chem. C* **2011**, *115*, 20225–20236. doi:10.1021/jp205476y.
18. Boll, W. Korrelation Zwischen Umsatzverhalten Und Katalytischer Oberfläche von Dieseloxidationskatalysatoren Unter Variation von Beladung Und Alterungszustand. Ph.D. Thesis, Karlsruhe Institute of Technology, Karlsruhe, 2011.
19. Cai, Y.; Stenger, Harvey G, J.; Lyman, C.E. Catalytic CO Oxidation over Pt–Rh/γ-Al2O3Catalysts. *J. Catal.* **1996**, *161*, 123–131. doi:10.1006/jcat.1996.0169.

20. Ito, S.I.; Fujimori, T.; Nagashima, K.; Yuzaki, K.; Kunimori, K. Strong Rhodium–Niobia Interaction in Rh/Nb$_2$O$_5$, Nb$_2$O$_5$–Rh/SiO$_2$ and RhNbO$_4$/SiO$_2$ Catalysts: Application to Selective CO Oxidation and CO Hydrogenation. *Catal. Today* **2000**, *57*, 247–254. doi:10.1016/S0920-5861(99)00333-8.

21. Zum Mallen, M.P.; Williams, W.R.; Schmidt, L.D. Steps in Hydrogen Oxidation on Rhodium: Hydroxyl Desorption at High Temperatures. *J. Phys. Chem.* **1993**, *97*, 625–632. doi:10.1021/j100105a016.

22. Hickman, D.A.; Schmidt, L.D. Steps in CH4 Oxidation on Pt and Rh Surfaces: High-Temperature Reactor Simulations. *AIChE J.* **1993**, *39*, 1164–1177. doi:10.1002/aic.690390708.

23. Appel, C.; Mantzaras, J.; Schaeren, R.; Bombach, R.; Inauen, A.; Tylli, N.; Wolf, M.; Griffin, T.; Winkler, D.; Carroni, R. Partial Catalytic Oxidation of Methane to Synthesis Gas over Rhodium: In Situ Raman Experiments and Detailed Simulations. *Proc. Combust. Inst.* **2005**, *30*, 2509–2517. doi:10.1016/j.proci.2004.08.055.

24. Sui, R.; Mantzaras, J.; Bombach, R. A Comparative Experimental and Numerical Investigation of the Heterogeneous and Homogeneous Combustion Characteristics of Fuel-Rich Methane Mixtures over Rhodium and Platinum. *Proc. Combust. Inst.* **2017**, *36*, 4313–4320. doi:10.1016/j.proci.2016.06.001.

25. Sui, R.; Mantzaras, J.; Bombach, R.; Denisov, A. Hetero-/Homogeneous Combustion of Fuel-Lean Methane/Oxygen/Nitrogen Mixtures over Rhodium at Pressures up to 12bar. *Proc. Combust. Inst.* **2017**, *36*, 4321–1328. doi:10.1016/j.proci.2016.06.003.

26. Maestri, M.; Beretta, A.; Faravelli, T.; Groppi, G.; Tronconi, E.; Vlachos, D.G. Two-Dimensional Detailed Modeling of Fuel-Rich H2 Combustion over Rh/Al$_2$O$_3$ Catalyst. *Chem. Eng. Sci.* **2008**, *63*, 2657–2669. doi:10.1016/j.ces.2008.02.024.

27. Karakaya, C.; Deutschmann, O. Kinetics of Hydrogen Oxidation on Rh/Al$_2$O$_3$ Catalysts Studied in a Stagnation-Flow Reactor. *Chem. Eng. Sci.* **2013**, *89*, 171–184. doi:10.1016/j.ces.2012.11.004.

28. Karadeniz, H.; Karakaya, C.; Tischer, S.; Deutschmann, O. Numerical Modeling of Stagnation-Flows on Porous Catalytic Surfaces: CO Oxidation on Rh/Al$_2$O$_3$. *Chem. Eng. Sci.* **2013**, *104*, 899–907. doi:10.1016/j.ces.2013.09.038.

29. Pery, T.; Schweitzer, M.G.; Volpp, H.R.; Wolfrum, J.; Ciossu, L.; Deutschmann, O.; Warnatz, J. Sum-Frequency Generation in Situ Study of CO Adsorption and Catalytic CO Oxidation on Rhodium at Elevated Pressures. *Proc. Combust. Inst.* **2002**, *29*, 973–980. doi:10.1016/S1540-7489(02)80123-7.

30. Karakaya, C.; Otterstätter, R.; Maier, L.; Deutschmann, O. Kinetics of the Water-Gas Shift Reaction over Rh/Al$_2$O$_3$ Catalysts. *Appl. Catal. A Gen.* **2014**, *470*, 31–44. doi:10.1016/j.apcata.2013.10.030.

31. McGuire, N.E.; Sullivan, N.P.; Deutschmann, O.; Zhu, H.; Kee, R.J. Dry Reforming of Methane in a Stagnation-Flow Reactor Using Rh Supported on Strontium-Substituted Hexaaluminate. *Appl. Catal. A Gen.* **2011**, *394*, 257–265. doi:10.1016/j.apcata.2011.01.009.

32. Behrendt, F.; Deutschmann, O.; Maas, U.; Warnatz, J. Simulation and Sensitivity Analysis of the Heterogeneous Oxidation of Methane on a Platinum Foil. *J. Vac. Sci. Technol. A* **1995**, *13*, 1373–1377. doi:10.1116/1.579566.

33. Kee, R.J.; Coltrin, M.E.; Glarborg, P. *Chemically Reacting Flow: Theory and Practice*; Wiley Interscience: Hoboken, NJ, USA, 2003.

34. Tischer, S.; Correa, C.; Deutschmann, O. Transient Three-Dimensional Simulations of a Catalytic Combustion Monolith Using Detailed Models for Heterogeneous and Homogeneous Reactions and Transport Phenomena. *Catal. Today* **2001**, *69*, 57–62. doi:10.1016/S0920-5861(01)00355-8.

35. Tavazzi, I.; Beretta, A.; Groppi, G.; Forzatti, P. Development of a Molecular Kinetic Scheme for Methane Partial Oxidation over a Rh/α-Al$_2$O$_3$ Catalyst. *J. Catal.* **2006**, *241*, 1–13. doi:10.1016/j.jcat.2006.03.018.

catalysts

MDPI

Article

Kinetic Study of the Selective Hydrogenation of Acetylene over Supported Palladium under Tail-End Conditions

Caroline Urmès [1,2], Jean-Marc Schweitzer [2], Amandine Cabiac [2] and Yves Schuurman [1,*]

[1] IRCELYON CNRS, UMR 5256, Univ Lyon, Université Claude Bernard Lyon 1, 2 avenue Albert Einstein, 69626 Villeurbanne Cedex, France; caroline.urmes@gmail.com

[2] IFP Energies nouvelles, Etablissement de Lyon, Rond-point de l'échangeur de Solaize, BP3, 69360 Solaize, France; jean-marc.schweitzer@ifpen.fr (J.-M.S.); amandine.cabiac@ifpen.fr (A.C.)

* Correspondence: yves.schuurman@ircelyon.univ-lyon1.fr; Tel.: +33-472445482

Received: 9 January 2019; Accepted: 31 January 2019; Published: 14 February 2019

Abstract: The kinetics of the selective hydrogenation of acetylene in the presence of an excess of ethylene has been studied over a 0.05 wt. % Pd/α-Al$_2$O$_3$ catalyst. The experimental reaction conditions were chosen to operate under intrinsic kinetic conditions, free from heat and mass transfer limitations. The data could be described adequately by a Langmuir–Hinshelwood rate-equation based on a series of sequential hydrogen additions according to the Horiuti–Polanyi mechanism. The mechanism involves a single active site on which both the conversion of acetylene and ethylene take place.

Keywords: power-law; Langmuir–Hinshelwood; kinetic modeling; Pd/α-Al$_2$O$_3$

1. Introduction

Ethylene is the largest of the basic chemical building blocks with a global market estimated at more than 140 million tons per year with an increasing growth rate. It is used mainly as precursor for polymers production, for instance polyethylene, vinyl chloride, ethylbenzene, or even ethylene oxide synthesis. New ways of production of ethylene are emerging, such as ethanol dehydration, but steam cracking of naphtha and gas remains the major producer of alkenes. The C2 fraction at the outlet of a steam cracker contains mainly ethane and ethylene, but also traces of acetylene. These trace amounts need to be removed as acetylene is known to poison the Ziegler–Natta catalyst that is used for the polymerization of ethylene. This important issue is done by selective hydrogenation of acetylene, important process in petrochemical industry. Thus, the initial acetylene content, approximately 0.8–1.6%, needs to be reduced to less than 5 ppmv for chemical grade and less than 1 ppmv for polymer grade ethylene. Depending on plant design, selective hydrogenation is carried in two different ways: front-end and tail-end [1]. In the tail-end configuration, which corresponds to 70% of all units worldwide, the process is placed after CH$_4$ and H$_2$ separation. The hydrogen is added in an amount slightly higher than the acetylene concentration and the majority of the stream is ethylene [1,2]. In front-end configuration, the selective hydrogenation unit is placed upstream of the demethanizer and a larger amount of hydrogen is available (around 20%).

Alumina-supported palladium or bimetallic palladium-silver catalysts are used for this process, assuring very high activity and selectivity for acetylene hydrogenation. The main goal is to reduce the acetylene content without the hydrogenation of ethylene to ethane. Catalyst deactivation by coke formation is very common under tail-end conditions as is the formation of C$_4$ byproducts.

The kinetics of this reaction have been the subject of several studies [3–9] and have been analyzed in detail by Borodziński and Bond [8]. The most elaborate models consist of 2 or 3 distinct sites, each catalyzing a specific reaction [8]. In the case of multiple active sites, a small and a large size site are considered. Small sites favour the adsorption and selective hydrogenation of acetylene to ethylene, whereas the adsorption of ethylene seems to be only possible on the large sites. Pachulski et al. [9], in a systematic study of acetylene hydrogenation over Pd-Ag/Al$_2$O$_3$, evaluated 77 different rate equations and found that the best rate equation was based on two different sites.

However, few studies discriminate between different models and in some cases, the need for a more elaborate mechanism might be due to additional factors such as the addition of carbon monoxide or the use of a catalyst promotor. For example, Bos et al. [6] discarded a single-site Langmuir–Hinshelwood mechanism because of its inability to predict the observed change of the ethane selectivity when carbon monoxide was added to the feed. An alternative explanation can be that addition of carbon monoxide actually creates an additional site by either an electronic or geometric effect.

In this study, we derived rate equations based on a sequence of elementary steps. Several rate equations were obtained depending on the assumption of the rate-determining step. A regression analysis was then performed to select the most appropriate mechanism based on our experimental data and to estimate the rate parameters.

2. Results

An analysis of the repeatability of the experiments has been performed. Several operating conditions have been tested at both temperatures:

- Operating conditions 1: $y_{C_2H_2} = 1.0\%$; $y_{H_2} = 4.3\%$; $y_{C_2H_4} = 70\%$ and $y_{Ar} = 24.7\%$
- Operating conditions 2, 51 °C: $y_{C_2H_2} = 0.6\%$; $y_{H_2} = 4.3\%$; $y_{C_2H_4} = 70\%$ and $y_{Ar} = 25.1\%$
- Operating conditions 2, 62 °C: $y_{C_2H_2} = 0.8\%$; $y_{H_2} = 4.3\%$; $y_{C_2H_4} = 70\%$ and $y_{Ar} = 24.9\%$

Each operating condition was tested three times per catalyst loading. Two catalyst loadings were used. All this data was used to calculate the relative standard deviations of both the acetylene conversion and the ethane exit molar flow rate. The relative standard deviations are given in Table 1. Rather large (10–20%) relative standard deviations were found for the molar exit flows of ethane.

Table 1. Relative standard deviation (rsd) of 6 repeated experiments (3 for each catalyst loading, 2 loadings) for both conversion and ethane outlet flow at 2 conversion levels.

	Operating Conditions 1		Operating Conditions 2		Operating Conditions 1		Operating Conditions 2	
T (°C)	X_{C2H2}	rsd (%)	X_{C2H2}	rsd (%)	F_{C2H6}	rsd (%)	F_{C2H6}	rsd (%)
51	0.13	9.3	0.36	6.2	0.07	19	0.13	11
62	0.23	7.8	0.37	8.5	0.11	16	0.16	15

The variation of the relative flows of acetylene, ethylene, hydrogen, and argon allowed determination of the apparent reaction orders with respect to acetylene, ethylene, and hydrogen. Apparent reaction orders are based on power law expressions for the rates as follows:

$$- r_{C_2H_2} = k_1 P_{C_2H_2}^{\alpha_1} P_{H_2}^{\beta_1} P_{C_2H_4}^{\gamma_1}$$

$$r_{C_2H_6} = k_2 P_{C_2H_2}^{\alpha_2} P_{H_2}^{\beta_2} P_{C_2H_4}^{\gamma_2}$$

As no C4 products were experimentally observed, the kinetic analysis is restricted to the hydrogenation of acetylene to ethylene and ethylene to ethane. Both rate equations were integrated numerically and the reaction orders were determined by regression analysis of the acetylene conversion and the molar exit flow rate of ethane of the data set at each temperature separately. The estimated values of the reaction orders are given in Table 2.

Table 2. Estimated values of the reaction orders with their 95% confidence intervals.

T (°C)	r_{C2H2}			r_{C2H6}		
	α_1 (C$_2$H$_2$)	β_1 (H$_2$)	γ_1 (C$_2$H$_4$)	α_2 (C$_2$H$_2$)	β_2 (H$_2$)	γ_2 (C$_2$H$_4$)
51	-0.88 ± 0.09	1.46 ± 0.16	-0.13 ± 0.07	-1.00 ± 0.20	1.42 ± 0.70	1.09 ± 0.50
62	-0.93 ± 0.07	1.49 ± 0.34	-0.19 ± 0.04	-1.56 ± 0.13	1.69 ± 0.88	0.30 ± 0.15

A negative reaction order for acetylene was observed, -0.9 with respect to acetylene consumption, and (-1)–(-1.5) with respect to ethane formation. This correspond to a strong adsorption of acetylene on the surface of the catalyst. This order is lower than the values reported in the literature for acetylene consumption, which are between 0–(-0.7) [7,10–14] depending on the conditions.

The order for hydrogen, approximately 1.5, is high and hard to explain mechanistically. However, the same range of magnitude was found by Aduriz and al.: 1.3–1.6 [10], but under front-end conditions. Most studies report an order of $+1$ [6,11,12]. Molero et al. observed a reaction order of hydrogen between 1–1.25, depending on the temperature for acetylene hydrogenation over a Pd foil [7]. From a careful analysis of the data, they derived that the hydrogen reaction order can vary between values of 1 and 1.5. Excess hydrogen can remove strongly adsorbed carbonaceous species from the catalyst surface and so creates free surface sites.

Regarding the rate of consumption of acetylene, no strong effect is observed for ethylene. The order is close to zero as shown by numerous studies [12]. However, some ethylene adsorption occurs as indicated by the small negative value of the reaction order. For ethane formation, the reaction order in ethylene is much higher, between 0.3 and 1. This is related to the fact that ethylene is the reactant for ethane production.

Inspection of the reaction orders give valuable insights into the reaction mechanism. However, this needs to be further validated by deriving the corresponding rate equation based on a sequence of elementary steps. Catalytic hydrogenation reactions of unsaturated hydrocarbons often follow a series of sequential hydrogen additions according to the Horiuti–Polanyi mechanism [13,15]. This mechanism is given for acetylene hydrogenation via ethylene to ethane in Table 3. A single site for all surface species has been assumed. This assumption will be discussed later on.

Table 3. Elementary steps for the reaction $C_2H_2 + H_2 \leftrightarrows C_2H_4$ and $C_2H_4 + H_2 \leftrightarrows C_2H_6$.

N°	Elementary Step	σ
1	$H_2 + 2^* \leftrightarrows 2H^*$	2
2	$C_2H_2 + {}^* \leftrightarrows C_2H_2^*$	1
3	$C_2H_2^* + H^* \leftrightarrows C_2H_3^* + {}^*$	1
4	$C_2H_3^* + H^* \leftrightarrows C_2H_4^* + {}^*$	1
5	$C_2H_4^* \leftrightarrows C_2H_4 + {}^*$	1
6	$C_2H_4^* + H^* \leftrightarrows C_2H_5^* + {}^*$	1
7	$C_2H_5^* + H^* \leftrightarrows C_2H_6 + 2^*$	1

Hydrogen adsorbs dissociatively on palladium, requiring two free neighboring surface sites (step (1)). The adsorbed hydrogen atom can react with adsorbed acetylene to form a vinyl intermediate in step (2). This intermediate can react with a second hydrogen atom to form adsorbed ethylene. Neurock and van Santen studied ethylene adsorption on a Pd(111) surface by DFT and found that ethylene adsorbs at low coverages as a di-σ species and at high coverage as a π-bonded species [16]. Ethylene can desorb or react with atomic hydrogen to form an ethyl intermediate. This intermediate can react again with a second hydrogen atom to form ethane, which has little interaction with the Pd surface and therefore desorbs instantaneously (step (7)).

Even though the reaction mechanism is still rather simple, the derivation of the corresponding rate-equation requires several assumptions. We assume that the adsorption take place according to the Langmuir isotherm [17]. The next assumption is with respect to the rate-determining step for the formation of ethylene and ethane. Since for both steps the reaction order in hydrogen was found to be larger than 1, the additions of the second hydrogen atom to the vinyl and ethyl intermediates are assumed to be rate-determining (steps (4) and (7), respectively). All other steps are assumed to be in quasi-equilibrium. To reduce the number of parameters in the rate equation only the most abundant reaction intermediates are kept in the site balance. The full site balance is given as:

$$\theta_* + \theta_H + \theta_{C_2H_2} + \theta_{C_2H_3} + \theta_{C_2H_4} + \theta_{C_2H_5} = 1$$

Here, we assume that the coverages of vinyl and ethyl intermediates are much smaller than those of adsorbed acetylene and ethylene and thus can be left out of the site balance. A combined DFT Monte-Carlo study for acetylene hydrogenation over Pd(111) at a hydrogen to acetylene ratio of 1, showed that this is indeed the case [13]. This same study indicates that the hydrogen coverage is not negligible and that it is larger than the ethylene coverage. Here, we take into account the ethylene coverage, because a negative reaction order in ethylene was observed. The last model assumption states that the rate-determining steps, (4) and (7), are irreversible (or one-way) under the given reaction conditions.

The rates for the two rate-determining steps can be written as:

$$r_4 = k_4 \theta_{C_2H_3} \theta_H = k_4 K_1 K_2 K_3 P_{C_2H_2} P_{H_2} \theta_*^2$$

$$r_7 = k_7 \theta_{C_2H_5} \theta_H = k_7 \frac{K_1}{K_5} K_6 P_{C_2H_2} P_{H_2} \theta_*^2$$

and the site balance is given by:

$$1 = \theta_* + \theta_H + \theta_{C_2H_2} + \theta_{C_2H_4}$$

or:

$$\theta_* = \frac{1}{\left(1 + K_2 P_{C_2H_2} + \frac{P_{C_2H_4}}{K_5} + \sqrt{K_1 P_{H_2}}\right)}$$

By introducing the number of palladium surface atoms per catalyst mass, N_S (mol Pd$_s$/kg$_{cat}$), and attributing all temperature effects to the change of the rate constant in the rate-determining steps, thus assuming that the adsorption equilibrium constants do not change between 51 and 62 °C, the following rate equations are obtained for the consumption of acetylene and the production of ethane, respectively:

$$-r_{C_2H_2} = \frac{N_S k_4^0 \exp\left(-\frac{E_4}{RT}\right) K_1 K_2 P_{C_2H_2} P_{H_2}}{\left(1 + \sqrt{K_1 P_{H_2}} + K_2 P_{C_2H_2} + \frac{P_{C_2H_4}}{K_5}\right)^2}$$

$$r_{C_2H_6} = \frac{N_S k_7^0 \exp\left(-\frac{E_7}{RT}\right) \left(\frac{K_1}{K_5}\right) P_{C_2H_4} P_{H_2}}{\left(1 + \sqrt{K_1 P_{H_2}} + K_2 P_{C_2H_2} + \frac{P_{C_2H_4}}{K_5}\right)^2}$$

Notice that k_4 and k_7 in the above equation are actually a combination of $k_4^*K_3$ and $k_7^*K_6$, respectively. Table 4 gives the correspondence between the reaction orders of the rate equations and the surface coverages as well as the range of reaction orders that are covered by the model.

Table 4. Reaction orders corresponding to the derived rate-equations.

Reaction Order	r_{C2H2}			r_{C2H6}		
	C_2H_2	H_2	C_2H_4	C_2H_2	H_2	C_2H_4
dependence on coverage	$1-2\theta_{C2H2}$	$1-\theta_{H2}$	$-2\theta_{C2H4}$	$-2\theta_{C2H2}$	$1-\theta_{H2}$	$1-2\theta_{C2H4}$
Min–max	(−1)–1	0–1	(−2)–0	(−2)–0	0–1	(−1)–1

The reaction order with respect to acetylene is smaller for the ethane production than for the acetylene consumption, whereas the reaction order with respect to ethylene is larger for the ethane production than for the acetylene consumption, in agreement with the experimental results. Taking the experimental reaction orders from Table 3, the mean coverage averaged over all tested reaction conditions of acetylene, hydrogen and ethylene can be estimated as $\theta_{C2H2} = 0.65\text{--}0.95$, $\theta_{H2} \approx 0$ and $\theta_{C2H4} = 0.06\text{--}0.15$.

In order to estimate the 7 parameters in the two rate equations, a multi-response regression analysis of the two data sets at 51 and 62 °C was carried out simultaneously. Initial results showed that all parameters were strongly correlated and no accurate estimates could be determined. Therefore, the hydrogen equilibrium constant was set at a fixed value, calculated from literature data. The following expression is used to calculate the adsorption equilibrium constant:

$$K_i = \frac{\sigma_s A_S}{\sqrt{2\pi M_w RT}} \frac{1}{10^{13} e^{\left(-\frac{E_d}{RT}\right)}} \quad \left(Pa^{-1}\right)$$

where σ_s is the sticking coefficient, A_S the surface area of Pd ($1.26 \; 10^4 \; m^2/mol$), M_w the molecular weight (kg/mol) and E_d the desorption activation energy (J/mol). A typical value for the pre-exponential factor for desorption of $10^{13} \; s^{-1}$ was assumed. Assuming a sticking coefficient for hydrogen adsorption of 0.16 (0.1–0.2 [18] and 0.17 [19]) and a desorption energy of 69 kJ/mol [20], a value of 611 Pa^{-1} was estimated for K_1. Fixing the adsorption equilibrium constant for hydrogen forces the model to account for the hydrogen coverage, else due to the hydrogen reaction order ≥ 1 the hydrogen coverage would be close to zero. Further regression analysis still showed an unacceptable correlation between the parameters. To get an accurate estimate of the adsorption equilibrium constant for acetylene the adsorption equilibrium constant for ethylene had to be set at a fixed value. Although it is not evident from the two rate equations, the correlation between these two equilibrium constants (K_2 and K_5) can be revealed by expressing the ethane selectivity as:

$$S_{C_2H_6} = \frac{r_{C_2H_6}}{r_{C_2H_2} + r_{C_2H_6}} \approx \frac{r_{C_2H_6}}{r_{C_2H_2}} = \frac{k_7}{k_4} \frac{P_{C_2H_4}}{P_{C_2H_2}} \frac{1}{K_2 K_5}$$

Apparently, the rate constants k_7 and k_4 can be decoupled by fitting the conversion and ethane production but not the term $K_2{}^*K_5$. However, the absolute values of k_7 and k_4 will depend on the values of K_2 and K_5.

As stated above, ethylene is adsorbed at low coverages as a di-σ species or at high coverage as a π-bonded species with adsorption enthalpies of −60 and −30 kJ/mol, respectively [16]. A TPD study gave a value of the adsorption enthalpy of −59 kJ/mol [21]. This value was used to estimate the equilibrium constant for ethylene at a value of 0.14 Pa^{-1}. Table 5 reports the values of the parameter estimates with their 95% confidence intervals from the regression analysis of all data. The parameters can be accurately estimated with 95% confidence intervals at the 10% level for those related the acetylene conversion and 20% for the ethane production, similar as the relative standard deviations on the repeated runs (Table 1). No strong parameter correlation was observed; the highest value of 0.85 was between k^0_4 and E_4. The value of the equilibrium adsorption constant for acetylene is approximately 3 orders of magnitude higher than that of ethylene. This corresponds to an enthalpy of adsorption for acetylene of (−80)–(−90) kJ/mol. Vattuone et al. [22] measured differential heat of adsorptions of acetylene over a single crystal of Pd(100) by calorimetry from 110–40 kJ/mol, decreasing with increasing acetylene coverage.

Table 5. Parameter estimates with their 95% confidence intervals.

Parameter	Estimated Value
K_1 (Pa^{-1})	13 (fixed)
K_2 (Pa^{-1})	107 ± 7
k^0_4 (mol/mol Pd$_s$/s)	$1.4 \pm 0.1 \times 10^8$
K_5 (Pa)	7.4 (fixed)
k^0_7 (mol/mol Pd$_s$/s)	$3.6 \pm 0.6 \times 10^9$
E_4 (kJ/mol)	48.5 ± 4.2
E_7 (kJ/mol)	54.8 ± 11
N_S (mol/kg)	2×10^{-3} (fixed)

An adequate fit of the data was obtained, as shown in Figures 1–6, organized per inlet flow of acetylene, ethane and hydrogen for both the acetylene conversion and ethane production. Although the model corresponds to a hydrogen reaction order of approximately 1, both the acetylene conversion and ethane production are well fitted by the model, especially at 51 °C. The power law model gave a significantly larger value for the hydrogen reaction order (~1.45 at 51 °C, Table 2). The cause of this discrepancy between the two models is not clear. The surface coverages as calculated by the model are in the range of θ_{C2H2} = 0.62–0.95, θ_{H2} = 0.002–0.01 and θ_{C2H4} = 0.04–0.36.

This is in good agreement with the surface coverages as calculated from the reaction orders. The apparent activation energy for the conversion of acetylene compares well with the value of 40 kJ/mol reported in the literature for acetylene hydrogenation over a Pd foil [7].

The proposed reaction mechanism based on a single site can adequately represent the experimental data, as shown in Figures 1–6. Neurock and coll. [13] could describe independent experimental data correctly by the same reaction mechanism as the one proposed in this study using also a single site. Bos et al. [6] could describe their experimental data by very similar rate equations as used here, in the absence of carbon monoxide. However, the addition of carbon monoxide to the feed resulted in a change of the ethane selectivity, which cannot be explained by the above model. As mentioned in the introduction, it could well be that the second site is related to the presence of carbon monoxide on the surface, which is known to have a strong electronic effect and can cause surface reconstruction.

Numerous other studies [2–4,8,9] and notably the model proposed by Borodziński and Cybulski [4] are based on two distinct active sites. We therefore performed a regression analysis of the model proposed by Borodziński and Cybulski to our data set (we used the equations given in the Appendix A of their article, which differ from the text) [4]. The regression analysis showed that the parameters associated with the second site were not statistically significant and the model was reduced to a single site model. The arguments of Borodziński and Cybulski [4] for a two-site model was that they observed a lack of the effect of the partial pressure of ethylene on r_{C2H2}/P_{H2}, indicating that the coverage of active sites by ethylene is negligible, while they observed an effect of the partial pressure of ethylene on r_{C2H4}/P_{H2}. The latter effect is also observed in this study and corresponds to the data in Figure 4. The rate of ethane production increases with the partial pressure of ethylene (the partial pressure of ethylene is proportional to the molar flow of ethylene in Figure 4, where the partial pressure of acetylene and the partial pressure of hydrogen are constant). Contrary to Borodziński and Cybulski, we do observe an effect of the partial pressure of ethylene on r_{C2H2}/P_{H2}. This is demonstrated in Figure 7 where we trace both the data of Borodziński and Cybulski [4] and our data for r_{C2H2}/P_{H2} as a function of the partial pressure of ethylene. Despite the rather different conditions of both studies, the r_{C2H2}/P_{H2} values are very similar. However, whereas r_{C2H2}/P_{H2} is independent of the ethylene partial pressure in the case of the data of Borodziński and Cybulski, in our case, r_{C2H2}/P_{H2} decreases with increasing ethylene partial pressure. A single site mechanism is thus validated for the conditions used in this study, or at least if two sites are present on the catalyst, the adsorption behavior of ethylene and acetylene on both sites is not different enough to be distinguished from the experiments under these conditions. Borodziński and Cybulski [4] attributed the different active sites on palladium to the deposit of carbonaceous species, creating pockets of different sizes that induce size dependent reactivity. In our study, the amount of carbonaceous species might be rather low, due to the short time on stream of the catalyst and the much higher stoichiometric ratio of hydrogen to acetylene. This is also consistent with the reaction order of hydrogen greater than 1, which was explained by Molero et al. [7] by excess hydrogen that removes adsorbed carbonaceous species from the catalyst surface and creates free palladium sites.

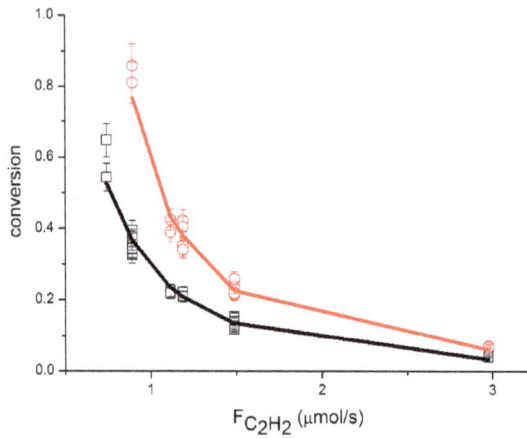

Figure 1. Evolution of acetylene conversion as a function of the acetylene inlet molar flow. Circles: data at 62 °C, squares: data at 51 °C.

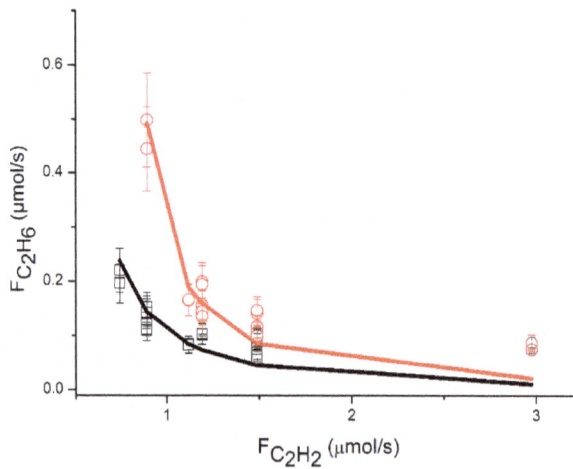

Figure 2. Ethane outlet molar flow as a function of the acetylene inlet molar flow. Circles: data at 62 °C, squares: data at 51 °C.

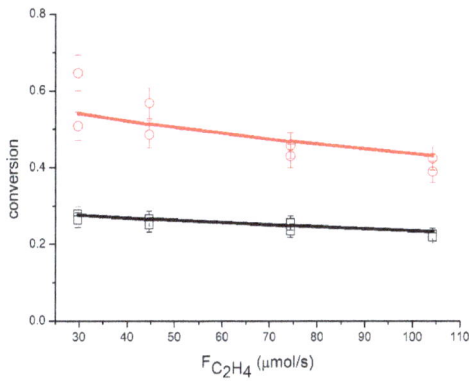

Figure 3. Evolution of acetylene conversion as a function of the ethylene inlet molar flow. Circles: data at 62 °C, squares: data at 51 °C.

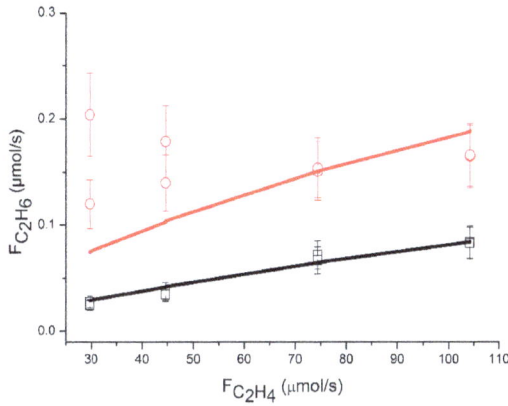

Figure 4. Ethane outlet molar flow as a function of the ethylene inlet molar flow. Circles: data at 62 °C, squares: data at 51 °C.

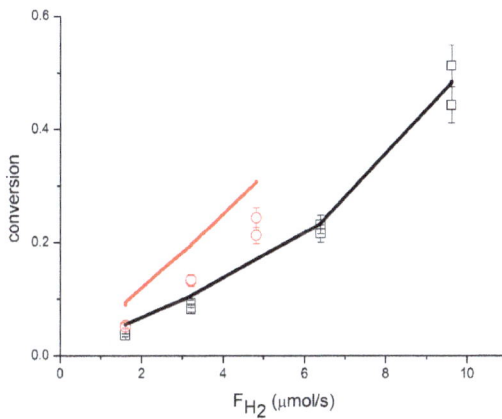

Figure 5. Evolution of acetylene conversion as a function of the hydrogen inlet molar flow. Circles: data at 62 °C, squares: data at 51 °C.

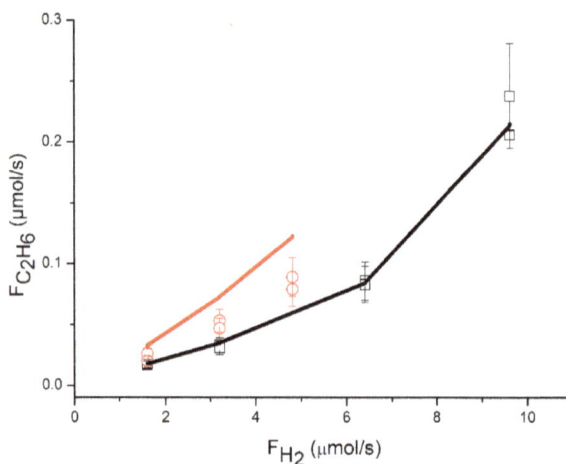

Figure 6. Ethane outlet molar flow as a function of the hydrogen inlet molar flow. Circles: data at 62 °C, squares: data at 51 °C.

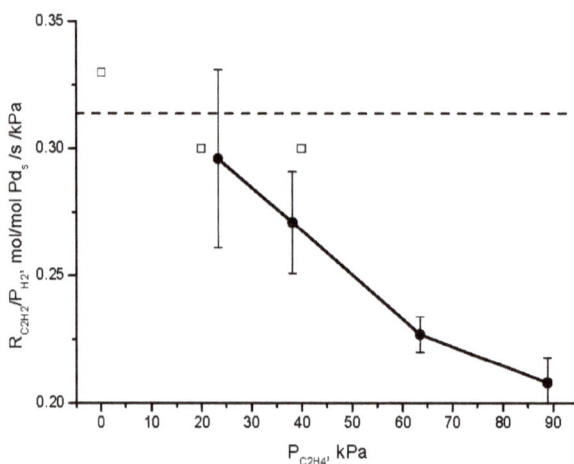

Figure 7. Plots of r_{C2H2}/P_{H2} vs. ethylene partial pressure. Open symbols: data from Borodziński and Cybulski [4]: T = 70 °C, P_{C2H2} = 0.02 kPa, P_{H2} = 0.64 kPa. Closed symbols: data from this study: T = 62 °C, P_{C2H2} = 0.95 kPa, P_{H2} = 5.5 kPa. The dotted horizontal line indicates the extrapolated average values of r_{C2H2}/P_{H2} for the data from Borodziński and Cybulski [4].

3. Materials and Methods

3.1. Experimental set-up and testing

Kinetic experiments for acetylene hydrogenation were carried out in an integral fixed bed reactor over 0.05 wt. % Pd/α-Al$_2$O$_3$ catalysts under typical tail-end conditions. During the experiment, 0.1 g of catalyst sieved between 100 and 200 μm was used, diluted with 0.5 g of the α-Al$_2$O$_3$ support, crushed and sieved identically, to improve the isothermicity of the catalytic bed. The fixed-bed reactor consisted of a quartz tube with an inner diameter of 4 mm, inserted in a cylindrical oven. A thermocouple was inserted into the catalyst bed. The reaction was carried out at 51 and 62 °C. A flow consisting of different ratios of acetylene, ethylene, hydrogen, and argon, using mass flow controllers, at a total flow rate of 200 mL/min was used and the reactor was operated at 1.26 bar. The experiment at standard conditions was performed regularly to check the stability of the catalyst and the data repeatability. The reactor was loaded twice with a fresh catalyst sample and the experiments were conducted twice. Before the experiments the catalyst was reduced under a flow of 100 mL/min of 50% hydrogen in argon at 150 °C for 2 h. The catalyst activity and selectivity were found to be stable during the kinetic runs.

Online analysis was done by passing the outlet gas flow through a small volume infrared gas cell connected to a FTIR spectrometer. Quantitative FTIR analysis of acetylene, ethylene and ethane was carried out. No C4 products were observed. The carbon balance was closed within 5%. Experimental conditions are given in Table 6. By using appropriate criteria, the kinetic measurements were shown to be free from heat and mass transfer limitations [23]. This can be easily verified using the tool developed by Eurokin [24]. The maximum observed rate of acetylene consumption of 9 mmol/kg$_{cat}$/s was used to check the criteria given in Table 7. As shown in Table 7 all criteria were met.

Table 6. Experimental conditions.

Variable	Values
T (°C)	51, 62
$F_{C_2H_2}$ (µmol/s)	0.7–3
F_{H_2} (µmol/s)	1.6–10
$F_{C_2H_4}$ (µmol/s)	29–105
F_{Ar} (µmol/s)	34–115
P (bar)	1.26
W_{cat} (g)	0.1
F_{tot} (µmol/s)	149

Table 7. Criteria to assess the conditions for the absence of mass and heat transfer limitations.

Physical Phenomenon	Criteria
plug flow (axial)	$h/d_p = 193 > 6$
plug flow (radial)	$d_R/d_p = 40 > 10$
degree of dilution (vol./vol.)	$0.89 < 0.97$
external mass transfer (Carberry number)	$3.3 \times 10^{-4} < 0.05$
internal mass transfer (Weisz-Prater)	$1.7 \times 10^{-2} < 0.33$
external heat transfer (film)	$0.14\,K < 1.1\,K$
external heat transfer (radial)	$0.17\,K < 1.1\,K$
internal heat transfer	$3 \times 10^{-3}\,K < 1.1\,K$

3.2. Catalyst

0.05 wt. % Pd supported on α-alumina was used as a catalyst. It was prepared by impregnation of the alumina by an aqueous palladium nitrate solution. The material was then dried at 120 °C under ambient air and calcined at 425 °C for 2 h. The detailed protocol has been previously described together with extensive XRD and HRTEM characterization [25,26]. The alumina support has a porous volume of 0.54 mL/g and a BET surface area of 10 m²/g. Electron Transmission Microscopy analysis showed a mean particle diameter of 2.4 ± 0.7 nm, corresponding to a dispersion of ca. 40%. From this and the wt. % of Pd the number of surface palladium atoms was calculated as $N_S = 2 \times 10^{-3}$ mol/kg. The catalyst was sieved between 100 and 200 µm to improve the heat transfer during the kinetic experiments. Runs under standard conditions using the original catalyst spheres in a single pellet reactor configuration showed similar performance as the crushed sample.

3.3. Modeling

An integral reactor operation was used and the rate equations have been integrated numerically using the ODEPACK library [27]. A one-dimensional homogeneous reactor model has been used, in agreement with the absence of mass and heat transfer limitations. A non-linear least-square multi-response regression analysis has been performed by a Levenberg–Marquardt minimization algorithm [28,29]. The fractional acetylene conversion and the molar exit flow of ethane (μmol/s) have been used for the objective function. No weighing factor has been applied. After regression analysis several statistical tests were performed, including the *t*-test, the 95% confidence intervals of the parameter estimates, the F-value, and binary correlation coefficients.

4. Conclusions

The kinetics of the selective hydrogenation of acetylene over a 0.05 wt. % Pd/α-Al$_2$O$_3$ catalyst was studied under intrinsic kinetic conditions. The operating conditions were chosen according to the tail-end process. Analysis of the experimental reaction orders gave good insight into the reaction mechanism. It allowed us to estimate the mean surface coverages of hydrogen, acetylene, and ethylene. The proposed reaction mechanism consists of a series of sequential hydrogen additions according to the Horiuti–Polanyi mechanism. The derived rate equations are based on the addition of the second hydrogen atom as rate-determining step for both acetylene and ethylene hydrogenation. Accurate parameter estimation was only possible after fixing the values of the equilibrium adsorption constant of ethylene. The relatively simple Langmuir–Hinshelwood-type rate-equations describe the experimental data adequately. It is based on physically meaningful parameters. The reaction mechanism involves only one active site, which was carefully verified.

Author Contributions: Conceptualization, J.-M.S., Y.S. and A.C.; methodology, J.-M.S., Y.S.; software, Y.S.; validation, Y.S., J.-M.S.; formal analysis, C.U., Y.S.; investigation, C.U.; resources, A.C.; data curation, Y.S.; writing—original draft preparation, C.U.; writing—review and editing, J.-M.S., Y.S. and A.C.; visualization, C.U.; supervision, J.-M.S., Y.S. and A.C.; project administration, J.-M.S.; funding acquisition, J.-M.S., A.C.

Funding: This research received no external funding.

Conflicts of Interest: The authors declare no conflict of interest.

Appendix A

Nomenclature

A_S	Surface area of Pd	1.26×10^4 m^2/mol
d_p	Catalyst particle diameter	m
d_R	Reactor diameter	m
E	Activation energy	kJ/mol
F	Molar flowrate	μmol s^{-1}
h	Bed height	m
k^0	Pre-exponential factor	(mol/mol Pd$_s$/s) or s^{-1}
k	Rate constant	(mol/mol Pd$_s$/s)
K	Adsorption equilibrium constant	Pa^{-1}
M_w	Molecular weight	kg/mol
N_S	Number of surface Pd atoms	mol Pd$_s$/kg
P	(Partial) pressure	bar, Pa or kPa
R	Gas constant	J/mol/K
r_i or R_i	Reaction rate for species i or step i	mol/kg/s
S_{C2H6}	Selectivity of ethane	-
T	Temperature	°C, K
W_{cat}	Catalyst mass	g
X	Conversion	%mol

Greek letters and symbols

*	Active site
α_i	Acetylene reaction order for reaction i
β_i	Hydrogen reaction order for reaction i
γ_i	Ethylene reaction order for reaction i
σ	Stoichiometric number
σ_s	Sticking coefficient
θ_i	Fractional surface coverage of species i

References

1. Derrien, M. Selective hydrogenation applied to refining of petrochemical raw materials produced by steam cracking. *Stud. Surf. Catal.* **1986**, *27*, 613–666.
2. Borodzinski, A.; Bond, G.C. Selective Hydrogenation of Ethyne in Ethene-Rich Streams on Palladium Catalysts. Part 1. Effect of Changes to the Catalyst During Reaction. *Catal. Rev.* **2006**, *48*, 91–144. [CrossRef]
3. Bos, A.N.R.; Westerterp, K.R. Mechanism and kinetics of the selective hydrogenation of ethyne and ethane. *Chem. Eng. Process.* **1993**, *32*, 1–7. [CrossRef]
4. Borodzinski, A.; Cybulski, A. The kinetic model of hydrogenation of acetylene-ethylene mixtures over palladium surface covered by carbonaceous deposits. *Appl. Catal. A Gen.* **2000**, *198*, 51–66. [CrossRef]
5. Cider, L.; Schöön, N.-H. Kinetics of cross-desorption of carbon monoxide by the influence of ethyne over palladium/alumina. *Appl. Catal.* **1991**, *68*, 207–216. [CrossRef]
6. Bos, A.N.R.; Bootsma, E.S.; Foeth, F.; Sleyster, H.W.J.; Westerterp, K.R. A kinetic study of the hydrogenation of ethyne and ethene on a commercial Pd/Al$_2$O$_3$ catalyst. *Chem. Eng. Process.* **1993**, *32*, 53–63. [CrossRef]
7. Molero, H.; Bartlett, B.F.; Tysoe, W.T. The Hydrogenation of Acetylene Catalyzed by Palladium: Hydrogen Pressure Dependence. *J. Catal.* **1999**, *181*, 49–56. [CrossRef]
8. Borodzinski, A.; Bond, G.C. Selective Hydrogenation of Ethyne in Ethene-Rich Streams on Palladium Catalysts, Part 2: Steady-State Kinetics and Effects of Palladium Particle Size, Carbon Monoxide, and Promoters. *Catal. Rev.* **2008**, *50*, 379–469. [CrossRef]
9. Pachulski, A.; Schödel, R.; Claus, P. Kinetics and reactor modeling of a Pd-Ag/Al$_2$O$_3$ catalyst during selective hydrogenation of ethyne. *Appl. Catal. A Gen.* **2012**, *445–446*, 107–120. [CrossRef]
10. Aduriz, H.R.; Bodnariuk, P.; Dennehy, M.; Gigola, C.E. Activity and Selectivity of Pd/Alpha-Al$_2$O$_3$ for Ethyne Hydrogenation in a large Excess of Ethene and Hydrogen. *Appl. Catal.* **1990**, *58*, 227–239. [CrossRef]
11. McGowm, W.T.; Kemball, C.; Whan, D.A. Hydrogenation of Acetylene in Excess Ethylene on a Alumina-Supported Palladium Catalyst at Atmospheric Pressure in a Spinning Basket Reactor. *J. Catal.* **1978**, *51*, 173–184. [CrossRef]
12. Vincent, M.J.; Gonzalez, R.D. A Langmuir–Hinshelwood model for a hydrogen transfer mechanism in the selective hydrogenation of acetylene over a Pd/γ-Al$_2$O$_3$ catalyst prepared by the sol-gel method. *Appl. Catal. A Gen.* **2001**, *217*, 143–156. [CrossRef]
13. Mei, D.; Sheth, P.A.; Neurock, M.; Smith, C.M. First-principles-based kinetic Monte Carlo simulation of the selective hydrogenation of acetylene over Pd(111). *J. Catal.* **2006**, *242*, 1–15. [CrossRef]
14. Bond, G.C.; Wells, P.B. The hydrogenation of acetylene: II. The reaction of acetylene with hydrogen catalyzed by alumina-supported palladium. *J. Catal.* **1965**, *5*, 65–67. [CrossRef]
15. Saeys, M.; Reyniers, M.-F.; Thybaut, J.W.; Neurock, M.; Marin, G.B. First-principles based kinetic model for the hydrogenation of toluene. *J Catal.* **2005**, *236*, 129–138. [CrossRef]
16. Neurock, M.; van Santen, R.A. A First Principles Analysis of C–H Bond Formation in Ethylene Hydrogenation. *J. Phys. Chem. B* **2000**, *104*, 11127–11145. [CrossRef]
17. Langmuir, I. The constitution and fundamental properties of solids and liquids. Part 1. Solids. *J. Am. Chem. Soc.* **1916**, *38*, 2221–2295. [CrossRef]
18. Conrad, H.; Ertl, G.; Latta, E.E. Adsorption of hydrogen on palladium single crystal surfaces. *Surf. Sci.* **1974**, *41*, 435–446. [CrossRef]
19. Michalakab, W.D.; Millerab, J.B.; Alfonsoa, D.; Gellmana, A.J. Uptake, transport, and release of hydrogen from Pd(100). *Surf. Sci.* **2012**, *606*, 146–155. [CrossRef]

20. Chou, P.; Vannice, M.A. Calorimetric heat of adsorption measurements on palladium: I. Influence of crystallite size and support on hydrogen adsorption. *J. Catal.* **1987**, *104*, 1–16. [CrossRef]
21. Tysoe, W.T.; Nyberg, G.L.; Lambert, R.M. Structural, kinetic, and reactive properties of the palladium(111)-ethylene system. *J. Phys. Chem.* **1984**, *88*, 1960–1963. [CrossRef]
22. Vattuone, L.; Yeo, Y.Y.; Kose, R.; King, D.A. Energetics and kinetics of the interaction of acetylene and ethylene with Pd{100} and Ni{100}. *Surf. Sci.* **2000**, *447*, 1–14. [CrossRef]
23. Schuurman, Y. Aspects of kinetic modeling of fixed bed reactors. *Catal. Today* **2008**, *138*, 15–20. [CrossRef]
24. Fixed-Bed Web Tool at the Eurokin Website. Available online: http://www.eurokin.org/ (accessed on 19 December 2018).
25. Ramos-Fernandez, M.; Normand, L.; Sorbier, L. Structural and Morphological Characterization of Alumina Supported Pd Nanoparticles Obtained by Colloidal Synthesis. *Oil Gas Sci. Technol.* **2007**, *62*, 101–113. [CrossRef]
26. Didillon, B.; Merlen, E.; Pagès, T.; Uzio, D. *Studies in Surface Science and Catalysis*; Delmon, B., Ed.; Elsevier: Amsterdam, The Netherlands, 1998; Volume 118, pp. 41–54.
27. Hindmarsh, A.C. *ODEPACK. A Systematized Collection of ODE Solvers*; Elsevier: Amsterdam, The Netherlands, 1983.
28. Levenberg, K. A Method for the Solution of Certain Non-Linear Problems in Least Squares. *Q. Appl. Math.* **1944**, *11*, 164–168. [CrossRef]
29. Marquardt, D.W. An algorithm for least-squares estimation of nonlinear parameters. *J. Soc. Ind. Appl. Math.* **1963**, *11*, 431–441. [CrossRef]

catalysts

MDPI

Article

Transient Kinetic Experiments within the High Conversion Domain: The Case of Ammonia Decomposition

Yixiao Wang [1], M. Ross Kunz [2], Skyler Siebers [3], Harry Rollins [1], John Gleaves [4], Gregory Yablonsky [4] and Rebecca Fushimi [1,*]

[1]   Department of Biological and Chemical Processing, Idaho National Laboratory, Idaho Falls, ID 83415, USA; yixiao.wang@inl.gov (Y.W.); harry.rollins@inl.gov (H.R.)
[2]   High Performance Computing and Data Analytics, Idaho National Laboratory, Idaho Falls, ID 83415, USA; ross.kunz@inl.gov
[3]   Department of Chemistry, Idaho State University, Pocatello, ID 83209, USA; siebskyl@isu.edu
[4]   Department of Energy, Environmental and Chemical Engineering, Washington University in Saint Louis, Saint Louis, MO 63130, USA; klatu_00@che.wustl.edu (J.G.); gregoryyablonsky@gmail.com (G.Y.)
*   Correspondence: rebecca.fushimi@inl.gov

Received: 24 December 2018; Accepted: 16 January 2019; Published: 19 January 2019

Abstract: In the development of catalytic materials, a set of standard conditions is needed where the kinetic performance of many samples can be compared. This can be challenging when a sample set covers a broad range of activity. Precise kinetic characterization requires uniformity in the gas and catalyst bed composition. This limits the range of convecting devices to low conversion (generally <20%). While steady-state kinetics offer a snapshot of conversion, yield and apparent rates of the slow reaction steps, transient techniques offer much greater detail of rate processes and hence more information as to why certain catalyst compositions offer better performance. In this work, transient experiments in two transport regimes are compared: an advecting differential plug flow reactor (PFR) and a pure-diffusion temporal analysis of products (TAP) reactor. The decomposition of ammonia was used as a model reaction to test three simple materials: polycrystalline iron, cobalt and a bimetallic preparation of the two. These materials presented a wide range of activity and it was not possible to capture transient information in the advecting device for all samples at the same conditions while ensuring uniformity. We push the boundary for the theoretical estimates of uniformity in the TAP device and find reliable kinetic measurement up to 90% conversion. However, what is more advantageous from this technique is the ability to observe the time-dependence of the reaction rate rather than just singular points of conversion and yield. For example, on the iron sample we observed reversible adsorption of ammonia and on cobalt materials we identify two routes for hydrogen production. From the time-dependence of reactants and product, the dynamic accumulation was calculated. This was used to understand the atomic distribution of H and N species regulated by the surface of different materials. When ammonia was pulsed at 550 °C, the surface hydrogen/nitrogen, (H/N), ratios that evolved for Fe, CoFe and Co were 2.4, 0.25 and 0.3 respectively. This indicates that iron will store a mixture of hydrogenated species while materials with cobalt will predominantly store NH and N. While much is already known about iron, cobalt and ammonia decomposition, the goal of this work was to demonstrate new tools for comparing materials over a wider window of conversion and with much greater kinetic detail. As such, this provides an approach for detailed kinetic discrimination of more complex industrial samples beyond conversion and yield.

Keywords: transient kinetics; TAP reactor; temporal analysis of products; ammonia decomposition

1. Introduction

An indispensable need in the design and development of catalytic materials is establishing a robust, broad-reaching yet precise basis for comparison of catalyst properties. A strong emphasis is based on unraveling the structure–activity relationship across singular points of fixed, static or steady-state conditions. Structural characterization techniques for properties of both the bulk and surface are numerous, widely used and well-developed: X-ray diffraction (XRD), X-ray photoelectron spectroscopy (XPS), Brunauer–Emmett–Teller (BET), infrared (IR), Raman, transmission electron microscopy (TEM), scanning tunneling microscopy (STM), atomic force microscopy (AFM), etc. There is an increasing drive to conduct structural characterization experiments in more 'kinetically relevant' environments, i.e., *in situ* and *operando* spectroscopy. The diversity of kinetic tools, however, is more limited and kinetic characterization primarily relies upon observation of reaction rates as a function of temperature and feed composition. There are two mainstream kinetic devices: the plug flow reactor (PFR) and the continuous stirred tank reactor (CSTR) (or differential PFR). These are typically operated at steady-state and yield useful global reaction conditions more similar to the industrial use setting.

Transient experiments that induce a change in temperature, pressure or concentration are more challenging to conduct and interpret but can offer greater detail of the fundamental kinetics that render the global performance differently from one material to the next [1–7]. Transient kinetic data can hence be a powerful tool for standardizing the comparison of active materials. One of the main requirements for non-steady state characterization is uniformity of chemical composition (both gas and solid) in the catalyst bed. The chemical composition in a PFR is non-uniform which limits its use in a transient mode. The CSTR offers uniformity only at low conversion (<15–20%); beyond this non-uniformity is proportional to conversion (*viz.* Figure 20 in Shekhtman et al. [8]).

The temporal analysis of products (TAP) reactor [8–10] is a third type of kinetic device that investigates materials at conditions far from equilibrium and has not been as widely adopted as PFR and CSTR. In contrast to the PFR and CSTR, which are reactors with convective transport, the TAP device is an example of a pure-diffusion reactor. In TAP, diffusion plays the role of an efficient 'impeller'. In contrast to the CSTR, conversion in TAP is proportional to the difference between inlet and outlet diffusional flux. These fluxes are proportional to the corresponding concentration *gradients*, not concentrations. Consequently, the thin-zone TAP reactor does not suffer from chemical non-uniformity even at high conversions (up to 75–80%) [8]. While the PFR and CSTR offer coarse kinetic screening near industrial conditions, the TAP approach is focused on capturing the precise kinetic properties of the catalyst at a well-defined catalyst state. By stripping away the complexity of the process these experiments can be used to obtain a more detailed kinetic characterization of materials.

For example, the Y-Procedure inverse-diffusion analysis method [11,12] enables time-dependent calculation of the reaction rate, R(t) ($mol/cm^2/s$ or $mol/cm^3/s$). The ability to observe the time-dependence of the reaction rate with millisecond time-resolution (well-matched for typical catalytic processes) is a unique feature of the TAP method. Using the time dependence, the integral of the rate can be calculated to determine the uptake (storage) of reactants or the release of gaseous products from the surface, U(t) (mol/cm^2). Temporal uptake/release information is extremely useful for estimating the surface composition of the catalyst.

In this paper we will compare the kinetic information available at steady-state with transient differential PFR and TAP pulse response experiments for the decomposition of ammonia. Following its original conception [12] the Y-Procedure has only been implemented in a limited number of experimental works [11,13–15]. Here we demonstrate this tool using the ammonia decomposition reaction over polycrystalline iron, cobalt and a bimetallic preparation of the two. Using these simple materials, our goal is to demonstrate the advanced information that can be gained from the time-dependence of the rate and the temporal change of the atomic surface composition.

Ammonia can play a key role in energy storage for both mobile and stationary applications [16,17]. Its high hydrogen storage capacity meets the targets set by the US Department of Energy and can be used as a CO_x-free H_2 source for fuel cells [18]. At the same time, catalytic decomposition of ammonia is the reverse of the Haber–Bosch process which has been studied for the past 100–150 years [19]. Generally, with the principle of microscopic reversibility, an understanding of catalyst properties that control the decomposition reaction informs the synthesis reaction as well. The ideal synthesis catalyst is, however, not the ideal decomposition catalyst as the two processes are carried out under different reaction conditions that call for very different optimal binding energies [20]. The decomposition of ammonia has been previously studied using the TAP technique where Ru and Ir materials were compared and a difference in reaction mechanism was suggested based on the time-dependence of exit flux [21]. Cobalt and iron were studied in conventional flow systems and synergistic effects for decomposition were detected in certain bimetallic compositions [22] similar to results found here.

The characterization of materials under working conditions is complicated by the confluence of reaction mechanism, surface complexity, and gas transport. The decomposition of ammonia is not an overwhelmingly complex multistep reaction (one reactant, two products) and we have used this reaction to compare three fairly simple materials, iron, cobalt and their bimetallic combination for the purpose of demonstrating methods. As a rule, from steady-state experiments we observe a singular rate at a singular gas concentration. In flow experiments we conduct a step-transient in gas concentration and observe the evolution of the rate at the time-scale of the device (>1 s). The pulse response experiment, with millisecond time-resolution, forces a gas concentration dynamic and the time-dependence of the rate is presented using the Y-Procedure method. We compare the kinetic information obtained in these two transient devices that operate in different transport regimes and on different time scales. The details of properly separating the transport and kinetic time-dependencies are discussed.

These results demonstrate the advantages of the TAP non-steady-state characterization method for discriminating subtleties of kinetic function in different materials *at high conversion*. In particular, we observe a reversibility in the ammonia adsorption process on the iron material and two distinct kinetic routes for hydrogen generation on cobalt and the bimetallic sample. The time-dependence of reaction rate and surface uptake/release provides a unique *kinetic fingerprint* for distinguishing one catalyst from the next. It enables a data-rich standardization for the comparison of materials based on their *intrinsic* chemical activity in terms of the ability to store, transform and release molecules from the gas phase. We find from observing the dynamic accumulation of H and N elements that iron supports hydrogenated surface species (NH_3 and NH_2) while materials with cobalt favor storage of NH and N species. From these initial results on simple materials the same transient methods in the future should be more interpretable on systems of greater complexity, for example to compare a collection of industrial catalyst with incremental changes in metal composition.

2. Results and Discussion

Three samples, referred to as Fe, CoFe and Co are the focus of this investigation. Both Fe and Co were high-purity polycrystalline materials. The CoFe material was prepared by wet impregnation of a cobalt precursor onto the polycrystalline iron; described in more detail in Section 3.1. In the first part of the Results section we present the conventional fixed-state ex situ characterization of structure and composition of prepared materials. Next, we compare kinetic measurements in a differential PFR at steady state and in response to a step-transient. Section 2.3 then introduces the pulse response characterization. First, we compare values calculated using conventional integral methods to the flow reactor results. Next, to study the time-dependence intact we present a result showing the importance of the Y-procedure method in accurately separating transport and kinetic effects in the pulse response experimental. The time-dependence of the reaction rate and uptake are compared for ammonia, nitrogen and hydrogen over our three samples. Finally, dynamic accumulation data of H and N species is presented to gain more insight into the composition of stored surface species.

2.1. Catalyst Structure and Chemical Composition

Measured by inductively coupled plasma optical emission spectrometry (ICP-OES), the mass loading of Co on the bimetallic sample was 3.16 wt.%. The BET surface areas of Fe and CoFe were both 4.5 m^2/g and 1 m^2/g was measured for Co. Figure 1A shows the TEM micrographs of the CoFe sample that was reduced in 10% H$_2$/Ar flow for 12 hours. The image indicates a general spherical morphology and an average crystallite size of 94 nm (Figure 1B). In order to investigate the chemical element distribution in the CoFe sample, elemental mapping was used. The EDS elemental mapping of Co and Fe (Figure 1C,D) showed a uniform distribution of Co throughout the whole of the iron surface. Similar homogeneity was observed in CoFe alloys supported on carbon nanotubes [22]. Following high-temperature reduction, a significant amount of cobalt in the bimetallic preparation remained as a dispersed surface alloy as would be expected based on surface free energy [23]. Similar bimetallic preparations have demonstrated a high degree of alloying [24,25]. Figure S1 in the Supplementary Information shows the pre- and post-reaction XRD patters of Fe, CoFe and Co. Following reaction, nitride phases were observed on all materials.

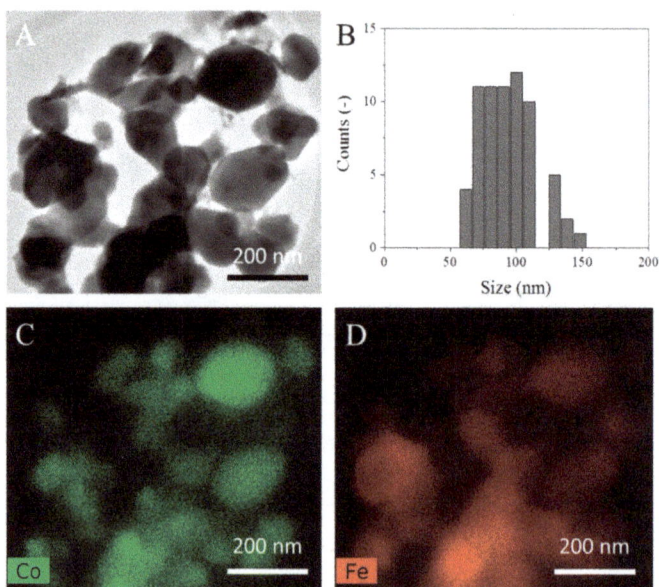

Figure 1. (**A**) Representative transmission electron microscope (TEM) images of the bimetallic catalyst, (**B**) particle size distribution, (**C**) Co mapping, (**D**) Fe mapping.

2.2. Differential Plug Flow Reactor (PFR): Steady-State and Step-Transient

Steady-state ammonia conversion is compared for three materials in the first series of Figure 2 (blue triangles). We find that the conversion of cobalt is around 10 times greater than that of pure iron. This is in agreement with the findings of Schlögl and coworkers at the same reaction temperature where cobalt supported on carbon nanotubes demonstrated conversion was 2.7 times that of iron on the same support [22]. In our experiment, the addition of cobalt to iron in the wet-impregnation method created a surface where both elements were accessible, Figure 1. Synergistic kinetic effects are indicated at steady-state with the bimetallic sample presenting the highest NH$_3$ conversion. Rather than a simple mixing of the properties of monometallic iron and cobalt, a 22-fold increase in conversion is found when only small amounts of cobalt are added to iron, Figure 2. Schlögl and coworkers similarly found a synergistic enhancement in the rate of ammonia decomposition for certain CoFe alloy compositions [22].

They offered the reasoning that since Co has one more d-band electron than Fe, the formation of a surface Co–Fe bond will transfer the electron from Co to Fe. The highest reaction barriers of Fe can be attributed to the strong binding energy of N with Fe. The modified electronic properties of Fe are expected to decrease the activation barrier and assist the desorption of N atoms from the surface. Moreover, it is interesting to note that Fe and Co fall on opposite sides of the volcano plot for ammonia synthesis [26]. From a synthesis point of view, iron is 'too reactive', binding NH_x too strongly while cobalt is 'too noble' and does not strongly bind synthesis intermediates. This corresponds well with our experimental observations from a decomposition perspective and suggests the improvements offered from the CoFe material arise from some 'interpolation' of binding strength.

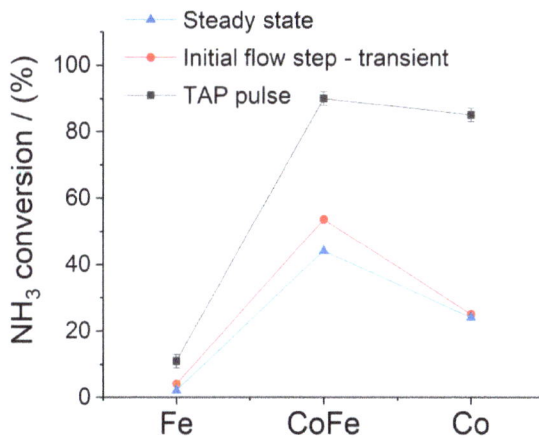

Figure 2. Ammonia % conversion in decomposition experiments at 550 °C in steady-state flow, the onset of the flow step-transient and temporal analysis of products (TAP) pulse response experiments for Fe, CoFe and Co.

This steady-state snapshot of kinetic performance is helpful to screen materials but does not offer much information as to why these materials perform differently. Under steady-state conditions the conversion equals the non-uniformity in chemical composition of the catalyst bed. For CoFe this was 44%; cobalt was 24%. In this case, rather than compare materials based on conversion at the same conditions, it makes sense to compare the conditions required to achieve the same conversion (<20%). Generally, the next step would be to compare the reaction rate at different ammonia concentrations, space velocities and temperatures to extract an apparent rate constant and activation energy. One could lower the temperature but then the conversion over iron would be unmeasurable.

At steady-state we can only compare singular points of performance on different materials. In an oxidation experiment we would also have gas phase selectivity as additional data to compare materials. However, for this simple decomposition N_2 and H_2 are the only products observed at steady-state. Often in the approach to steady-state, an induction period in performance is observed. Such a transient regime offers more data as shown in Figure 3. In this experiment the catalyst was brought to reaction temperature with an inert gas flow before switching to the ammonia feed at time zero.

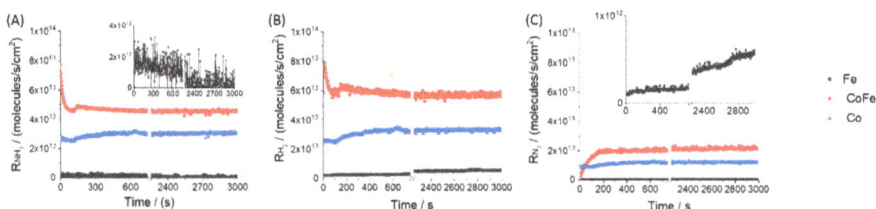

Figure 3. Flow regime step-transient data during ammonia decomposition at 550 °C. Ammonia feed is switched on at time = 0 s. Gas phase reaction rate data for (**A**) ammonia, (**B**) hydrogen and (**C**) nitrogen as a function of time. Note the break in the time scale to distinguish the transient regime from the approach to steady-state. The inset shows trends in performance on the iron sample which was generally an order or magnitude lower.

The CoFe sample shows the clearest induction period with large changes in ammonia conversion and product formation observed before steady-state performance is achieved. The initial conversion detected following the step-transient is compared to steady-state conversion in Figure 2. For Co and CoFe, the ammonia conversion rate generally matches the hydrogen production rate over the entire experiment. This indicates that there is little accumulation of H on the catalyst surface. For iron, the ammonia conversion rate decreases during the experiment while hydrogen production increases slightly before assuming a constant value. For the CoFe sample at the start of the feed switch we observe that all ammonia converted is retained by the sample. The rate of gas phase nitrogen production slowly increases as the surface accumulates N but even at the end of the experiment (nearly 1 h), the nitrogen rate is still slightly less than the conversion rates, indicating a true steady-state has not been achieved. Following the flow experiments, XRD analysis confirmed the accumulation of N with nitridation peaks, Supplementary Information Figure S1; Fe_4N, Fe_3N, and Fe_2N over the Fe and CoFe samples, in agreement with Feyen et al. [27]; CoN and Co_2N were observed over the Co sample.

The break-in period can be interpreted as resulting from dynamic surface and bulk processes that change the chemical composition of the material. In the flow mode the CoFe sample had the highest NH_3 conversion (44%) and hence non-uniformity in catalyst bed composition can also be the key reason for the observed dynamics. As mentioned previously, for an advecting device, non-uniformity is equal to conversion. Furthermore, the final steady-state performance can be greatly influenced by the history of these induction processes. More colloquially, how you approach steady-state can influence where you end up. We might have compared these materials at lower temperature or higher space velocity but considering the disparity in the performance of the iron sample (conversion would not be detectable), it is difficult to find a uniform basis for comparison of all three samples at the same conditions. The non-uniformity of gas and solid chemical composition is the main source of uncertainly in kinetic parameters obtained from advecting devices.

2.3. Temporal Analysis of Products (TAP) Pulse Response

While TAP experiments can be challenging in terms of instrumentation and analysis of data, some advantages of the technique are (a) well-defined transport that can be separated from chemical reaction kinetics, (b) uniformity of the chemical composition at high conversion, (c) an insignificant change in the material during the experiment. In any steady-state experiment the observed kinetics reflect the slow reaction steps. Dynamics are often observed in flow experiments (*viz.* Figure 3) and can certainly be forced (e.g., frequency response) but generally the time-scale for reactor transport and product detection does not enable sufficient resolution to observe fast kinetic processes. A key difference between atmospheric flow and low-pressure TAP experiments is surface coverage. One might draw a comparison with the initial data points of the flow experiment; however, all processes that commence when 1.1 mol/s are introduced in the flow mode cannot be resolved as finely as the 10 nanomol pulse. In other words, the material has already significantly changed by the time the first data point in the flow mode can be collected. Thus, it is the insignificant perturbation of the material in a TAP experiment that enables a uniform basis of comparison of *intrinsic* kinetic properties of a collection of catalysts.

2.3.1. Integral Analysis of Transient Data

Ammonia Conversion

Each catalyst was exposed to a long series of NH_3 pulses to incrementally titrate the surface. In fact, even with 2000 pulses (6.8×10^3 nmol NH_3 total) there was still significant conversion of ammonia; multipulse series are presented in the Supplementary Information, Figure S2. In similar TAP experiments of ammonia decomposition on ruthenium catalysts, Garcia-Garcia et al. observed constant conversion over even longer pulse series [21]. Over the course of the pulsed experiment ammonia conversion moderately declined but comparison of height normalized pulse intensities at select pulse intervals indicated that the shape of the pulse response does not change significantly, Figure S3 in the Supplementary Information. This indicates that while active sites were incrementally consumed, the overall kinetic properties of the available sites did not change. Conversion data are calculated in the conventional fashion using integral quantities of the pulse response curve (see Supplementary Information).

Conversion under TAP conditions is compared to flow reactor conversion in Figure 2. The ammonia conversion over iron was very similar under both TAP and flow conditions. The cobalt materials showed significantly higher conversion (90 ± 2 and $85 \pm 2\%$ for CoFe and Co, respectively) in the low-pressure pulse response experiment. This occurs due to the unique experimental configuration of the pure-diffusion reactor. While plug flow reactors can suffer from bed-bypassing, hot-spots and inhomogeneity in the gas phase the experimental conditions of the TAP experiment, as described in the work of Schuurman [28], ensure that each active site of the catalyst surface receives roughly 1000 collisions from gas phase reactants. Generally, a sticking coefficient of 10^{-4} corresponds to a conversion of more than 99%. Thus, the higher conversion of the Co and CoFe samples under TAP conditions are understood as a consequence of the high sticking coefficient at low coverage. A similar example was demonstrated by Zheng et al. for CO oxidation over platinum. In this case, only a single catalyst particle was added to the inert reactor packing. Under flow conditions conversion was near 20% while under TAP conditions conversion exceeded 90% [29].

Where is the Boundary of Non-Uniformity?

One of the main advantages of the TAP experiment is the ability to characterize kinetics at high conversion while maintaining uniformity in both the catalyst bed and gas phase composition. In previous work, the upper limit for conversion in the thin zone TAP reactor was described as 80% (where nonuniformity would not exceed 20%) based on a linear analytical estimate [30]. Our conversion measurements for the cobalt samples in this work exceed this guideline so the reliability of the data may be questioned. The topic of non-uniformity has been previously addressed in great detail theoretically [30,31]. Figure 4A from Shekhtman et al. [30] compares non-uniformity as a function of conversion. For the thin-zone TAP reactor, they provide the calculation of non-uniformity according to:

$$\frac{C_{in} - C_{out}}{C_{in}} \approx 2\frac{L_c}{L_r} + \frac{X}{1 + (1 - X)\frac{L_r}{L_c}} \tag{1}$$

where C represents concentration, L_c is the length of the catalyst zone, L_r is the length of the whole TAP reactor and X is conversion. Figure 4B demonstrates the reactor concentration profile and highlights the non-uniformity in the catalyst zone. Non-uniformity is a function of concentration (or conversion). The important distinction between the differential-PFR and TAP arises from the way rate is calculated. In differential-PFR the rate is proportional to the difference in concentrations while in TAP rate is proportional to the difference in concentration *gradients*. Figure 4B illustrates how in TAP the difference in concentration at the boundaries of the thin zone is not as dramatic as the difference in gradients. As a result, TAP can access uniform rates at much higher conversion.

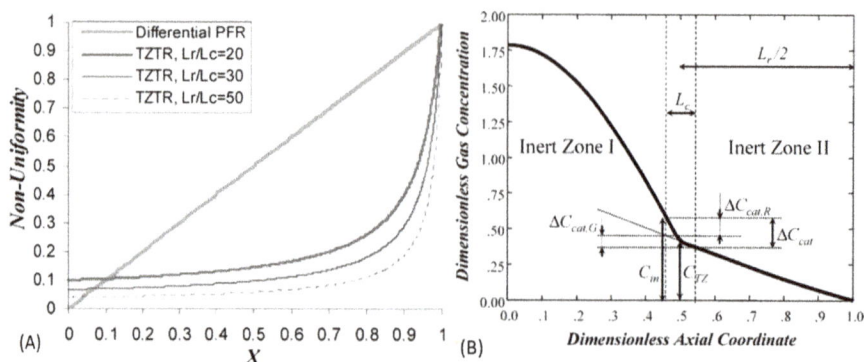

Figure 4. (A) Gas concentration nonuniformity versus conversion for the differential plug flow reactor (PFR) and thin-zone TAP reactor at different values of the geometric parameter L_r/L_c where L_c is the length of the catalyst zone, L_r is the length of the whole TAP reactor and X is conversion; **(B)** thin-zone concentration profile with dashed lines indicating nonuniformity in the catalyst zone; reprinted with permission from Shekhtman et al. [30] Copyright 2005 American Chemical Society.

Using Equation (1), we determine the concentration nonuniformity for our reactor configuration can be as high as 24% at 90% conversion. While this estimate is high, closer examination of the data does not indicate any sign of deviations caused by non-uniformity. The same experiments were repeated at lower temperatures for the CoFe and Co sample where conversions were lower. Figure 5 shows the Arrhenius plot of the apparent rate constant along with conversion. We find that the high conversion data falls in line with lower temperature data where non-uniformity estimates are within acceptable levels. Calculation of the rate constant in Figure 5 assumes first-order kinetics of the gas phase reactant which is applicable in the TAP device where one reactant interacts with a well-defined surface under conditions far from equilibrium (similar to molecular beam scattering experiments). TAP is rooted in the concept of insignificant perturbation and the simplicity of the measurement brings it closer to that of an elementary process. Furthermore, the expression for the apparent rate constant is analogous to that observed in a CSTR and derivation from the diffusion-reaction equations was demonstrated explicitly by Phanawadee et al. [31].

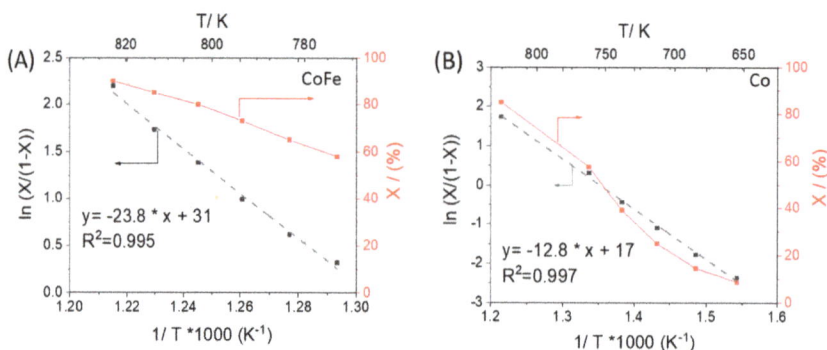

Figure 5. Arrhenius plot (assuming first order reaction) and conversion when ammonia is pulsed over **(A)** CoFe and **(B)** Co samples at different temperatures; X is NH_3 conversion, the apparent rate constant is proportional to $X/(1-X)$; see derivation in [31].

In Figure S4 of the Supplementary Information, the exit flux of the hydrogen produced when ammonia is pulsed over CoFe and Co is compared at 500 and 550 °C. By height normalizing to compare shape, we do not observe any significant deviation in the kinetic response. Even over the course of 2000 pulses, Figure S3 in the Supplementary Information does not show a significant change in the pulse response shape of ammonia. From these observations we can conclude that the gas composition and catalyst composition are not significantly impacted by non-uniformity even at 90% conversion in our experiments. Previously, the upper limit for conversion was set at 80% based on a linear analytical estimate [30]. However, it was an *estimate* in which only the difference of gas concentrations was taken into account as the main component of the non-uniformity. Rigorously speaking, although the two go hand-in-hand, the explicit non-uniformity in surface composition should be considered as well. It should slightly decrease upon lengthening the thin catalytic zone (averaging over a longer length). Also, there is significant experimental uncertainty in measuring the actual thin zone thickness. In addition to the support from lower temperature observations, we consider the measurements at 90% conversion to be reliably uniform although at the upper boundary. With any conversion higher than this it would be difficult to capture the reactant pulse shape with a significant signal to noise ratio.

With conventional integral analysis, the TAP pulse response is reduced to one data point for conversion. There is no additional information from what is obtained under flow conditions. The key difference in the experiments is surface coverage and pressure. While TAP offers the ability to incrementally control surface coverage with high precision, the data presented here is just a snapshot of performance at two different conditions. What is more important is to preserve the time-dependence of the pulse response whereby greater information can be gained. However, there are special considerations to take into account in order to properly separate the time-dependence of the transport and kinetic information.

2.3.2. Decoupling Transport and Kinetic Information

Figure 6A demonstrates the experimentally observed exit flux of H_2 and N_2 products when NH_3 is pulsed over the cobalt catalyst; the responses have been height-normalized for shape comparison. Hydrogen is observed to leave the reactor before nitrogen. The information we need however is the time-dependence of hydrogen production in the catalyst zone. Hydrogen has a significantly higher diffusivity compared to nitrogen and this transport consideration should be accounted for. One approach [32] is to normalize the time scale according to the effective diffusivity of each molecule based on Graham's law:

$$D \sim \sqrt{\frac{T}{M}} \qquad (2)$$

In Figure 6B, the time vectors for hydrogen and nitrogen have been separately corrected according to molecular weight and the interpretation of the temporal characteristics is significantly different from Figure 6A. This simple scaling successfully corrects the transport time dependence however, the exit flux contains both transport and kinetic information. For adsorbing/desorbing and reacting molecules, this method is unreliable since it also operates on the kinetic dependence which does not scale accordingly with molecular weight. However, this method can be used to correct the exit flux of an inert molecule for comparison to the kinetic dependence at the same molecular weight.

Figure 6C presents the rate time-dependence of hydrogen and nitrogen calculated by the Y-Procedure. This method accounts for transport through the inert zones before and after the catalyst. The arrival of reactant molecules from the input pulse into the thin zone presents the time-dependent 'initial condition' for reaction. Using Fourier analysis, diffusion of the inlet pulse is 'fed forward' and the exit flux is 'dragged backward' to the catalyst zone; the reader may refer to the original publication for greater detail [12]. The rate is calculated as the difference between fluxes in and out of the thin catalyst zone. In contrast to the PFR, these fluxes are diffusional in nature which provides effective mixing. A thin-zone approach could also be utilized in a plug flow reactor but in this case convection is dominant and mixing is not as efficient.

Application of the Y-Procedure to exit flux data decouples diffusional transport and the rate time-dependence only reflects the kinetic process taking place in the catalyst zone. As a result, the interpretation of TAP pulse response curves can be markedly different from exit flux and diffusion corrected exit flux. Figure 6C demonstrates fast hydrogen formation rate followed by slower evolution of nitrogen from the surface. In addition, the log-based time scale emphasizes a small shoulder on the hydrogen peak which coincides with the timing of the nitrogen peak. These two distributions in the hydrogen concentration will be deconvoluted and discussed more in the next section.

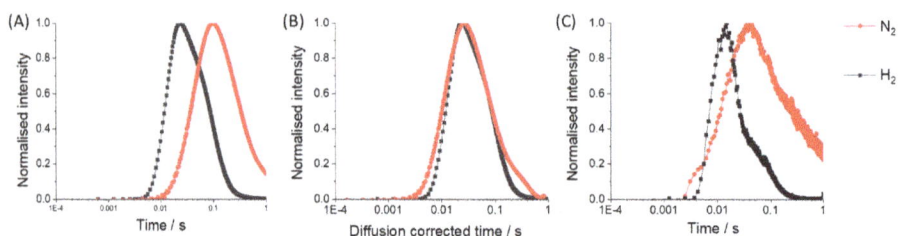

Figure 6. Transient product time-dependencies for H_2 and N_2 when NH_3/Argon is pulsed over the Cobalt catalyst; height normalized for comparison, (**A**) experimentally observed exit flux, (**B**) N_2 flux was corrected uniformly to match the effective diffusivity of H_2, (**C**) product rate calculated using the Y-Procedure.

2.3.3. Time-Dependent Reaction Rate and Uptake/Release

In a steady-state experiment the initial concentration of reactant is fixed, C_{A0}. In a pulse response experiment the inlet gas pulse arriving in the catalyst zone sets the initial conditions for reactions, $C_{A0}(t)$. Gas concentration passes through a maximum and the catalyst responds with dynamic accumulation of reactant molecules on the surface and release of products back to the gas phase. The time-dependence of the molecular reaction rate calculated via the Y-procedure is demonstrated in Figure 7 along with the surface total uptake for ammonia and total release of hydrogen and nitrogen for each sample. Again, the uptake/release is simply the integral of the rate data. The reaction rates under TAP conditions are generally 2 or more orders of magnitude less than that observed under flow. This information is captured under unique conditions in order to provide a uniform basis of comparison. The pressure may be significantly less in the TAP experiment but as such it enables an intrinsic perspective for how each material functions.

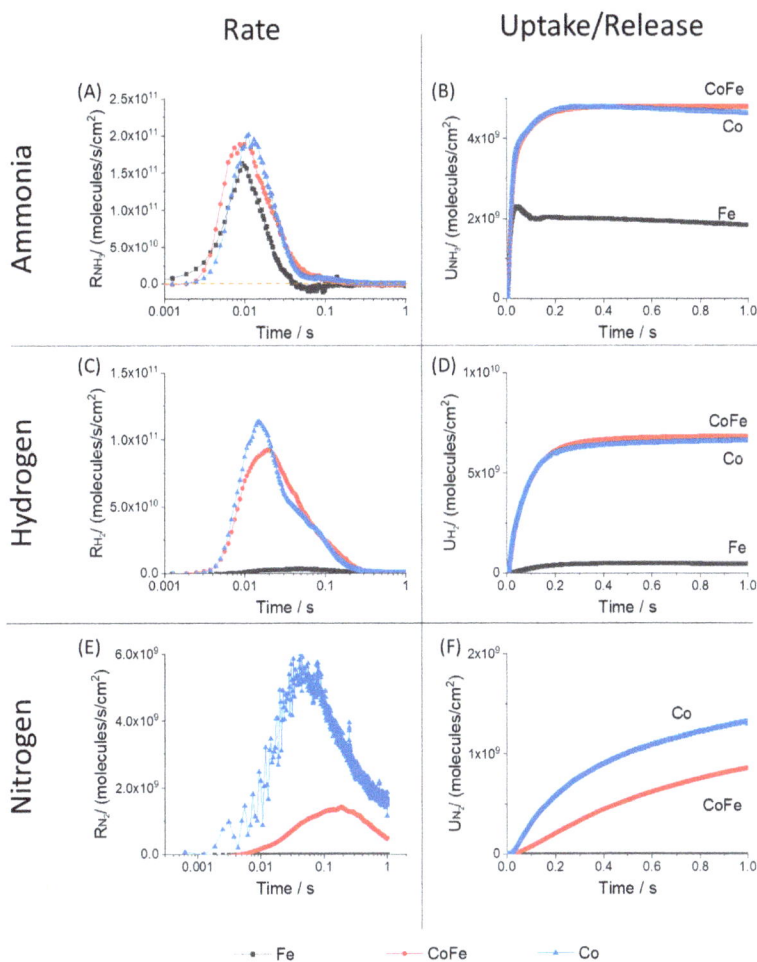

Figure 7. Molecular reaction rate calculated from pulse response data when ammonia is pulsed at 550 °C for (**A**) ammonia, (**C**) hydrogen and (**E**) nitrogen; catalyst uptake calculated from the integral of the rate for (**B**) ammonia; catalyst release calculated from the integral of the rate for (**D**) hydrogen and (**F**) nitrogen on different samples (Fe, CoFe and Co). Note: log-based time scale on rate data, linear time scale on uptake data.

For all samples the rate of ammonia conversion, Figure 7A, passes through a singular maximum in response to the inlet pulse. The highest reaction rates are observed for cobalt and CoFe. The iron sample achieves an instantaneous rate maximum that is nearly as high as the cobalt materials. This is surprising since conversions over iron are much lower (in both low and ambient pressure conditions). This indicates that while iron may have a population of significantly active sites they are generally far fewer in number. Only the iron sample demonstrates a small negative reaction rate which indicates that NH_3 is sequentially released from the surface of iron. Simultaneous adsorption/desorption is a possibility for all samples; this would simply lower the total rate observed and is better distinguished using isotopic studies (a subject of current work). The sequential desorption indicates a stronger dependence on coverage that lowers the cumulative uptake, Figure 7B, detected over the duration of the pulse. Figure 7B is simply the integral of the rate and shows the balance of ammonia both consumed by the surface and released back to the gas phase. Note that rate in Figure 7A is on a log-based time scale while uptake in Figure 7B is on a linear time scale. It is important to note that the data shown here is an average of 2000 pulses. The total uptake from one pulse to the next is similar; adding to the total accumulation over the pulse series but a change in pulse shape is not observed over the pulse series (Supplementary Information, Figure S3). The total ammonia consumption is highest for the CoFe and Co samples which is in agreement with the high conversion activity of this material. Although there is a minor decline in the ammonia uptake for cobalt the iron sample demonstrates an obvious maximum early in the pulse period which is due to the sequential release. This is an important point for demonstrating the greater detail derived from transient experiments. The iron catalyst has a higher ammonia desorption rate that contributes to an overall lower conversion. This effect would not be detected under steady-state conditions and the timing of the process is too fast for observation under flow transient conditions.

Hydrogen and nitrogen were the only gas phase products detected during ammonia pulsing. Figure 7C,D show the time-dependent production rate and cumulative product release calculated for hydrogen. Both the maximum and cumulative hydrogen generation rate from the cobalt catalysts are significantly higher than the iron sample. The cumulative release of hydrogen shows a fast increase and then a plateau. This verifies that there is no backward consumption and these experiments are conducted far from equilibrium. In separate experiments where hydrogen was directly pulsed under the same conditions we observe a kinetic response on all materials (Figure S5). To more closely compare the time-dependence of the reaction rate on iron, the height normalized rate is presented in Figure 8. Closer inspection indicates two separate hydrogen release processes for materials with cobalt; one narrow peak centered near 0.02 s with a slower distribution centered near 0.04 s. The slower distributions demonstrate similar timing to the rate observed on the iron sample. This indicates that while cobalt materials may have more active sites (summary from conversion data) a primary driver for the increase in the hydrogen production rate on cobalt is access to a faster reaction pathway.

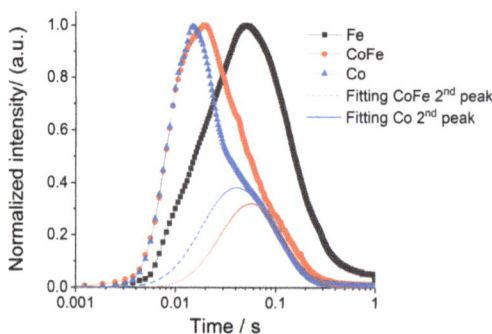

Figure 8. Height normalized hydrogen production rates collected for Fe, CoFe and Co during ammonia pulsing at 550 °C, deconvolution of secondary peak on CoFe and Co samples to emphasize timing of slower process.

The nitrogen production rate and cumulative release data are found in Figure 7E,F. While a very small nitrogen pulse response was detected over the iron sample the signal intensity was too low for analysis using the Y-Procedure. This is an important point, because due to the low chemical potential the TAP technique does have limitations on what reactions can be observed. Ammonia synthesis from N_2 and H_2 is unlikely but synthesis with N atoms added to the surface with other means is possible (such work using isotopic pump/probe experiments is in preparation for publication). The nitrogen rate detected over cobalt is higher than the CoFe sample. This may be a reflection of the low N surface coverage as an inversion in the dominant material for gas phase N_2 release was also observed early in the flow transient experiment, Figure 3.

A more extensive pulse response or ex situ nitridation experiment could be used in future work to better understand the N_2 rate in TAP experiments. In general, the magnitude of the N_2 rates were significantly less than the rate of H_2 production; by more than an order of magnitude. Also, the timing of the rate maximum is shifted to the right and coincided more with the slow hydrogen generation process. The cumulative nitrogen release was much slower than the ammonia uptake and hydrogen release processes. While only 1 s of data is shown for clarity, the total experiment duration was 3.3 s.

2.3.4. Surface Mass Balance and Temporal Atomic Accumulation

The molecular uptake reported for reactants and products in Figure 7B,D,E can be used in mass balance to determine the atomic accumulation of H and N species by considering the reaction stoichiometry:

$$H^{Surface} = 3NH_3{}^{Gas} - 2H_2{}^{Gas}$$

$$N^{Surface} = NH_3{}^{Gas} - 2N_2{}^{Gas}$$

This takes into account the uptake of reactant and release of product molecules in order to identify the balance of species that remain on the surface. The temporal atomic accumulation of both H and N pass through a maximum for all catalysts, Figure 9, which indicates the conversion of reactants to gas phase products. Following the peak, all materials show a slow decay in the amount of H and N on the surface. In each pulse, there is incomplete release and consequently an accumulation of species. Only the cobalt sample demonstrates a significant decrease in the hydrogen and nitrogen surface concentrations during the pulse. For iron, the accumulation remains high over the pulse duration which indicates much of the converted ammonia is stored on the surface. From Figure 9 we can surmise that iron is a more effective material for surface storage while cobalt will push products back to the gas phase. These features are well-known for iron and cobalt materials. For example the Co(111) surface was reported to have much higher ammonia decomposition activity than the Fe(110) surface. Also, Co has a lower activation energy for N recombination than Fe [33]. Thus, this 'simple' and well-documented system serves as a good model for demonstrating a unique characterization method that can be applied to more complex systems, e.g., to compare a set of ill-defined industrial catalysts with incremental changes in composition. One might use this method to understand the role of different metals of a complex multi-component formulation in regulating surface and gas phase species.

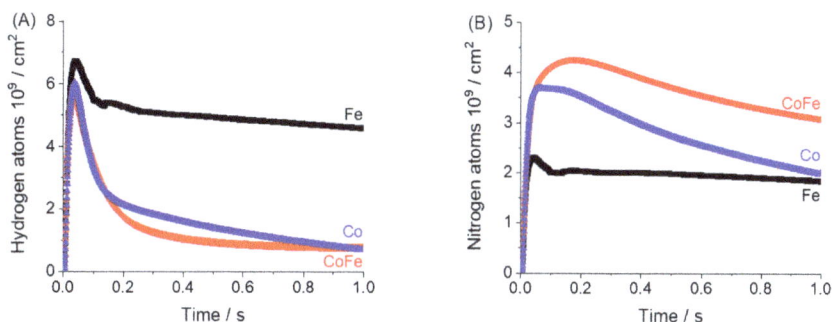

Figure 9. Dynamic atomic accumulation calculated from pulse response data for (**A**) hydrogen and (**B**) nitrogen when ammonia is pulsed over Fe, CoFe and Co at 550 °C.

By examining the ratio of hydrogen to nitrogen accumulation, we can better understand the surface stoichiometry. Figure 10 shows the time dependence of the hydrogen/nitrogen, (H/N), ratio.

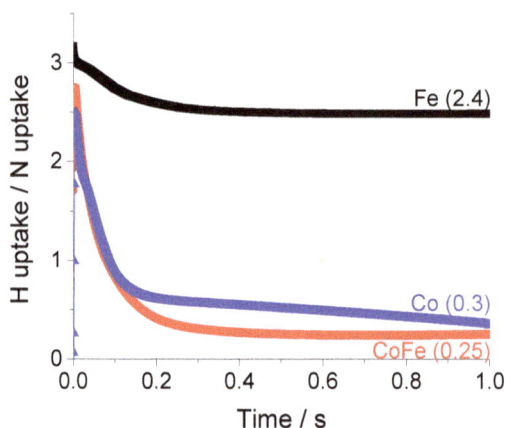

Figure 10. Dynamic hydrogen/nitrogen, (H/N), ratio determined from atomic accumulation data when ammonia is pulsed over Fe, CoFe and Co at 550 °C.

Following a transition period which represents the non-steady state behavior of the experiment (0–0.2 s), the H/N ratio assumes a constant value for all samples. Although the surface concentrations of H and N are always declining in the experiment, they reach a point where they decrease coherently at the same rate. For iron, the final H/N ratio of 2.4 is an indication that a mixture of NH_3 and NH_2 species are stored on the surface. The addition of cobalt to iron lowers the ratio to 0.25 while the ratio for pure cobalt is slightly higher; 0.35. This indicates a mixture of NH and N species are more likely to dominate the surface. Of course, this test does not provide an unequivocal indication of the distribution of surface species but is a useful basis for comparison. The addition of cobalt can be understood to destabilize the surface NH_3 species, promote deprotonation and release of N_2 and H_2 products to the gas phase.

From the surface mass balance, we observe iron to favor nitrogen storage while cobalt accelerates recombination and release. These observations are in agreement with DFT (density functional theory) modeling results that indicate greater stability of nitrogen species on Fe(110) surfaces compared to Co(111) [33]. In alloy species the electron transfer from cobalt to iron is expected to promote the desorption of nitrogen atoms from the surface [22]. The main conclusions stemming from the time-dependence of the rate and atomic accumulation data form the basis of mechanistic analysis which is the topic of a separate manuscript presently in preparation.

3. Materials, Experiment and Analysis Methods

3.1. Catalyst Preparation

Polycrystalline iron (99.99% purity, particle size 450 μm) and cobalt (99.9% purity, particle size 600 μm) were purchased from Good Fellow and Alfa Aesar, respectively. Samples were sieved and a 250–300 μm size range was used for analysis. A bimetallic cobalt–iron sample was prepared by wet impregnation of the polycrystalline iron using cobalt (II) nitrate hexahydrate, $Co(NO_3)_2 \cdot 6H_2O$ (99.999% purity, Sigma Aldrich, St. Louis, MO, USA). Nominally, 1 ML of cobalt was deposited over the Fe support. A 3.15 wt.% loading of cobalt was determined by inductively coupled plasma mass spectroscopy (ICP-MS). All samples were reduced with a 10% H_2/Argon flow at 550 °C for 17 h. Samples were cooled and then exposed to ambient conditions before further analysis. These three samples, referred to as Fe, CoFe and Co are the focus of this investigation.

3.2. Catalyst Structural Characterization

The BET surface area of Fe and Co were determined by N_2 adsorption at -196 °C (NOVA 2200e Micromeritics, Norcross, GA, USA). The catalyst samples were pre-treated by degassing at 200 °C for 16 h in vacuum (0.05 mbar). The amount of Co loading of CoFe sample was measured by ICP-MS) (Thermo iCAP Qc, Thermo Fisher Scientific, Waltham, MA, USA). The morphology of Fe, Co and CoFe were measured by transmission electron microscopy (TEM) on a JEM-2100P electron microscope operating at 200 kV (JEOL Ltd., Tokyo, Japan). The powder X-ray diffraction (XRD) was recorded by a Bruker-AXS D5005 (Bruker, Billerica, MA, USA) with a Cu Kα source to obtain the structure of materials used in this study. Prior to reaction XRD of Co and Fe were were reduced in situ overnight using H_2 before diffraction was recorded at 550 °C under vacuum. XRD of the pre-reaction CoFe sample and the post-reaction Fe, CoFe and Co samples were recorded at room temperature.

3.3. Differential PFR Kinetic Characterization

100 mg of sample (250–300 μm particle size) was loaded in a quartz tube fixed bed reactor. Prior to NH_3 flow experiments, the catalyst was reduced with a 10% H_2/Ar (Airgas, Radnor, PA, USA) flow at 550 °C for 12 h, then cooled down to room temperature, and flushed with He. Subsequently, the sample was heated in He to 550 °C before the feed gas was switched to 15% NH_3/He (Airgas, Radnor, PA, USA). The effluent gas was analyzed by an on-line MS (Micromeritics Cirrus2 quadrupole mass spectrometer). Different atomic mass units (AMU) of m/e = 2, 4, 17, 28 were recorded every 2.8 s.

3.4. TAP Pulse Response Experiment

The defining feature of a TAP pulse response experiment is the use of a sufficiently small gas pulse injected into an evacuated microreactor to ensure gas transport in the Knudsen diffusion regime. Well-defined transport enables decoupling of intrinsic kinetic phenomena on the catalyst surface from reactor transport. For reactants, any deviation from Knudsen diffusion is evidence of a gas-solid interaction, reaction, including the elementary steps of adsorption, surface diffusion, interparticle diffusion, surface reaction, desorption, etc. For the formed products, the response curve contains information about the product formation rate and product desorption.

TAP pulse response experiments were carried out in a commercial TAP-3 reactor system, Mithra Technologies, Foley, Missouri. 30 mg of catalyst sample was loaded in the thin zone reactor configuration [30,34,35] using ground quartz in the inert zones (also sieved to the 250–300 μm size range). The quartz was ground from larger pieces prior to sieving and was washed in 3% nitric acid followed by triple rinsing with water (Millipore-Q). The quartz was calcined overnight at 1000 °C prior to use. A quartz reactor measuring 4 mm internal diameter (ID), 8 mm outer diameter (OD) by 38.3 mm in length was used. O-rings at the top and bottom of the quartz tube were water cooled and maintained near 70 °C. Prior to pulsed experiments the catalyst was again reduced with a 10% H_2/Ar flow at 550 °C. Following *in situ* reduction the reactor was cooled to room temperature and evacuated.

The reduced sample was then heated at 10 °C/min to 550 °C while monitoring the desorption spectra using a RGA200 mass spectrometer, Stanford Research Systems, Sunnyvale, California. At 550 °C a blend of ammonia/argon (1:1 by volume, 99.999%, Matheson Gas) was pulsed into the packed bed reactor using a magnetically operated solenoid valve. An electronic pulse width of 135 μs was sent to the valve every 3.3 s for a total of 2000 pulses. The pulse injection size was calibrated separately, approximately 10^{-8} mol per pulse to ensure operation in the Knudsen flow regime where gas-gas collisions are insignificant. At each pulse a different atomic mass unit (AMU) was monitored for 3.0 s cycling between 17, 2, 28 and 40 to record the pulse response for ammonia, hydrogen, nitrogen and argon, respectively (0.3 s of dead time before the next pulse allows time for the mass spectrometer to adjust to the next mass in the sequence).

3.5. Y-Procedure Analysis Method

While the experimental observable in the TAP pulse response experiment is reactor exit flux the desired information for any kinetic experiment is rate and concentration. The Y-Procedure analysis method [11,12] is an inverse-diffusion technique that reconstructs observed exit flux to gas concentration and reaction rate in the reaction zone while maintaining the time-dependence of the transient experiment. This method does not require a priori assumption of any kinetic model. The chemical transformation rate (conversion of reactants, generation of products) is determined as the difference in diffusive flux at either side of the differentially thin active zone. The uptake/release of a gas is the integral of the rate. For example, for a simple adsorption reaction, the uptake is equal to the increase of the adsorbing gas surface concentration, or the decrease of empty sites.

The general data analysis procedure has been described previously [13] and will only be summarized here. Pulse response data collected from the mass spectrometer were first baseline corrected and 2000 pulses were averaged to improve the signal-to-noise ratio. Concentration and rate calculations for hydrogen and nitrogen were conducted by scaling the argon response according to molecular weight. Similar calculations for ammonia required the use of ammonia pulse response data collected over a one-zone inert reactor in order to account for minor reversible adsorption of ammonia on the silica packing material. To suppress the amplification of higher frequencies in the Fourier domain analysis a smoothing factor of 1.0 was used for NH_3 and H_2 and 3.0 was used for N_2.

One drawback of the Y-Procedure method is the noise induced from sinh and cosh terms of the solution to the transport model in the Fourier domain. More sophisticated methods for noise analysis, filtering and minimization have been described [36–38] and are currently in development. In this work, since the time-dependence remains constant from pulse-to-pulse over the course of the experiment a simple average of 2000 pulses was used for the presentation of trends.

4. Conclusions

It has been well established that transient methods offer significantly more information than steady-state [1–7]. More specifically, steady-state characteristics (rates of transformation) reflect the slow, limiting steps while transient techniques, depending on the time resolution of the device, can be used to distinguish faster and slower steps. For all cases, the necessary condition for precise kinetic characterization is uniformity of the catalyst zone. The comparison of transient experiments in an advecting versus pure diffusion device demonstrated the greater window of opportunity when diffusion is the dominant transport mode and ensures efficient mixing. In the differential-PFR (CSTR) reactor, non-uniformity is linearly proportional to conversion while in TAP non-uniformity is a non-linear function of conversion, i.e., $x/(1 - x)$. Thus, the TAP pulse response experiment enables kinetic characterization under conditions of much higher conversion. Previously the limit was estimated at 80% [30]; by corroborating the observed pulse response shapes with data collected at lower conversion we push this estimate towards 90%.

What is more important is the ability to observe the time-dependence of product rates in response to the reactant pulse. Conventionally, dynamic experiments such as TAP, TPD (temperature programmed desorption) and SSITKA (steady-state isotopic transient kinetic analysis) have used integral analysis for interpretation of transient data which, unfortunately, averages-out the rich kinetic information. By implementing the Y-Procedure tool in the TAP methodology to preserve the time dependence, the amount of data available from one pulse is increased at least 100-fold; 1000-fold or more for examining a series of pulses [39]. Moreover, the time resolution of TAP is 1000-fold greater than most advecting devices and enables the observation of fast kinetic processes.

From steady-state and flow step-transient experiments we observed the addition of cobalt to iron creates higher activity that either material alone; in agreement with previous reports [22]. From the time-dependence of pulse response experiments we extracted the following **new observations and facts** which support greater understanding for why these materials perform differently:

- Iron can generate instantaneous reaction rates on a par with cobalt and CoFe but, since conversions are much lower for iron, this indicates the site density is lower.
- Iron demonstrates sequential reversible adsorption of ammonia which leads to lower global activity. Reversibility of adsorption is coverage dependent. For iron the cumulative rate is lower because of desorption of unconverted molecules.
- The rate of hydrogen production on cobalt materials shows two time distributions representing fast and slow processes. The presence of cobalt indicated an additional fast pathway for hydrogen production that contributes to higher activity. The timing of the slow process coincides with the rate observed on iron. Thus, a primary driver for increasing hydrogen production is the access to two routes, one that is faster than iron.
- Nitrogen production on cobalt coincides with the slow hydrogen process indicating they may be formed in the same step.
- Surface mass balances demonstrate how iron stores converted ammonia on the surface while cobalt accelerates recombination and releases N_2 into the gas phase.
- The dynamic atomic accumulation indicates that Fe predominantly stores more hydrogenated NH_3 and NH_2 species while cobalt favors deprotonation predominantly to store NH and N.

This information forms the basis of work currently in progress to describe a detailed mechanism and corresponding kinetic model on each material.

In conclusion, the time dependence of the rate offers greater information (both quantity and quality) and enables the calculation of dynamic uptake. Each of these transient kinetic observations points to a distinct feature of the catalyst that could be 'microkinetically' optimized towards the desired performance. This approach was demonstrated using simple materials and the ammonia decomposition reaction as a model system. However, this characterization provides a unique vantage point that can be applied to *complex* active materials via two methods:

(a) Comparison of materials with incremental changes in composition;
(b) 'Smart' design of complex active materials assisted by knowledge of the storage properties of different components.

Moreover, these transient methods conducted in the pure diffusion regime provide a uniform basis where we can compare these features at the same experimental conditions on materials that vary widely in conversion.

Supplementary Materials: The following are available online at http://www.mdpi.com/2073-4344/9/1/104/s1: Figure S1. (A) Pre-reaction XRD patterns for Fe and Co at 550 °C, CoFe at room temperature; post-reaction XRD patterns following exposure to 15%NH₃/He at 50 mL/min for 1 h at 550 °C, (B) Fe, (C) CoFe, (D) Co. Figure S2. NH₃ conversion over the course of 2000 pulses NH₃ pulses at 550 °C over Fe, CoFe and Co materials in the TAP reactor. Figure S3. Height normalized NH₃ reaction rate at different pulses over the 2000 pulse sequence for (A) Fe, (B) CoFe and (C) Co. Figure S4. Height normalized hydrogen pulse response for (A) CoFe and (B) Co during ammonia decomposition experiments at 500 and 550 °C. Figure S5. Height normalized hydrogen and argon (internal standard) pulse response during H2/Ar direct pulsing at 550 °C over Fe, CoFe and Co. Argon diffusion time was corrected by Graham's law.

Author Contributions: Conceptualization, R.F.; Data curation, Y.W., M.R.K., S.S. and H.R.; Formal analysis, Y.W., M.R.K., J.G., G.Y. and R.F.; Funding acquisition, R.F.; Investigation, Y.W., M.R.K., S.S., H.R., G.Y. and R.F.; Methodology, M.R.K., J.G., G.Y. and R.F.; Project administration, R.F.; Software, M.R.K.; Validation, Y.W.; Writing—original draft, R.F., Y.W.; Writing—review and editing, Y.W., M.R.K., G.Y. and R.F.

Funding: This work was supported by U.S. Department of Energy (USDOE), Office of Energy Efficiency and Renewable Energy (EERE), Advanced Manufacturing Office Next Generation R&D Projects under contract no. DE-AC07-05ID14517. Accordingly, the U.S. Government retains a non-exclusive, royalty-free license to publish or reproduce the published form of this contribution, or allow others to do so, for U.S. Government purposes.

Acknowledgments: The authors are grateful to A.J. Medford for reviewing the manuscript and providing useful discussion. The authors acknowledge J.M. Yoda for many fruitful discussions.

Conflicts of Interest: The authors declare no conflict of interest.

References

1. Bennett, C.O. A dynamic method for the study of heterogeneous catalytic kinetics. *AIChE J.* **1967**, *13*, 890–895. [CrossRef]
2. Kobayashi, H.; Masayoshikobayashi. Transient response method in heterogeneous catalysis. *Catal. Rev. Sci. Eng.* **1974**, *10*, 139–176. [CrossRef]
3. Bennett, C.O. *Understanding Heterogeneous Catalysis through the Transient Method*; American Chemical Society: Washington, DC, USA, 1982; Volume 178.
4. Dekker, F.; Bliek, A.; Kapteijn, F.; Moulijn, J. Analysis of heat and mass transfer in transient experiments over heterogeneous catalysts. *Chem. Eng. Sci.* **1995**, *50*, 3573–3580. [CrossRef]
5. Pérez-Ramírez, J.; Berger, R.J.; Mul, G.; Kapteijn, F.; Moulijn, J.A. The six-flow reactor technology: A review on fast catalyst screening and kinetic studies. *Catal. Today* **2000**, *60*, 93–109. [CrossRef]
6. Van, V.A.C.; Farrusseng, D.; Rebeilleau, M.; Decamp, T.; Holzwarth, A.; Schuurman, Y.; Mirodatos, C. Acceleration in catalyst development by fast transient kinetic investigation. *J. Catal.* **2003**, *216*, 135–143. [CrossRef]
7. Berger, R.J.; Kapteijn, F.; Moulijn, J.A.; Marin, G.B.; De Wilde, J.; Olea, M.; Chen, D.; Holmen, A.; Lietti, L.; Tronconi, E. Dynamic methods for catalytic kinetics. *Appl. Catal. A Gen.* **2008**, *342*, 3–28. [CrossRef]
8. Gleaves, J.T.; Yablonsky, G.; Zheng, X.; Fushimi, R.; Mills, P.L. Temporal analysis of products (TAP)—recent advances in technology for kinetic analysis of multi-component catalysts. *J. Mol. Catal. A Chem.* **2010**, *315*, 108–134. [CrossRef]
9. Morgan, K.; Maguire, N.; Fushimi, R.; Gleaves, J.; Goguet, A.; Harold, M.; Kondratenko, E.; Menon, U.; Schuurman, Y.; Yablonsky, G. Forty years of temporal analysis of products. *Catal. Sci. Technol.* **2017**, *7*, 2416–2439. [CrossRef]
10. Perez-Ramirez, J.; Kondratenko, E.V. Evolution, achievements, and perspectives of the TAP technique. *Catal. Today* **2007**, *121*, 160–169. [CrossRef]
11. Redekop, E.A.; Yablonsky, G.S.; Constales, D.; Ramachandran, P.A.; Pherigo, C.; Gleaves, J.T. The Y-Procedure methodology for the interpretation of transient kinetic data: Analysis of irreversible adsorption. *Chem. Eng. Sci.* **2011**, *66*, 6441–6452. [CrossRef]
12. Yablonsky, G.S.; Constales, D.; Shekhtman, S.O.; Gleaves, J.T. The Y-procedure: How to extract the chemical transformation rate from reaction–diffusion data with no assumption on the kinetic model. *Chem. Eng. Sci.* **2007**, *62*, 6754–6767. [CrossRef]
13. Kunz, R.; Redekop, E.A.; Borders, T.; Wang, L.; Yablonsky, G.S.; Fushimi, R. Pulse Response Analysis Using the Y-Procedure Computational Method. *Chem. Eng. Sci.* **2018**, *192*, 46–60. [CrossRef]
14. Redekop, E.A.; Yablonsky, G.S.; Constales, D.; Ramachandran, P.A.; Gleaves, J.T.; Marin, G.B. Elucidating complex catalytic mechanisms based on transient pulse-response kinetic data. *Chem. Eng. Sci.* **2014**, *110*, 20–30. [CrossRef]
15. Redekop, E.A.; Yablonsky, G.S.; Galvita, V.V.; Constales, D.; Fushimi, R.; Gleaves, J.T.; Marin, G.B. Momentary Equilibrium (ME) in transient kinetics and its application for estimating the concentration of catalytic sites. *Ind. Eng. Chem. Res.* **2013**, *52*, 15417–15427. [CrossRef]
16. Schüth, F.; Palkovits, R.; Schlögl, R.; Su, D.S. Ammonia as a possible element in an energy infrastructure: Catalysts for ammonia decomposition. *Energy Environ. Sci.* **2012**, *5*, 6278–6289. [CrossRef]
17. Bell, T.; Torrente-Murciano, L. H2 production via ammonia decomposition using non-noble metal catalysts: A review. *Top. Catal.* **2016**, *59*, 1438–1457. [CrossRef]
18. Satyapal, S.; Petrovic, J.; Read, C.; Thomas, G.; Ordaz, G. The US Department of Energy's National Hydrogen Storage Project: Progress towards meeting hydrogen-powered vehicle requirements. *Catal. Today* **2007**, *120*, 246–256. [CrossRef]
19. Schlögl, R. Catalytic Synthesis of Ammonia—A "Never–Ending Story"? *Angew. Chem. Int. Ed.* **2003**, *42*, 2002–2008.
20. Boisen, A.; Dahl, S.; Nørskov, J.K.; Christensen, C.H. Why the optimal ammonia synthesis catalyst is not the optimal ammonia decomposition catalyst. *J. Catal.* **2005**, *230*, 309–312. [CrossRef]
21. García-García, F.; Guerrero-Ruiz, A.; Rodríguez-Ramos, I.; Goguet, A.; Shekhtman, S.; Hardacre, C. TAP studies of ammonia decomposition over Ru and Ir catalysts. *Phys. Chem. Chem. Phys.* **2011**, *13*, 12892–12899. [CrossRef]

22. Zhang, J.; Müller, J.-O.; Zheng, W.; Wang, D.; Su, D.; Schlögl, R. Individual Fe−Co alloy nanoparticles on carbon nanotubes: Structural and catalytic properties. *Nano Lett.* **2008**, *8*, 2738–2743. [CrossRef] [PubMed]

23. Christensen, A.; Ruban, A.; Stoltze, P.; Jacobsen, K.W.; Skriver, H.L.; Nørskov, J.K.; Besenbacher, F. Phase diagrams for surface alloys. *Phys. Rev. B* **1997**, *56*, 5822. [CrossRef]

24. Raróg-Pilecka, W.; Jedynak-Koczuk, A.; Petryk, J.; Miśkiewicz, E.; Jodzis, S.; Kaszkur, Z.; Kowalczyk, Z. Carbon-supported cobalt–iron catalysts for ammonia synthesis. *Appl. Catal. A Gen.* **2006**, *300*, 181–185. [CrossRef]

25. Hagen, S.; Barfod, R.; Fehrmann, R.; Jacobsen, C.J.; Teunissen, H.T.; Chorkendorff, I. Ammonia synthesis with barium-promoted iron–cobalt alloys supported on carbon. *J. Catal.* **2003**, *214*, 327–335. [CrossRef]

26. Jacobsen, C.J.; Dahl, S.; Clausen, B.S.; Bahn, S.; Logadottir, A.; Nørskov, J.K. Catalyst design by interpolation in the periodic table: Bimetallic ammonia synthesis catalysts. *J. Am. Chem. Soc.* **2001**, *123*, 8404–8405. [CrossRef] [PubMed]

27. Feyen, M.; Weidenthaler, C.; Güttel, R.; Schlichte, K.; Holle, U.; Lu, A.H.; Schüth, F. High-Temperature Stable, Iron-Based Core–Shell Catalysts for Ammonia Decomposition. *Chem. A Eur. J.* **2011**, *17*, 598–605. [CrossRef] [PubMed]

28. Schuurman, Y. Assessment of kinetic modeling procedures of TAP experiments. *Catal. Today* **2007**, *121*, 187–196. [CrossRef]

29. Zheng, X.; Gleaves, J.T.; Yablonsky, G.S.; Brownscombe, T.; Gaffney, A.; Clark, M.; Han, S. Needle in a haystack catalysis. *Appl. Catal. A* **2008**, *341*, 86–92. [CrossRef]

30. Shekhtman, S.O.; Yablonsky, G.S. Thin-Zone TAP Reactor versus Differential PFR: Analysis of Concentration Nonuniformity for Gas-Solid Systems. *Ind. Eng. Chem. Res.* **2005**, *44*, 6518–6522. [CrossRef]

31. Phanawadee, P.; Shekhtman, S.O.; Jarungmanorom, C.; Yablonsky, G.S.; Gleaves, J.T. Uniformity in a thin-zone multi-pulse TAP experiment: Numerical analysis. *Chem. Eng. Sci.* **2003**, *58*, 2215–2227. [CrossRef]

32. Kondratenko, E.V.; Kondratenko, V.A.; Santiago, M.; Pérez-Ramírez, J. Mechanistic origin of the different activity of Rh-ZSM-5 and Fe-ZSM-5 in N_2O decomposition. *J. Catal.* **2008**, *256*, 248–258. [CrossRef]

33. Duan, X.; Ji, J.; Qian, G.; Fan, C.; Zhu, Y.; Zhou, X.; Chen, D.; Yuan, W. Ammonia decomposition on Fe (1 1 0), Co (1 1 1) and Ni (1 1 1) surfaces: A density functional theory study. *J. Mol. Catal. A Chem.* **2012**, *357*, 81–86. [CrossRef]

34. Shekhtman, S.O.; Yablonsky, G.S.; Chen, S.; Gleaves, J.T. Thin-zone TAP-reactor—Theory and application. *Chem. Eng. Sci.* **1999**, *54*, 4371–4378. [CrossRef]

35. Shekhtman, S.O.; Yablonsky, G.S.; Gleaves, J.T.; Fushimi, R.R. Thin-zone TAP reactor as a basis of "state-by-state transient screening". *Chem. Eng. Sci.* **2004**, *59*, 5493–5500. [CrossRef]

36. Reece, C. Kinetic Analysis and Modelling in Heterogeneous Catalysis. Ph.D. Thesis, Cardiff University, Cardiff, Wales, 2017.

37. Roelant, R. Mathematical Determination of Reaction Networks from Transient Kinetic Experiments. Ph.D. Thesis, Ghent University, Ghent, Belgium, 2011.

38. Roelant, R.; Constales, D.; Yablonsky, G.S.; Van Keer, R.; Rude, M.A.; Marin, G.B. Noise in temporal analysis of products (TAP) pulse responses. *Catal. Today* **2007**, *121*, 269–281. [CrossRef]

39. Medford, A.J.; Ewing, S.; Kunz, M.R.; Borders, T.; Fushimi, R. Extracting knowledge from data through catalysis informatics. *ACS Catal.* **2018**, *8*, 7403–7429. [CrossRef]

catalysts

MDPI

Article

Gaseous Nitric Acid Activated Graphite Felts as Hierarchical Metal-Free Catalyst for Selective Oxidation of H₂S

Zhenxin Xu [1,*], Cuong Duong-Viet [2,*], Housseinou Ba [1], Bing Li [1], Tri Truong-Huu [3], Lam Nguyen-Dinh [3] and Cuong Pham-Huu [1,*]

[1]   Institute of Chemistry and Processes for Energy, Environment and Health (ICPEES), ECPM, UMR 7515 CNRS-University of Strasbourg, 25 rue Becquerel, 67087 Strasbourg CEDEX 02, France; h.ba@unistra.fr (H.B.); bing.li@etu.unistra.fr (B.L.)
[2]   Ha Noi University of Mining and Geology, 18 Pho Vien, Duc Thang, Bac Tu Liem, Ha Noi, Viet-Nam
[3]   The University of Da-Nang, University of Science and Technology, 54 Nguyen Luong Bang, Da-Nang, Viet-Nam; thtri@dut.udn.vn (T.T.-H.); ndlam@dut.udn.vn (L.N.-D.)
*   Correspondence: zhenxin.xu@etu.unistra.fr (Z.X.); duongviet@unistra.fr (C.D.-V.); cuong.pham-huu@unistra.fr (C.P.-H.)

Received: 5 March 2018; Accepted: 30 March 2018; Published: 4 April 2018

Abstract: In this study, we reported on the influence of gaseous HNO₃ treatment on the formation of defects decorated with oxygenated functional groups on commercial graphite felts (GFs). The gaseous acid treatment also leads to a remarkable increase of the specific as well as effective surface area through the formation of a highly porous graphite structure from dense graphite filamentous. The as-synthesized catalyst was further used as a metal-free catalyst in the selective oxidation of H₂S in industrial waste effluents. According to the results, the defects decorated with oxygenated groups were highly active for performing selective oxidation of H₂S into elemental sulfur. The desulfurization activity was relatively high and extremely stable as a function of time on stream which indicated the high efficiency of these oxidized un-doped GFs as metal-free catalysts for the selective oxidation process. The high catalytic performance was attributed to both the presence of structural defects on the filamentous carbon wall, which acting as a dissociative adsorption center for the oxygen, and the oxygenated functional groups, which could play the role of active sites for the selective oxidation process.

Keywords: gas-phase oxidation; HNO₃; hierarchical graphite felts; selective oxidation; H₂S

---

1. Introduction

Nanocarbon-based metal-free catalysts consisting of a nitrogen-doped carbon matrix have received an ever increasing scientific and industrial interest in the field of heterogeneous catalysis over the last decade for several potential processes [1–6]. The introduction of hetero-element atoms, i.e., N, S, or P, inside the carbon matrix leads to the formation of metal-free catalysts which can activate oxygen bonding to generate reactive intermediates in different catalytic reactions according to the first report from Dai and co-workers [7]. The most studied form of these metal-free catalysts consisted of carbon nanotubes doped with nitrogen atoms, which has been extensively used in several catalytic processes [8–13]. Recently, work reported by Pham-Huu and Gambastiani [14,15] has shown that nitrogen-doped mesoporous carbon film, synthesized from food stuff raw materials, displays a high performance for different catalytic processes such as oxygen reduction reaction (ORR), direct dehydrogenation of ethylbenzene and selective oxidation of H₂S. Such nitrogen-doped metal-free catalysts display an extremely high stability as a function of time on stream or cycling tests which could be directly attributed to the complete lack of sintering consecutively to the direct incorporation of the nitrogen atoms inside the carbon matrix.

Nitrogen sites could also be efficiently replaced by carbon nanotubes containing surface defects decorated with oxygenated functional groups for the selective oxidation of H_2S into elemental sulfur [16]. Such a carbon metal-free catalyst displays a high stability as a function of the test duration under severe reaction conditions, i.e., high space velocity, low O_2-to-H_2S ratio. In the literature, the incorporation of these oxygenated functional groups has generally been carried out through oxidation treatments of the pristine carbon materials with different oxidants such as liquid HNO_3 [17,18], H_2SO_4 [19], $KMnO_4$ [20], and H_2O_2 [21] or through gaseous reactants like oxygen plasma [22], ozone [23], or CO_2 [24]. The main drawbacks of the liquid-phase treatments are the generation of a large amount of acid waste and the need for a subsequent washing step to remove the residual acid adsorbed on the sample surface. The gas-phase treatments seem to be the most appropriate ones for generating oxygenated functional groups on the carbon-based surface. Recent work by Su and co-workers [25] has shown that catalyst consisting of carbon nanotubes treated under ozone displays a high catalytic performance for different catalytic processes. The main drawback of such nano-catalyst is its nanoscopic dimension which renders difficult handling and transport and induces a high pressure drop in an industrial fixed-bed configuration. The catalyst recovery also represents a problem of health concerns due to its high ability to be breathed. In addition, the carbon nanotubes synthesis also requires the use of explosive and toxic organic compounds and hydrogen which induce a high cost operation due to the post-synthesis treatment of the by-products [26,27]. The purification process to remove the growth catalyst also leads to the generation of a large amount of wastewater, consecutive to the acid and basic treatment of the as-synthesized samples, which represents an environmental concern as well. It is thus of high interest to develop new metal-free carbon-based catalysts with high porosity, low cost, environmental benign, controlled macroscopic shape, and easy to scale up to replace the metal-free based carbon nanomaterials.

In the present article, we report on the use of gaseous oxidative HNO_3 to create surface defects, with exposed prismatic planes and decorated with oxygen functionalized groups, on the commercially available macroscopic carbon filamentous surface. The oxidized graphite felts (OGFs) will be directly tested as metal-free catalyst for the selective oxidation of H_2S issued from the refinery stream effluents to prevent the problem of air pollution [28–33]. Indeed, sulfur recovery from H_2S containing industrial effluents, mostly generated from oil refineries and natural gas plants, has become an increasingly important topic as H_2S is a highly toxic compound and represents a major air pollutant, which enters the atmosphere and causes acid rain [34,35]. The general process is to selectively transform H_2S into elemental sulfur by the equilibrated Claus process: $2\,H_2S + SO_2 \rightarrow (3/n)\,S_n + 2\,H_2O$ [36]. However, because of the thermodynamic limitations of the Claus equilibrium reaction, a residual concentration of H_2S of up to 3 vol. % is still present in the off-gas. To remove this residual H_2S in the effluent gas before releasing into atmosphere, a new process called super-Claus has been developed, which is a single-step catalytic selective oxidation of H_2S to elemental sulfur by using oxygen gas: $H_2S + 1/2\,O_2 \rightarrow (1/n)\,S_n + H_2O$. The super-Claus process is a direct oxidation process and thus is not limited by thermodynamic equilibrium. In the present work, the as-treated metal-free catalyst exhibits an extremely high catalytic performance as well as stability compared to the untreated. The catalytic sites could be attributed to the presence of oxygen species such as carbonyl, anhydride, and carboxyl groups decorating the structural defects present on the GFs surface defects upon treating under gaseous HNO_3. It is worthy to note that as far as the literature results are concerned, no such catalytic study using oxidized commercial filamentous GFs directly as metal-free catalyst with controlled macroscopic shape has been reported so far. The GFs also avoids the use of nanoscopic carbon with uncertainty about health concerns along with a validated industrial production and competitive production cost compared to the carbon nanotubes or carbon nanofibers.

2. Results and Discussion

2.1. Characteristics of the Acid Treated Graphite Felts

The macroscopic shape of the filamentous GFs was completely retained after the gaseous HNO$_3$ treatment according to scanning electron microscopy (SEM) micrographs with difference magnifications presented in Figure 1A–D. The HNO$_3$ treated filamentous GFs was decorated with evenly distributed carbon nodules as evidenced by the SEM analysis (Figure 1D,E). High magnification SEM micrographs (Figure 1E,F) also evidence the formation of cracks and holes on the cross section of the OGFs sample (indicated by arrows). Such cracks could be attributed to the degradation of the graphite structure during the acid treatment. It is worthy to note that the acid treatment also leads to the formation of a rougher graphite surface (Figure 1F) compared to the smooth one for the pristine graphite material. Such roughness could be attributed to the formation of defects on the surface of the treated sample.

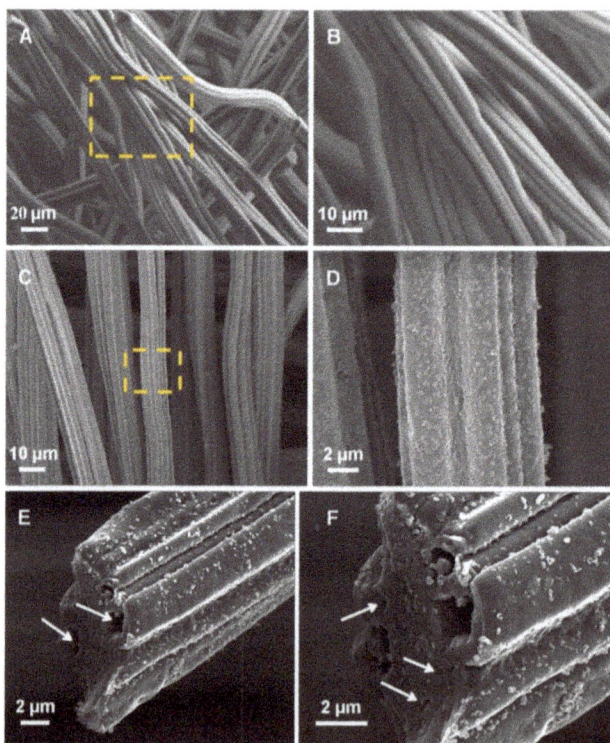

Figure 1. Scanning electron microscopy (SEM) micrographs of the (**A,B**) pristine commercial graphite felts (GFs), and (**C,D**) the same after treatment under gaseous HNO$_3$ at 250 °C for 24 h, noted oxidized graphite felts (OGFs)-24 showing the formation of nanoscopic nodules on its surface. SEM micrographs with medium and high resolution (**E,F**) reveal the formation of cracks on the cross section of the OGFs-24 (indicated by arrows) as well as a rougher surface after acid treatment.

High-resolution SEM image (Figure 2A) evidences the formation of defects on the whole surface of the carbon filamentous and some carbon extrusion in the form of discrete nodules. The Energy Dispersive X-ray analysis (EDX) carried out on the sample evidences the presence of oxygen intimately linked with carbon on the surface of the acid treated sample surface (Figure 2B–D) which confirms the high concentration of oxygenated functional groups decorating the surface defects.

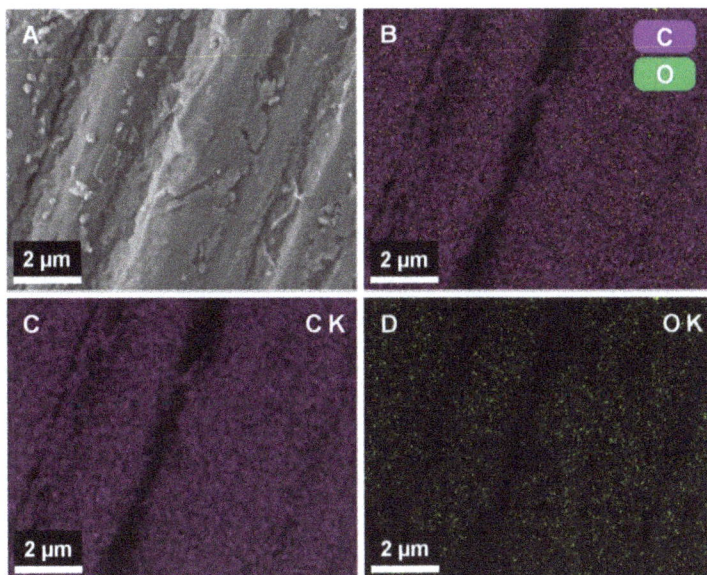

Figure 2. (**A**) SEM micrograph of the acid treated graphite felts, OGFs-24, (**B**) Elemental mapping showing the presence of C and O on the sample surface. (**C,D**) Elemental maps of carbon and oxygen elements on the OGFs-24 surface.

For industrial applications, the catalyst should be prepared in a controlled macroscopic shape in order to avoid problems of handling and transport and also to prevent excessive pressure drop within the catalyst bed. In the present synthesis method, the GFs can be prepared with different macroscopic shapes, i.e., pellets, disk with different holes, depending to the downstream applications as shown in Figure 3. The as-synthesized GFs could be directly used as metal-free catalyst, see catalytic application below, or also as catalyst support where the defective surface could lead to a high metal dispersion and stability. The macroscopic shape allows easy catalyst/products separation for liquid-phase catalytic processes which represents a costly process in the case of powdered catalysts.

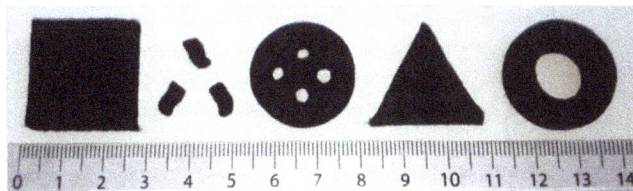

Figure 3. Graphite felts with different macroscopic shapes for various catalytic applications both in gas- and liquid-phase processes.

According to our previous work on carbon nanotubes, the acid treatment lead to the formation of defects on the surface of the carbon material which was decorated by oxygenated functional groups. Such defects are expected to be formed through oxidative reaction between the gaseous nitric acid vapors and the graphite sample. The functionalization of the formed defects is expected to take place by the partial decomposition of the oxygen in the gas-phase medium. The formation of defects along the graphite microfilamentous surface during the acid treatment step also significantly increases the overall specific surface area (SSA) of the as-treated materials. The specific surface area of the acid treated filamentous GFs steadily increased as a function of the acid treatment duration as evidenced in Figure 4A,B. According to the results, the SSA of the filamentous GFs was stepwise raised from 10 m^2/g to more than 300 m^2/g after the gaseous HNO$_3$ treatment for 30 h. It should be noted that such high SSA has never been accounted for chemical treated commercial microfilamentous carbon fibers. The increase of the SSA was attributed to the formation of a more porous graphite structure with higher effective surface area consecutive to the removal of carbon during the treatment from the sample (see transmission electron microscopy (TEM) analysis below). It is expected that such porosity was formed in or close to the surface of the carbon filaments and thus, allows the complete maintenance of the macroscopic shape of the material. One cannot exclude some porosity network which could be generated inside the pristine graphite microfilamentous.

The gaseous acid treatment also induces an overall oxidation of carbon matrix leading to a weight loss of the treated material compared to that of the pristine one. Such phenomenon has already been reported by Xia and co-workers with a similar treatment [37] and also by several groups in the literature [18]. The weight loss during the acid treatment process is accounted for the corrosion of the filaments where part of the carbon with low degree of graphitization was removed leaving behind the carbon nodules or porosity as observed by SEM. The weight loss calculated on the basis of the initial weight and the one after acid treatment as a function of the treatment duration is presented in Figure 4B.

It is expected that the corrosion phenomenon which occurring during the treatment was responsible for the increase of the SSA of the treated samples similarly to that reported for the carbon nanotubes treated with gaseous HNO$_3$ or O$_3$-H$_2$O mixture [16,25]. The treatment induces the formation of surface pores along the carbon fiber axis, i.e., corrosion, which significantly contribute to the improvement of the overall SSA of the treated samples. These defects are also the place for oxygen insertion to generate oxygenated functional groups on the surface of the OGFs samples as evidenced by X-ray photoelectron spectroscopy (XPS) presented below.

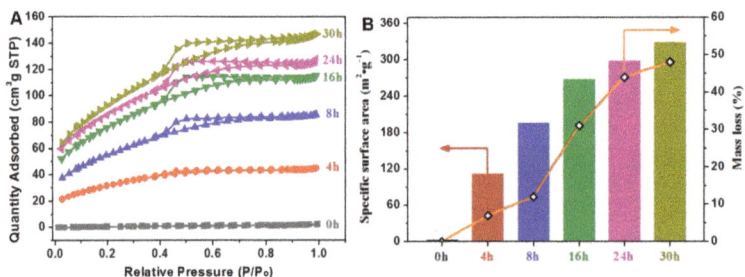

Figure 4. N$_2$ adsorption-desorption isotherms (**A**) and weight loss (open circles) and specific surface area (columns) modification of the acid treated carbon-based materials as a function of the treatment duration (**B**).

The weight loss increases with the treatment duration, especially for durations longer than 8 h. It is expected that the low temperature weight loss could be assigned to the removal of pore amorphous carbon, since they are considered to be more reactive than the graphitic carbon filamentous while at high temperature, weight loss is linked with the removal of carbon in the graphitic structure consecutive to the formation of structural defects and nodules on the remained filament wall. Indeed, under a more severe treatment, i.e., longer duration, the weight loss becomes significantly, i.e., the weight loss recorded for the sample after being treated at 250 °C for 8 h and 24 h are 12% and 44%, respectively. According to the results the graphite displayed three distinct weight loss regions: (i) at treatment duration <8 h, the oxidation process is relatively slow which could be attributed to the low reactivity of the graphite felt surface; (ii) at treatment ranged between 8 to 24 h, the oxidation rate is significantly increased and could be due to the depth oxidation of graphite matter through the surface defects generated previously; and (iii) at duration >24 h, the oxidation process becomes almost flat which could be attributed to the fact that depth oxidation process could be hinder due to some diffusion problem.

The oxygen incorporation into the acid treated samples can be clearly observed through XPS survey spectra recorded on the fresh and HNO_3 treated samples (Figure 5A). It is worthy to note that XPS analysis allows one to map out elements concentration at a depth of ca. 6 nm from the surface and thus part of the oxygenated functional groups localized at a distance >6 nm cannot be accurately detected. The deconvoluted O1s spectrum in Figure 5B shows the presence of three peaks which can be assigned to the C=O (ketone, aldehyde, quinone . . .), –C–OH, –C–O–C– (alcohol, ether), and –O–C=O (carboxylic, ester) oxygen species [25].

Raman spectroscopy was performed to investigate the change in the graphitic structure of the GFs after treatment with gaseous nitric acid at different durations. As shown in Figure 5C, every sample displayed three bands corresponding to the different carbonaceous structures: the G band attributed to an ideal graphitic lattice at around 1580 cm^{-1} [38]; the D band (~1350 cm^{-1}) associated with the structural defects [39]; and D' corresponding to the disordered graphitic fragments at ~1620 cm^{-1} [40], respectively. The I_D/I_G ratio increases as increasing the acid treatment duration (Figure 5D). After the treatment of 24 h, the I_D/I_G increased from 0.77 for GFs to 1.86. Meanwhile $I_{D'}/I_G$ increased more than three times from 0.26 to 0.80, which indicated the strong acidic oxidant etched the graphene lattice of GFs and created more defects and disordered graphitic fragments. Furthermore, the greater duration of the treatment on samples, the more structural defects and disordered fragments that were obtained. Consistent with the morphology from SEM, the etching effect of the treatment made the GFs with an extremely rough surface and much more macroscopic carbon fragments. Moreover, the G band shifted to the higher wavenumber by about 8 cm^{-1}, which may be attributed to the oxygen-containing functional groups generated on the surface of treated samples and confirmed by the results of XPS and temperature-programmed desorption (TPD), such as the O–H bending and C=O stretching [41].

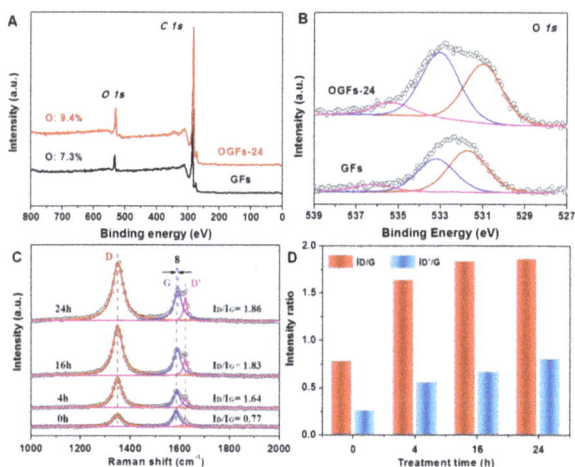

Figure 5. (**A**) Survey X-ray photoelectron spectroscopy (XPS) of OGFs-24 in comparison with pristine GFs, (**B**) deconvolution O1s present the oxygen species on the surface of the samples, (**C**) Raman spectra, and (**D**) I_D/I_G and $I_{D'}/I_G$ ratios of the pristine GFs and the OGFs after acid treatment with various durations.

The physical characteristics of the samples after acid treatment with different durations are summarized in Table 1.

Table 1. The physical chemistry properties of samples as a function of the acid treatment duration.

Treatment Duration	SSA [a] m^2/g	Mass Loss [b] %	I_D/I_G [c]	$I_{D'}/I_G$ [c]	O at % [d]	T_{WL} [e] °C
0 h	10	0	0.77	0.26	7.3	802
4 h	112	7	1.64	0.55	-	-
8 h	196	12	-	-	-	-
16 h	268	31	1.83	0.66	-	685
24 h	298	44	1.86	0.80	9.4	612
30 h	329	48	-	-	-	-

[a] BET specific areas. [b] The mass loss of samples after acid treatment. [c] The I_D/I_G and $I_{D'}/I_G$ ratio calculated from Raman spectra. [d] The atom percent of surface oxygen elemental from XPS analysis. [e] The temperature of weight loss peak determined by TG/DTG profiles. - Not detected.

TEM analysis is also used to investigate the influences of the HNO$_3$ treatment on the microstructure of the GFs (Figure 6). Compared with parallel graphitic layers on pristine GFs (Figure 6A,B), there are porous structures with disordered graphitic fragments formed on the outer region of the filamentous carbon of the acid treated GFs (pointed out by arrows in Figure 6C,D), which are consistent with the analysis of SSA and by Raman. TEM analysis reveals the formation of a less dense graphite structure in the OGFs sample (Figure 6C) compared to that observed for the pristine GFs (Figure 6B). High resolution TEM micrographs (Figure 6D) clearly evidence the porous structure of the treated sample. Such phenomenon can be attributed to the oxidation of a weakly graphitized carbon by the gaseous HNO$_3$ during the treatment, which forms entangled carbon sidewalls with high defect density.

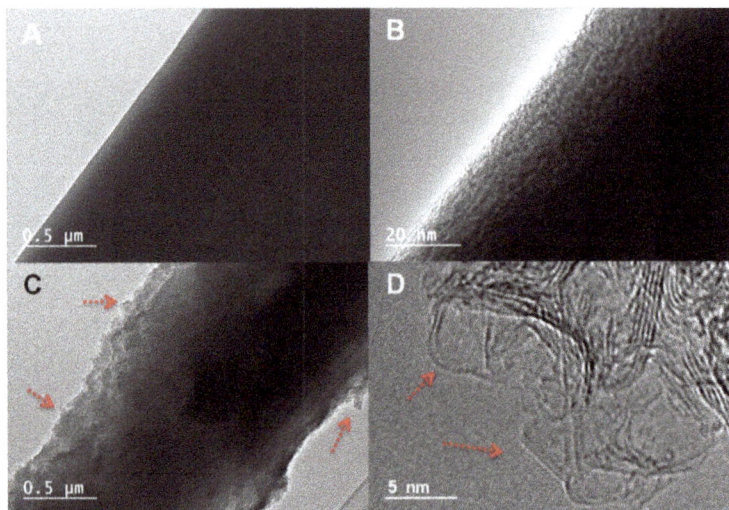

Figure 6. Transmission electron microscopy (TEM) images of GFs (**A,B**) and OGFs-24 (**C,D**). Surface defects generated on the OGFs are indicated by arrows.

According to the TEM results one could expected that during the acid treatment process part of the graphitic structure is slowly attacked, leading to a weight loss as a function of time of treatment, leaving behind porous structure with defects which contributes to an increase of the material SSA with time. Such corrosion process explains the formation of cracks and holes within the pristine graphite filamentous as observed by SEM. The as generated porosity with a highly defective surface decorated with oxygenated functional groups is expected to be of great interest for being used either as metal-free catalyst but also as catalyst support with high density of anchorage sites for hosting metal or oxide nanoparticles. Recent work has pointed out the high efficiency of defects decorated carbon nanotubes after treatment in the presence of ozone and water for anchoring gold nanoparticles [42]. The porous structure of the OGFs will be investigated in detail by mean of transmission electron microscopy tomography (TEM-3D) technique [43] to map out the porosity of the OGFs material and its influence on the metal nanoparticles dispersion.

The characterization of the different oxygenated groups present on the graphite surface was investigated by temperature-programmed desorption coupled with mass spectroscopy (TPD-MS). The surface oxygen groups can be assessed by the type of released molecules with their relevant peak areas and decomposition temperatures [44]. The amount of CO (m/e = 28) and CO_2 (m/e = 44) generated during the TPD process is presented in Figure 7 as a function of the desorption temperature. The evolution of CO_2 was ascribed to the decomposition of carboxylic acids, anhydrides, and lactones (Figure 7A), whereas the CO evolution was resulted from the decomposition of anhydrides, phenols, and carbonyls (Figure 7B). The amounts of corresponding groups determined by TPD with the deconvolution of evolved CO_2 and CO peaks are summarized in Table 2. The CO and CO_2 concentration increases as increasing the treatment duration, confirming that the formation of oxygenated functional groups is directly depending to the acid treatment, which is in accordance with the XPS analysis (Figure 5A,B).

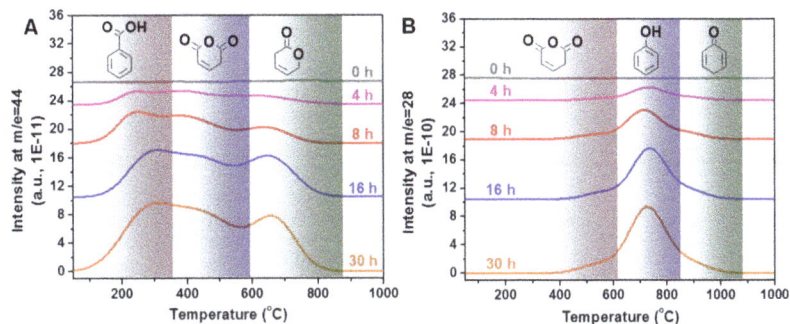

Figure 7. Temperature-programmed desorption (TPD) profiles showing the evolvement of CO_2 (**A**) and CO (**B**) as a function of the desorption temperature on the different oxidized samples at 250 °C and with different treatment duration.

It is expected that the acid treatment duration increases the defect density which in turn increases the amount of oxygenated functional groups directly linked to such defects. Furthermore, the total amount of oxygen on the sample increased from 165 μmol/g, on the pristine GFs, to 9116 μmol/g, on the OGFs-30, after 30 h of gas-phase oxidation treatment. Such results have already been reported by other research groups in the literature during acid activation process for the synthesis of highly reactive carbon-based catalyst supports [45]. However, data deals with the use of such defective macroscopic OGF materials with high effective surface area as metal-free catalyst has never been reported so far.

Table 2. The content of oxygen functional groups as a function of the acid treatment duration at 250 °C on the samples determined from TPD profiles.

Treatment Duration	CO_2 Desorption µmol/g			CO Desorption µmol/g			Total O µmol/g
	Carboxylic Acids	Anhydrides	Lactones	Anhydrides	Phenols	Carbonyls	
0 h	-	-	-	-	-	-	165
4 h	53	187	128	141	951	230	1691
8 h	134	486	148	332	2199	498	3797
16 h	304	626	412	527	3865	660	6393
30 h	422	1186	424	735	5462	887	9116

Thermogravimetry analysis (TG/DTG) has been generally used to study the oxidative stability of carbon materials [46–48]. As shown in Figure 8, all the samples exhibited a weight loss step during heat-treatment process, resulted from the combustion of carbon at high temperature. Obviously, the temperature of weight loss peak for OGFs-16 on the DTG curve (685 °C) was lower than that of the pristine GFs (802 °C), indicating the formation of highly reactive graphite species on the sample after the oxidation process. Meanwhile, a further decline of the oxidation temperature was observed for the OGF-24 (612 °C) with the longer treatment duration, i.e., 24 h instead of 16 h. On one hand, these reactive carbon species may be linked with the presence of structural defects or from the disordered graphitic structure, according to the Raman and TEM results (Figures 5C and 6), which display lower thermal stability than the pristine graphite material [49]. On the other hand, higher SSA (Figure 4) of OGFs provides higher surface contact between the sample and reactant gas which could favor the oxidation process. Moreover, the abundant oxygen functional groups derived from acid treatment (Table 2) could be active sites for the dissociative adsorption of O_2 [40]. It is expected that all those parameters will actively contribute to the lower oxidative resistance of the acid treated graphite felt.

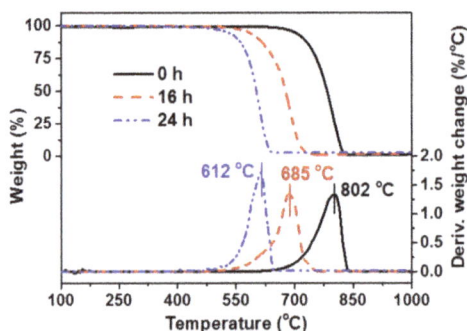

Figure 8. Thermogravimetry analysis (TG/DTG) profiles of pristine GFs, OGFs-16 and OGFs-24.

2.2. OGFs as Metal-Free Catalyst for Selective Oxidation of H_2S

The GFs and OGFs-16 catalysts were tested in the gas-phase selective oxidation of H_2S into elemental sulfur under realistic reaction conditions. The OGFs-16 catalyst has been chosen among the other treated samples according to the following facts: (i) the OGFs-16 displays a relatively high SSA along with a lower weight loss during the acid treatment and (ii) the OGFs-16 also displays higher oxidative resistance compared to the samples treated with longer duration. The pristine GFs catalyst shows no noticeable desulfurization activity under the operated reaction conditions (not reported) and confirms its inactivity for the reaction considered. The desulfurization performance obtained on the OGFs-16 catalyst at reaction temperature of 230 °C and 250 °C is presented in Figure 9A as a function of time on stream. The OGFs-16 catalyst displays a relative high H_2S conversion at a Weight Hourly Space Velocity (WHSV) of 0.05 h^{-1} with a total H_2S conversion of 87% and a sulfur selectivity of 83%. In addition, it is worthy to note that the desulfurization activity remains stable as a function of time-on-stream for more than several dozen hours on stream, which indicates that no deactivation occurred on the catalyst. Such relatively high desulfurization activity could be attributed to the high stability of the defect decorated with oxygenated functional groups on the OGFs-16 catalyst which are generated at a relatively high treatment temperature, i.e., 250 °C. Similar results have also been reported by the HNO$_3$ treated carbon nanotubes (CNTs) which is expected to bear the same active center [16]. Jiang et al. [50] have reported that pure carbon nanocages, which possess abundant holes, edges, and positive topological disclinations display a relatively good oxygen reductive performance which is even better than those reported for undoped CNTs. In the present work, the control of the macroscopic shape of the OGFs catalyst represents also a net advantage as the catalyst shape can be modified in a large range depending to the downstream applications.

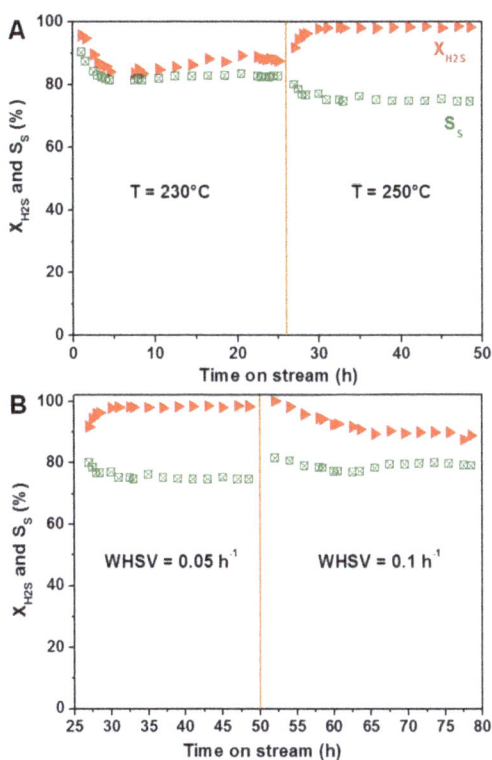

Figure 9. Desulfurization performance as a function of time on stream on the OGFs-16 catalyst. (**A**) Reaction conditions: [H$_2$S] = 1 vol. %, [O$_2$] = 2.5 vol. %, [H$_2$O] = 30 vol. %, Weight Hourly Space Velocity (WHSV) = 0.05 h^{-1}; (**B**) Reaction conditions: [H$_2$S] = 1 vol. %, [O$_2$] = 2.5 vol. %, [H$_2$O] = 30 vol. %, reaction temperature = 250 °C.

Increasing the reaction temperature from 230 °C to 250 °C leads to an improvement of the H$_2$S conversion, i.e., 99% instead of 87%, along with a slight decrease of the sulfur selectivity from 83% to 77% due to the fact that high reaction temperature favors the complete oxidation of S to SO$_2$ in the presence of excess oxygen in the feed [51]. However, after a period of induction where both H$_2$S conversion and sulfur selectivity are modified the catalyst reached a steady-state for the rest of the test which again confirm its high stability.

Increasing the WHSV from 0.05 h^{-1} to 0.1 h^{-1}, keeping the reaction temperature at 250 °C, leads to a slight decrease of the H$_2$S conversion from 100% to 90% followed by a steady-state (Figure 9B).

The sulfur selectivity slightly increases from 75% to 82% at high space velocity and remains stable for the test. Such results are in good agreement with literature results as increasing the WHSV leads to a shorter sojourn time of the reactant and thus, reduce the rate of reactant dissociation for the reaction. The specific activity calculated is 0.56 $mol_{H2S}/g_{catalyst}/h$ which is relatively close to those reported for other metal-free catalysts [52]. Further improvement of the catalytic performance could be done by using compressed OGFs with higher specific weight in order to reduce the empty space inside the catalyst bed.

The results obtained indicate that the OGF metal-free catalyst displays relatively high sulfur selectivity, i.e., >70%, even at a relatively high reaction temperature, i.e., 250 °C, and in the presence of a high H_2S concentration. It is expected that such sulfur selectivity is linked with the high thermal conductivity of the graphite felt support which could efficiently disperse the reaction heat through the catalyst matrix to avoid local hot spots formation which is detrimental for the sulfur selectivity. Similar results have also been reported on the medium thermal conductive silicon carbide carrier where the lack of local hot spots leads to a significant improvement of the reaction selectivity for selective oxidation of H_2S [10] and also in other exothermal reactions such as Fisher-Tropsch synthesis [53,54], dimethyl ether [55,56], and propylene synthesis [57,58]. The relatively high sulfur selectivity observed in the present work could also be attributed to the presence of large voids inside the sample which could favor the rapid evacuation of the sulfur intermediate species before complete oxidation.

The effluents containing H_2S could be originated from different sources with various steam concentrations ranged from few to several percent, i.e., effluent from biogas plant or from the Claus reactor. It is expected that the steam concentration could have a significant influence on the desulfurization performance, i.e., H_2S conversion and sulfur selectivity, due to the problem of competitive adsorption. The influence of steam concentration was investigated and the results are presented in Figure 10. Decreasing the steam concentration in the reactant feed from 30% to 10% leads to a similar H_2S conversion, i.e., 99%, but to a significant increase of the sulfur selectivity from 75% to 85% keeping other reaction conditions similar. Such result indicates that steam could condense to yield water film inside the catalyst porosity, even at a relatively high reaction temperature, which favors oxygen dissociation leading to a higher complete oxidation to yield SO_2. Decreasing the steam concentration leads to a lower oxygen atoms available on the catalyst surface which in turn, reduces the selectivity towards SO_2. Such hypothesis can be confirmed by decreasing the O_2-to-H_2S ratio keeping the other reaction conditions similar. Decreasing the O_2-to-H_2S ratio seems to have hardly affected the sulfur selectivity while a slight decrease of the H_2S conversion is observed. Such result could be attributed to the fact that the dissociated oxygen on the catalyst surface reacts with H_2S to yield both S and SO_2 in a parallel reaction pathway but with a different reaction rate. Decreasing the O_2-to-H_2S ratio leads to a decrease of the available oxygen for the reaction and, as a consequence lowers the H_2S conversion. Finally, regarding the influence of steam on the H_2S conversion it should be noted that under the reaction conditions used the H_2S conversion is almost complete which could render the investigation of small effect, such as steam concentration, difficult as at such high conversion level one cannot rule out the fact that some active sites remain unemployed.

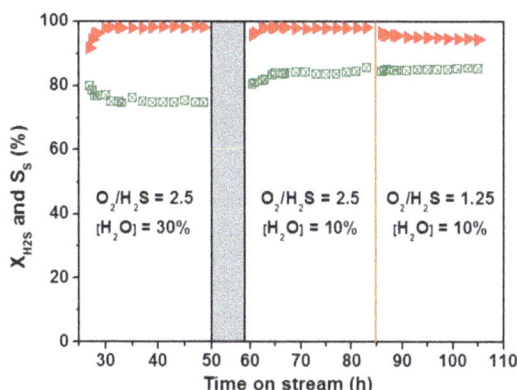

Figure 10. Influence of the steam concentration and O_2-to-H_2S ratio on the desulfurization performance on the pristine GFs and the one after treatment under gaseous HNO_3 at 250 °C for 16 h (OGFs-16) catalysts. Reaction conditions: $[H_2S]$ = 1 vol. %, reaction temperature – 250 °C and WHSV = 0.05 h^{-1}.

It is worthy to note that the reaction temperature of 250 °C is slightly higher than that usually used for the selective oxidation of H_2S into elemental sulfur on other metal-free carbon-based catalysts. However, on the OGFs-16 metal-free catalyst the sulfur selectivity remains relatively high, i.e., >70% and even ≥85% at low steam concentration (Figure 10), at almost complete conversion of H_2S. Such results could be attributed to the large open porosity of the catalyst, i.e., 90% of empty space, which allow the formed sulfur to rapidly escape the catalyst before secondary reaction with excess oxygen to yield SO_2.

It is also noted that the catalyst displays also a relatively high stability, both in terms of H_2S conversion and sulfur selectivity under severe reaction conditions, i.e., 0.1 h^{-1} of WHSV and high reaction temperature where the H_2S conversion is not complete indicating that all the active sites were involved in the reaction, which confirm again the advantage of the oxygenated functional groups on the defect sites. Such results are of high interest as usually on oxides or metals containing catalysts sintering is the main cause of deactivation with time on stream under severe reaction conditions.

The results obtained indicate that acid treated graphite felts could be efficiently used as metal-free catalyst for the selective oxidation of H_2S into elemental sulfur. The different catalytic results are summarized in Table 3 and compared with those reported on other metal-free catalysts. The OGFs-16 catalyst displays a relatively high sulfur selectivity compared to the other metal-free catalysts operated at higher space velocity and temperature. Such results pointed out the high efficiency of the OGFs-16 catalyst to perform selective oxidation of H_2S which could be attributed to the large open porosity of the catalyst providing high rate of intermediate sulfur escaping.

Table 3. Selective oxidation of H_2S to sulfur over different metal-free catalysts.

Catalysts	T °C	$[H_2S]$ vol. %	$[O_2]$ vol. %	$[H_2O]$ vol. %	WHSV h^{-1}	X_{H2S} [a] %	S_S [b] %	Y_S [c] %	Ref.
OGFs-16	250	1	2.5	10	0.1	98	86	84	This work
O-CNT-250-24	230	1	2.5	30	0.6	95	76	72	16
N-CNT/SiC-750	190	1	2.5	30	0.6	97	75	73	52

[a] Maximum H_2S conversion. [b] The corresponding sulfur selectivity. [c] The corresponding yield of sulfur. ($Y_S = X_{H2S} \times S_S$).

3. Materials and Methods

3.1. Graphite Felts

The commercial filamentous GFs, ex-polyacrinonitrile (PAN), with a dimension of 1×3 m^2 (thickness of 6 mm) was supplied by Carbone Lorraine (MERSEN, La Défense, France). The as-received GFs was shaped in the form of disks (Ø × thickness of 4 × 6 mm) for the experiment. The GFs displays a relatively low specific surface area, 10 ± 2 m^2/g measured by means of N$_2$ adsorption, which is mostly linked with the geometric surface area in good agreement with the extremely low porosity of the material. It is worthy to note that the GFs can be also shaped in various dimensions including pellets, disks or complexes structures, depending to the downstream applications, i.e., gas- or liquid-phase reactions, which represent a net advantage compared to other metal-free catalysts where low-dimensional shape is a main concern for further industrial development.

3.2. Gaseous HNO$_3$ Treatment of GFs

For the gaseous acid treatment, as shown in Figure 11, the GFs was first oxidized at moderate temperature (500 °C) in air for 1 h in order to remove as much as possible residue from it surface. The as-treated pre-shaped GFs was loaded inside a tubular reactor (15 × 100 mm) and heated to the desired temperature of 250 °C by an external electrical furnace. The treatment temperature was fixed at 250 °C according to the previous investigation results reported by Duong-Viet et al. [16]. The treatment temperature was controlled by a thermocouple inserted inside the furnace. The reactor containing GFs in the form of disk (Ø × thickness of 4 × 6 mm) was connected to a round bottom flask filled with 150 mL of HNO$_3$ with a concentration of 65% (SIGMA-ALDRICH, Saint-Quentin Fallavier, France). The temperature of the round bottom flask was fixed at 125 °C and the HNO$_3$ solution was keep under magnetic stirring. The gaseous acid passed through the GFs bed was further condensed in another flask which can be re-used for the process and thus, reducing in a significant manner the problem linked with liquid waste recycling and recovery. The sample was treated with different durations in order to rule out the influence of these treatments on its final microstructure and chemical properties and to correlate these physical properties with the catalytic activity. The sample was washed with deionized water after the acid treatment and oven dried at 130 °C for overnight. In this work, the GFs after acid treated are noted as follows: OGFs-X, for the treatment duration in hour, for example: OGFs-24 indicates that the raw GFs were treated for 24 h under gaseous HNO$_3$. It is noted that the treatment is not only limited to small amount of catalyst as higher amount of GFs can be also prepared by changing the reactor size.

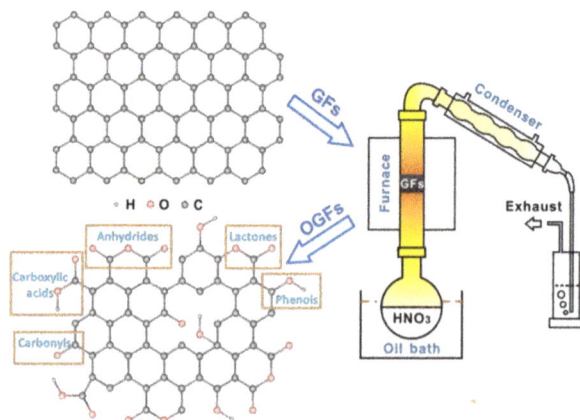

Figure 11. Schematic illustration of the OGFs prepared by the gaseous HNO$_3$ treatment and the different oxygenated functional groups generated on the graphite surface after the treatment.

3.3. Characterization Techniques

The scanning electron microscopy (SEM) and elemental mapping were carried out on a ZEISS GeminiSEM 500 microscope (ZEISS, Oberkochen, Germany) with a resolution of 5 nm. The sample was deposited onto a double face graphite tape in order to avoid the problem of charging effect during the analysis.

The Raman analysis was carried out using a LabRAM ARAMIS confocal microscope spectrometer (HORIBA, Kyoto, Japan) equipped with CCD detector. A laser line with the following characteristics was used to excite sample, 532 nm/100 mW (YAG) with Laser Quantum MPC600 PSU (Novanta, Bedford, OH, USA).

The specific surface area of the sample was determined in a Micromeritics sorptometer (Micromeritics, Norcross, GA, USA). The sample was outgassed at 250 °C under vacuum for 14 h in order to desorb moisture and adsorbed species on its surface.

The X-ray photoelectron spectroscopy (XPS) measurements of the support and catalyst were performed by using a MULTILAB 2000 (THERMO) spectrometer (Thermo Fisher Scientific, Waltham, MA, USA) equipped with an AlKα anode ($h\nu$ = 1486.6 eV) with 10 min of acquisition to achieve a good signal to noise ratio. Peak deconvolution was performed with the "Avantage" program from the Thermoelectron Company. The C1s photoelectron binding energy was set at 284.6 $\pm$ 0.2 eV relative to the Fermi level and used as reference to calibrate the other peak positions.

The transmission electron microscopy (TEM) analysis was performed on a JEOL 2100F instrument (JEOL, Tokyo, Japan) working at a 200 kV accelerated voltage, equipped with a probe corrector for spherical aberrations, and with a point-to-point resolution of 0.2 nm. The sample was ground and dispersed by ultrasound in an acetone solution for 5 min and then a drop of the solution was deposited on a copper grid covered with a holey carbon membrane for observation.

The temperature-programmed desorption coupled with mass spectroscopy (TPD-MS) (Micromeritics, Norcross, GA, USA) experiments were carried out under helium flow at atmospheric pressure. The sample was flushed under helium for 30 min at room-temperature and then the temperature was raised from room-temperature to 1000 °C with a heating rate of 10 °C /min. The resulting gases, mostly CO (m/z signal at 28) and CO_2 (m/z signal at 44) were continuously monitored with a time interval of 0.1 s.

3.4. Selective Oxidation Process

The catalytic selective oxidation of H_2S by oxygen (Equation (1)) was carried out in an all glass microreactor working isothermally at atmospheric pressure. During the reaction other secondary reactions could also take place: consecutive oxidation of the formed sulfur with an excess of oxygen or direct oxidation of H_2S to yield SO_2 (Equations (2) and (3)). The temperature was controlled by a K-type thermocouple and a Minicor regulator. The gas mixture was passed downward through the catalyst bed. Before the test, the reactor was flushed with helium at room temperature until no trace of oxygen was detected at the outlet. The helium flow was replaced by the one containing steam. The catalyst was slowly heated up to the reaction temperature, and then the wet helium flow was replaced by the reactant mixture. The gases (H_2S, O_2, He) flow rate was monitored by Brooks 5850TR mass flow controllers (Brooks Instrument, Hatfield, PA, USA) linked to a control unit. The composition of the reactant feed was H_2S (1 vol. %), O_2 (1.25 vol. % or 2.5 vol. %), H_2O (10 vol. % or 30 vol. %) and He (balance). The use of a relatively high concentration of steam in the feed is motivated by the will to be as close as possible to the industrial working conditions as the steam formed during the former Claus units is not removed before the oxidation step and remains in the treated tail gas. The steam (10 vol. % or 30 vol. %) was fed to the gas mixture by bubbling a helium flow through a liquid tank containing water maintained at 56 °C or 81 °C. The O_2-to-H_2S molar ratio was varied from 1.25 to 2.5 with a Weight Hourly Space Velocity (WHSV) at 0.05 h^{-1} or 0.1 h^{-1}. It is worth to note that the WHSV used in the present work is close to the usual WHSV used in the industrial process for this kind of reaction, i.e., 0.09 h^{-1} [59].

$$H_2S + 1/2\,O_2 \rightarrow 1/n\,S_n + H_2O \qquad \Delta H = -222\ \text{kJ·mol}^{-1} \qquad (1)$$

$$1/n\,S_n + O_2 \rightarrow SO_2 \qquad \Delta H = -297\ \text{kJ·mol}^{-1} \qquad (2)$$

$$H_2S + 3/2\,O_2 \rightarrow SO_2 + H_2O \qquad \Delta H = -519\ \text{kJ·mol}^{-1} \qquad (3)$$

The reaction was conducted in a continuous mode and the sulfur formed during the reaction was vaporized, due to the relatively high partial pressure of sulfur at these reaction temperatures, and was further condensed at the exit of the reactor in a trap maintained at room temperature.

The analysis of the inlet and outlet gases was performed on-line using a Varian CP-3800 gas chromatography (GC) (Agilent, Santa Clara, CA, USA) equipped with a Chrompack CP-SilicaPLOT capillary column coupled with a thermal conductivity detector (TCD), allowing the detection of O_2, H_2S, H_2O, and SO_2. The limit detection of the H_2S and SO_2 is about 10 ppm. The results are reported in terms of H_2S conversion and sulfur selectivity in percent. The percent of sulfur and SO_2 selectivity is calculated on a basis of 100%. The sulfur balance is calculated on the basis of the sum of sulfur detected in the solid sulfur recovered in the cold trap and the selectivity towards SO_2 and the theoretical sulfur calculated from the H_2S conversion on the catalyst. The sulfur balance is about 92% and the difference could be attributed to (i) some dissolution of SO_2 into the condensed water at the exit of the reactor and (ii) the incomplete recovery of the solid sulfur in the cold trap.

4. Conclusions

In summary, we have shown that oxidation with gaseous HNO_3 can be an efficient and elegant pre-activation step to generate active metal-free carbon-based catalysts decorated with surface defects containing oxygenated functional groups from available and low cost commercial filamentous graphite felts. The gaseous acid treatment leads to the formation of a high surface area carbon-based material which can find use in several catalytic processes as either metal-free catalyst or as catalyst support. It is worthy to note that it is the first time that such results are reported as literature only reports metal-free catalysts based on nanocarbons, whose synthesis requires harsh reaction conditions along with the problem linked with waste treatment. According to the obtained results the defects created on the filamentous carbon wall and the formation of oxygenated functional groups during the gaseous acid treatment provide active sites for H_2S and oxygen adsorption which contribute to the selective oxidation of H_2S into elemental sulfur under similar reaction conditions with those operated in the industrial plants. The catalyst displays a relatively high sulfur selectivity as well as relatively high stability as a function of time-on-stream, under severe reaction conditions, indicating that deactivation by surface fouling or oxygen groups removing is unlikely to occur. The high specific surface area as well as surface porous structure could be extremely helpful for developing new catalytic systems. Such hierarchical metal-free catalysts can be prepared with different macroscopic shapes for subsequence downstream applications. Work is ongoing to evaluate such carbon-based materials as hierarchical macroscopic support for metal nanoparticles which could find use in other catalytic processes where high dispersion and strong anchorage of the active phase are required in order to prevent long term deactivation through catalyst sintering (gas-phase reaction) or leaching (liquid-phase reaction) and also in terms of recovery.

Acknowledgments: Zhenxin Xu and Bing Li would like to thank the Chinese Scholarship Council (CSC) for the PhD grant for their stay at the ICPEES. The project is partly supported by the National Foundation for Science and Technology Development of Vietnam (Nafosted) program of research. The SEM experiments were carried out at the facilities of the ICPEES-IPCMS platform. Vasiliki Papaefthimiou, Thierry Romero, and Sécou Sall (ICPEES, UMR 7515) are gratefully acknowledged for performing XPS, TPD-MS, and SEM experiments. Jean-Mario Nhut (ICPEES) is gratefully acknowledged for technical and scientific help during the project. Loïc Vidal (IS2M, UMR 7361) is gratefully acknowledged for TEM experiments.

Author Contributions: Z.X., C.D.-V., and C.P.-H. conceived and design the experiments; H.B., Z.X., and C.D.-V. participated in the analytical experiments; T.T.-H., H.B., Z.X., C.D.-V., B.L., C.P.-H., and L.N.-D. participated in the discussion and interpretation of the data. Z.X., C.D.-V., and C.P.-H. wrote the article.

Conflicts of Interest: The authors declare no conflict of interest.

References

1. Su, D.S.; Perathoner, S.; Centi, G. Nanocarbons for the development of advanced catalysts. *Chem. Rev.* **2013**, *113*, 5782–5816. [CrossRef] [PubMed]
2. Tessonnier, J.P.; Su, D.S. Recent progress on the growth mechanism of carbon nanotubes: A review. *ChemSusChem* **2011**, *4*, 824–847. [CrossRef] [PubMed]
3. Duong-Viet, C.; Ba, H.; Truong-Phuoc, L.; Liu, Y.; Tessonnier, J.P.; Nhut, J.M.; Granger, P.; Pham-Huu, C. *Nitrogen-Doped Carbon Composites as Metal-Free Catalysts*; Elsevier Series Book; Parvulescu, V., Kemnitz, E., Eds.; Elsevier: Amsterdam, The Netherlands, 2016; pp. 273–312, ISBN 978-0-444-63587-7.
4. Liang, J.; Du, X.; Gibson, C.; Du, X.W.; Qiao, S.Z. N-doped graphene natively grown on hierarchical ordered porous carbon for enhanced oxygen reduction. *Adv. Mater.* **2013**, *25*, 6226–6231. [CrossRef] [PubMed]
5. Wei, W.; Liang, H.; Parvez, K.; Zhuang, X.; Feng, X.; Mullen, K. Nitrogen-doped carbon nanosheets with size-defined mesopores as highly efficient metal-free catalyst for the oxygen reduction reaction. *Angew. Chem. Int. Ed.* **2014**, *53*, 1570–1574. [CrossRef] [PubMed]
6. Tang, Y.; Allen, B.L.; Kauffman, D.R.; Star, A. Electrocatalytic activity of nitrogen-doped carbon nanotube cups. *J. Am. Chem. Soc.* **2009**, *131*, 13200–13201. [CrossRef] [PubMed]
7. Gong, K.; Du, F.; Xia, Z.; Durstock, M.; Dai, L. Nitrogen-doped carbon nanotube arrays with high electrocatalytic activity for oxygen reduction. *Science* **2009**, *323*, 760–764. [CrossRef] [PubMed]
8. Liu, Y.; Jin, Z.; Wang, J.; Cui, R.; Sun, H.; Peng, F.; Wei, L.; Wang, Z.; Liang, X.; Peng, L.; et al. Nitrogen-doped single-walled carbon nanotubes grown on substrates: Evidence for framework doping and their enhanced properties. *Adv. Funct. Mater.* **2011**, *21*, 986–992. [CrossRef]
9. Tuci, G.; Pilaski, M.; Ba, H.; Rossin, A.; Luconi, L.; Caporali, S.; Pham-Huu, C.; Palkovits, R.; Giambastiani, G. Unraveling surface basicity and bulk morphology relationship on covalent triazine frameworks with unique catalytic and gas adsorption properties. *Adv. Funct. Mater.* **2017**, *27*, 1605672. [CrossRef]
10. Chizari, K.; Deneuve, A.; Ersen, O.; Florea, I.; Liu, Y.; Edouard, D.; Janowska, I.; Begin, D.; Pham-Huu, C. Nitrogen-doped carbon nanotubes as a highly active metal-free catalyst for selective oxidation. *ChemSusChem* **2012**, *5*, 102–108. [CrossRef] [PubMed]
11. Zhang, J.; Qu, L.; Shi, G.; Liu, J.; Chen, J.; Dai, L. N, P-codoped carbon networks as efficient metal-free bifunctional catalysts for oxygen reduction and hydrogen evolution reactions. *Angew. Chem. Int. Ed.* **2016**, *55*, 2230–2234. [CrossRef] [PubMed]
12. Tuci, G.; Zafferoni, C.; Rossin, A.; Luconi, L.; Milella, A.; Ceppatelli, M.; Innocenti, M.; Liu, Y.; Pham-Huu, C.; Giambastiani, G. Chemical functionalization of N-doped carbon nanotubes: A powerful approach to cast light on the electrochemical role of specific N-functionalities in the oxygen reduction reaction. *Catal. Sci. Technol.* **2016**, *6*, 6226–6236. [CrossRef]
13. Lv, R.; Cui, T.; Jun, M.-S.; Zhang, Q.; Cao, A.; Su, D.S.; Zhang, Z.; Yoon, S.-H.; Miyawaki, J.; Mochida, I.; et al. Open-ended, N-doped carbon nanotube-graphene hybrid nanostructures as high-performance catalyst support. *Adv. Funct. Mater.* **2011**, *21*, 999–1006. [CrossRef]
14. Ba, H.; Liu, Y.; Truong-Phuoc, L.; Duong-Viet, C.; Nhut, J.-M.; Nguyen, D.L.; Ersen, O.; Tuci, G.; Giambastiani, G.; Pham-Huu, C. N-doped food-grade-derived 3D mesoporous foams as metal-free systems for catalysis. *ACS Catal.* **2016**, *6*, 1408–1419. [CrossRef]
15. Tuci, G.; Luconi, L.; Rossin, A.; Berretti, E.; Ba, H.; Innocenti, M.; Yakhvarov, D.; Caporali, S.; Pham-Huu, C.; Giambastiani, G. Aziridine-functionalized multiwalled carbon nanotubes: Robust and versatile catalysts for the oxygen reduction reaction and knoevenagel condensation. *ACS Appl. Mater. Interfaces* **2016**, *8*, 30099–30106. [CrossRef] [PubMed]
16. Duong-Viet, C.; Liu, Y.; Ba, H.; Truong-Phuoc, L.; Baaziz, W.; Nguyen-Dinh, L.; Nhut, J.-M.; Pham-Huu, C. Carbon nanotubes containing oxygenated decorating defects as metal-free catalyst for selective oxidation of H_2S. *Appl. Catal. B* **2016**, *191*, 29–41. [CrossRef]
17. Qi, W.; Liu, W.; Guo, X.; Schlogl, R.; Su, D. Oxidative dehydrogenation on nanocarbon: Intrinsic catalytic activity and structure-function relationships. *Angew. Chem. Int. Ed.* **2015**, *54*, 13682–13685. [CrossRef] [PubMed]
18. Pereira, M.F.R.; Órfão, J.J.M.; Figueiredo, J.L. Oxidative dehydrogenation of ethylbenzene on activated carbon catalysts. I. Influence of surface chemical groups. *Appl. Catal. A* **1999**, *184*, 153–160. [CrossRef]

19. Avilés, F.; Cauich-Rodríguez, J.V.; Moo-Tah, L.; May-Pat, A.; Vargas-Coronado, R. Evaluation of mild acid oxidation treatments for mwcnt functionalization. *Carbon* **2009**, *47*, 2970–2975. [CrossRef]

20. Hiura, H.; Ebbesen, T.W.; Tanigaki, K. Opening and purification of carbon nanotubes in high yields. *Adv. Mater.* **1995**, *7*, 275–276. [CrossRef]

21. Qui, N.V.; Scholz, P.; Krech, T.; Keller, T.F.; Pollok, K.; Ondruschka, B. Multiwalled carbon nanotubes oxidized by UV/H_2O_2 as catalyst for oxidative dehydrogenation of ethylbenzene. *Catal. Commun.* **2011**, *12*, 464–469. [CrossRef]

22. Mahata, N.; Pereira, M.F.; Suarez-Garcia, F.; Martinez-Alonso, A.; Tascon, J.M.; Figueiredo, J.L. Tuning of texture and surface chemistry of carbon xerogels. *J. Colloid Interface Sci.* **2008**, *324*, 150–155. [CrossRef] [PubMed]

23. Simmons, J.M.; Nichols, B.M.; Baker, S.E.; Marcus, M.S.; Castellini, O.M.; Lee, C.-S.; Hamers, R.J.; Eriksson, M.A. Effect of ozone oxidation on single-walled carbon nanotubes. *J. Phys. Chem. B* **2006**, *110*, 7113–7118. [CrossRef] [PubMed]

24. Huang, R.; Xu, J.; Wang, J.; Sun, X.; Qi, W.; Liang, C.; Su, D.S. Oxygen breaks into carbon nanotubes and abstracts hydrogen from propane. *Carbon* **2016**, *96*, 631–640. [CrossRef]

25. Luo, J.; Liu, Y.; Wei, H.; Wang, B.; Wu, K.-H.; Zhang, B.; Su, D.S. A green and economical vapor-assisted ozone treatment process for surface functionalization of carbon nanotubes. *Green Chem.* **2017**, *19*, 1052–1062. [CrossRef]

26. Hu, J.; Guo, Z.; Chu, W.; Li, L.; Lin, T. Carbon dioxide catalytic conversion to nano carbon material on the iron–nickel catalysts using CVD-IP method. *J. Energy Chem.* **2015**, *24*, 620–625. [CrossRef]

27. Liu, Y.; Dintzer, T.; Ersen, O.; Pham-Huu, C. Carbon nanotubes decorated α-Al_2O_3 containing cobalt nanoparticles for fischer-tropsch reaction. *J. Energy Chem.* **2013**, *22*, 279–289. [CrossRef]

28. Więckowska, J. Catalytic and adsorptive desulphurization of gases. *Catal. Today* **1995**, *24*, 405–465. [CrossRef]

29. Zhang, X.; Tang, Y.; Qu, S.; Da, J.; Hao, Z. H_2S-selective catalytic oxidation: Catalysts and processes. *ACS Catal.* **2015**, *5*, 1053–1067. [CrossRef]

30. Bashkova, S.; Baker, F.S.; Wu, X.; Armstrong, T.R.; Schwartz, V. Activated carbon catalyst for selective oxidation of hydrogen sulphide: On the influence of pore structure, surface characteristics, and catalytically-active nitrogen. *Carbon* **2007**, *45*, 1354–1363. [CrossRef]

31. Shinkarev, V.V.; Glushenkov, A.M.; Kuvshinov, D.G.; Kuvshinov, G.G. Nanofibrous carbon with herringbone structure as an effective catalyst of the H_2S selective oxidation. *Carbon* **2010**, *48*, 2004–2012. [CrossRef]

32. PiéPlu, A.; Saur, O.; Lavalley, J.-C.; Legendre, O.; NéDez, C. Claus catalysis and H_2S selective oxidation. *Catal. Rev. Sci. Eng.* **1998**, *40*, 409–450. [CrossRef]

33. Liu, Y.; Duong-Viet, C.; Luo, J.; Hébraud, A.; Schlatter, G.; Ersen, O.; Nhut, J.-M.; Pham-Huu, C. One-pot synthesis of a nitrogen-doped carbon composite by electrospinning as a metal-free catalyst for oxidation of H_2S to sulfur. *ChemCatChem* **2015**, *7*, 2957–2964. [CrossRef]

34. Keller, N.; Pham-Huu, C.; Ledoux, M.J. Continuous process for selective oxidation of H_2S over SiC-supported iron catalysts into elemental sulfur above its dewpoint. *Appl. Catal. A* **2001**, *217*, 205–217. [CrossRef]

35. Lee, E.-K.; Jung, K.-D.; Joo, O.-S.; Shul, Y.-G. Catalytic wet oxidation of H_2S to sulfur on V/MgO catalyst. *Catal. Lett.* **2004**, *98*, 259–263. [CrossRef]

36. Claus Process. Available online: https://en.wikipedia.org/wiki/Claus_process (accessed on 3 April 2018).

37. Xia, W.; Jin, C.; Kundu, S.; Muhler, M. A highly efficient gas-phase route for the oxygen functionalization of carbon nanotubes based on nitric acid vapor. *Carbon* **2009**, *47*, 919–922. [CrossRef]

38. Cancado, L.G.; Jorio, A.; Ferreira, E.H.; Stavale, F.; Achete, C.A.; Capaz, R.B.; Moutinho, M.V.; Lombardo, A.; Kulmala, T.S.; Ferrari, A.C. Quantifying defects in graphene via raman spectroscopy at different excitation energies. *Nano Lett.* **2011**, *11*, 3190–3196. [CrossRef] [PubMed]

39. Sadezky, A.; Muckenhuber, H.; Grothe, H.; Niessner, R.; Pöschl, U. Raman microspectroscopy of soot and related carbonaceous materials: Spectral analysis and structural information. *Carbon* **2005**, *43*, 1731–1742. [CrossRef]

40. Datsyuk, V.; Kalyva, M.; Papagelis, K.; Parthenios, J.; Tasis, D.; Siokou, A.; Kallitsis, I.; Galiotis, C. Chemical oxidation of multiwalled carbon nanotubes. *Carbon* **2008**, *46*, 833–840. [CrossRef]

41. Thanh, T.T.; Ba, H.; Truong-Phuoc, L.; Nhut, J.-M.; Ersen, O.; Begin, D.; Janowska, I.; Nguyen, D.L.; Granger, P.; Pham-Huu, C. A few-layer graphene–graphene oxide composite containing nanodiamonds as metal-free catalysts. *J. Mater. Chem. A* **2014**, *2*, 11349–11357. [CrossRef]

42. Luo, J.; Wei, H.; Liu, Y.; Zhang, D.; Zhang, B.; Chu, W.; Pham-Huu, C.; Su, D.S. Oxygenated group and structural defect enriched carbon nanotubes for immobilizing gold nanoparticles. *Chem. Commun.* **2017**, *53*, 12750–12753. [CrossRef] [PubMed]

43. Ersen, O.; Hirlimann, C.; Drillon, M.; Werckmann, J.; Tihay, F.; Pham-Huu, C.; Crucifix, C.; Schultz, P. 3D-TEM characterization of nanometric objects. *Solid State Sci.* **2007**, *9*, 1088–1098. [CrossRef]

44. Figueiredo, J.L. Functionalization of porous carbons for catalytic applications. *J. Mater. Chem. A* **2013**, *1*, 9351–9364. [CrossRef]

45. Shi, W.; Zhang, B.; Lin, Y.; Wang, Q.; Zhang, Q.; Su, D.S. Enhanced chemoselective hydrogenation through tuning the interaction between pt nanoparticles and carbon supports: Insights from identical location transmission electron microscopy and X-ray photoelectron spectroscopy. *ACS Catal.* **2016**, *6*, 7844–7854. [CrossRef]

46. Scheibe, B.; Borowiak-Palen, E.; Kalenczuk, R.J. Oxidation and reduction of multiwalled carbon nanotubes—Preparation and characterization. *Mater. Charact.* **2010**, *61*, 185–191. [CrossRef]

47. Chu, W.; Ran, M.; Zhang, X.; Wang, N.; Wang, Y.; Xie, H.; Zhao, X. Remarkable carbon dioxide catalytic capture (CDCC) leading to solid-form carbon material via a new cvd integrated process (CVD-IP): An alternative route for CO_2 sequestration. *J. Energy Chem.* **2013**, *22*, 136–144. [CrossRef]

48. Bom, D.; Andrews, R.; Jacques, D.; Anthony, J.; Chen, B.; Meier, M.S.; Selegue, J.P. Thermogravimetric analysis of the oxidation of multiwalled carbon nanotubes: Evidence for the role of defect sites in carbon nanotube chemistry. *Nano Lett.* **2002**, *2*, 615–619. [CrossRef]

49. Ran, M.; Sun, W.; Liu, Y.; Chu, W.; Jiang, C. Functionalization of multi-walled carbon nanotubes using water-assisted chemical vapor deposition. *J. Solid State Chem.* **2013**, *197*, 517–522. [CrossRef]

50. Jiang, Y.; Yang, L.; Sun, T.; Zhao, J.; Lyu, Z.; Zhuo, O.; Wang, X.; Wu, Q.; Ma, J.; Hu, Z. Significant contribution of intrinsic carbon defects to oxygen reduction activity. *ACS Catal.* **2015**, *5*, 6707–6712. [CrossRef]

51. Shinkarev, V.V.; Glushenkov, A.M.; Kuvshinov, D.G.; Kuvshinov, G.G. New effective catalysts based on mesoporous nanofibrous carbon for selective oxidation of hydrogen sulfide. *Appl. Catal. B* **2009**, *85*, 180–191. [CrossRef]

52. Duong-Viet, C.; Truong-Phuoc, L.; Tran-Thanh, T.; Nhut, J.-M.; Nguyen-Dinh, L.; Janowska, I.; Begin, D.; Pham-Huu, C. Nitrogen-doped carbon nanotubes decorated silicon carbide as a metal-free catalyst for partial oxidation of H_2S. *Appl. Catal. A* **2014**, *482*, 397–406.

53. Liu, Y.; Ersen, O.; Meny, C.; Luck, F.; Pham-Huu, C. Fischer–tropsch reaction on a thermally conductive and reusable silicon carbide support. *ChemSusChem* **2014**, *7*, 1218–1239. [CrossRef] [PubMed]

54. Lacroix, M.; Dreibine, L.; de Tymowski, B.; Vigneron, F.; Edouard, D.; Bégin, D.; Nguyen, P.; Pham, C.; Savin-Poncet, S.; Luck, F.; et al. Silicon carbide foam composite containing cobalt as a highly selective and re-usable Fischer–Tropsch synthesis catalyst. *Appl. Catal. A* **2011**, *397*, 62–72. [CrossRef]

55. Elamin, M.M.; Muraza, O.; Malaibari, Z.; Ba, H.; Nhut, J.-M.; Pham-Huu, C. Microwave assisted growth of SAPO-34 on β-SiC foams for methanol dehydration to dimethyl ether. *Chem. Eng. J.* **2015**, *274*, 113–122. [CrossRef]

56. Liu, Y.; Podila, S.; Nguyen, D.L.; Edouard, D.; Nguyen, P.; Pham, C.; Ledoux, M.J.; Pham-Huu, C. Methanol dehydration to dimethyl ether in a platelet milli-reactor filled with H-ZSM5/SiC foam catalyst. *Appl. Catal. A* **2011**, *409–410*, 113–121. [CrossRef]

57. Jiao, Y.; Yang, X.; Jiang, C.; Tian, C.; Yang, Z.; Zhang, J. Hierarchical ZSM-5/SiC nano-whisker/SiC foam composites: Preparation and application in mtp reactions. *J. Catal.* **2015**, *332*, 70–76. [CrossRef]

58. Duong-Viet, C.; Ba, H.; El-Berrichi, Z.; Nhut, J.-M.; Ledoux, M.J.; Liu, Y.; Pham-Huu, C. Silicon carbide foam as a porous support platform for catalytic applications. *New J. Chem.* **2016**, *40*, 4285–4299. [CrossRef]

59. Mares, B.; Prosernat Com. Personal communication, 2018.

catalysts

Article

Must the Best Laboratory Prepared Catalyst Also Be the Best in an Operational Application?

Lucie Obalová [1,*], Anna Klegova [1], Lenka Matějová [1], Kateřina Pacultová [1]
and Dagmar Fridrichová [1,2]

[1] Institute of Environmental Technology, VŠB – Technical University of Ostrava, 17. listopadu 15/2172,
    708 00 Ostrava, Czech Republic; anna.klegova@vsb.cz (A.K.); lenka.matejova@vsb.cz (L.M.);
    katerina.pacultova@vsb.cz (K.P.); dagmar.fridrichova@vsb.cz (D.F.)
[2] Centre ENET, VŠB – Technical University of Ostrava, 17. listopadu 15/2172, 708 00 Ostrava, Czech Republic
*   Correspondence: lucie.obalova@vsb.cz

Received: 23 December 2018; Accepted: 30 January 2019; Published: 7 February 2019

Abstract: Three cobalt mixed oxide deN$_2$O catalysts, with optimal content of alkali metals (K, Cs), were prepared on a large scale, shaped into tablets, and tested in a pilot plant reactor connected to the bypassed tail gas from the nitric production plant, downstream from the selective catalytic reduction of NO$_x$ by ammonia (SCR NO$_x$/NH$_3$) catalyst. High efficiency in N$_2$O removal (N$_2$O conversion of 75–90% at 450 °C, VHSV = 11,000 m^3 m$_{bed}^{-3}$ h^{-1}) was achieved. However, a different activity order of the commercially prepared catalyst tablets compared to the laboratory prepared catalyst grains was observed. Catalytic experiments in the kinetic regime using laboratory and commercial prepared catalysts and characterization methods (XRD, TPR-H$_2$, physisorption, and chemical analysis) were utilized to explain this phenomenon. Experimentally determined internal effectiveness factors and their general dependency on kinetic constants were evaluated to discuss the relationship between the catalyst activity in the kinetic regime and the internal diffusion limitation in catalyst tablets as well as their morphology. The theoretical N$_2$O conversion as a function of the intrinsic kinetic constants and diffusion rate, expressed as effective diffusion coefficients, was evaluated to estimate the final catalyst performance on a large scale and to answer the question of the above article title.

Keywords: internal effectiveness factor; effective diffusion coefficient; N$_2$O; catalytic decomposition; cobalt mixed oxide; alkali metal; promoter

---

1. Introduction

Research of new catalysts for industrial application is a time-consuming and costly process. It is usually based on a large number of laboratory catalytic experiments, which provide feedback for the optimization of catalyst preparation procedure, its chemical and phase composition, morphology, dispersion, and a number of other physicochemical characteristics which are necessary to obtain the desired catalytic properties. The result of laboratory research is the recipe for preparation of the optimized catalyst on a larger scale (approx. 100 kg), its production, and pilot plant testing.

In the presented work, based on previous extensive laboratory screening tests of N$_2$O catalytic decomposition [1–3], three cobalt mixed oxide deN$_2$O catalysts with optimal content of alkali metals K/Co$_4$MnAlO$_x$, Cs/Co$_4$MnAlO$_x$, and Cs/Co$_3$O$_4$ were prepared on a large scale, shaped into tablets and tested in the pilot plant reactor connected to the bypassed tail gas from the nitric production plant. Results of pilot plant testing of K/Co$_4$MnAlO$_x$ for decreasing N$_2$O emissions from a nitric acid plant were published in our previous work [4]. Although high efficiency in N$_2$O removal (N$_2$O conversion of 75–90% at 450 °C, VHSV = 11,000 m^3 m$_{bed}^{-3}$ h^{-1}) were reached; the question is if the prepared tablet catalyst achieved similar catalytic performance to the laboratory catalyst or if it was lower and in the latter case optimization of the pilot catalyst's preparation procedure would then be beneficial.

The same question applies in the case of the remaining two catalysts. For Cs/Co$_4$MnAlO$_x$, the results of pilot plant testing were published in [5] and for Cs/Co$_3$O$_4$ catalyst, the pilot-testing results have not yet been published.

The aim of this paper is to answer the question already posed in the title. Catalytic experiments in kinetic regime using laboratory and commercial prepared catalysts and characterization methods (XRD, TPR-H$_2$, physisorption and chemical analysis) were utilized to answer this question. The procedure for determination of (i) the effective diffusion coefficients from catalytic experiments and (ii) the theoretical limits of catalyst performance on a large scale is shown. The presented approach is generally valid and can be used also for other reactions of 1st order kinetics.

2. Results

2.1. N$_2$O Catalytic Decomposition

Figure 1 compares the laboratory and pilot-plant results of N$_2$O catalytic decomposition. Laboratory experiments were performed using laboratory-prepared catalysts in the form of grains and the conversion of N$_2$O was observed in a gaseous mixture simulating real waste gas conditions. Pilot-plant experiments were performed on commercially prepared tablets in real tail-gas from a nitric acid plant. High efficiencies in N$_2$O removal in both laboratory and pilot-plant conditions were reached, comparable with the best results from the literature [6,7].

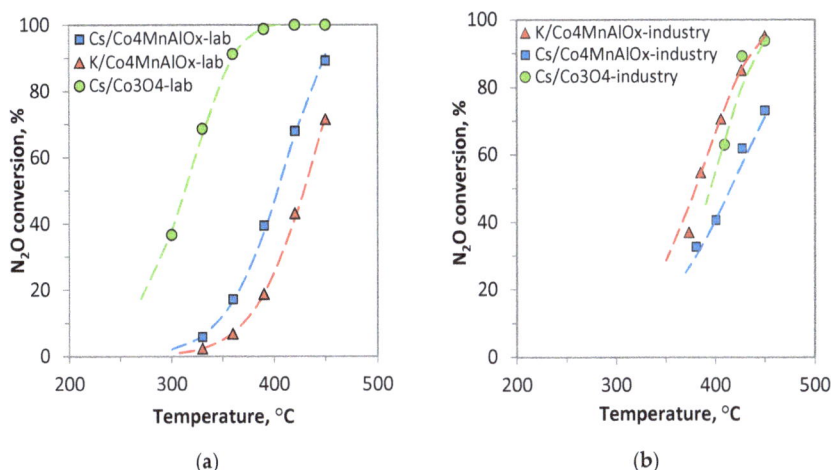

(a) (b)

Figure 1. N$_2$O catalytic decomposition over: **(a)** laboratory prepared catalyst grains in simulated process condition (1000 ppm N$_2$O + 5 mol% O$_2$ + 2 mol% H$_2$O in N$_2$, GHSV = 60,000 l kg^{-1} h^{-1}); **(b)** commercially prepared catalyst tablets in real waste gas (VHSV = 11,000 m^3 m$_{bed}^{-3}$ h^{-1}, real waste gas (512 ± 135 ppm N$_2$O, 26 ± 16 ppm NO$_x$, 9 ± 3 ppm NH$_3$, *p* = 0.6 MPa). Points—experimental data; dashed line—model (Equation (2)).

As expected, the conversions achieved under laboratory and pilot-plant conditions vary due to different composition of the gaseous mixture, space velocities, pressure, and the effect of internal diffusion inherent to the use of commercial tablets. However, the order of catalysts' activities reached under laboratory and pilot-plant conditions also changed, which was unexpected. The highest N$_2$O conversions under laboratory conditions were achieved using the Cs/Co$_3$O$_4$ catalyst, but under pilot-plant conditions, comparable N$_2$O conversions were obtained over the K/Co$_4$MnAlO$_x$ catalyst, which was the worst performing under laboratory conditions.

Three questions arise:

1. What are the reasons for the different activity order of commercially prepared catalyst tablets compared to laboratory-prepared samples?
2. Could our pilot-plant catalysts achieve higher activity (N_2O conversions)?
3. What are the theoretical limits of N_2O conversions over catalyst tablets and how to approach them?

To answer the first question, catalytic experiments with commercially prepared tablets were carried out under laboratory conditions in simulated tail-gas (Figure 2). It is evident that the order of catalytic activities of commercial tablets was quite different to that in real waste gas (Figure 1b) but the same as for laboratory prepared catalysts tested in kinetic regime (Figure 1a). It is therefore implied that the different order of catalytic activities in the pilot-plant experiment (Figure 1b) was caused by the different conditions of the said experiment, particularly by the presence of NO_x gases, although their concentration was quite low (max 26 ppm). Significant inhibition of N_2O decomposition caused by NO_x in the presence of cobalt catalysts modified by alkali promoters was previously reported [2,8,9]. The extent of the inhibiting effect depends on the alkali promoter type and its amount. The real tail gas also contained residual NH_3. The effect of NH_3 on the decomposition of N_2O in the presence of cobalt catalysts was studied only sporadically; mainly on zeolites and to a lesser extent on oxidic catalysts [10,11]. The influence of NH_3 in the feed gas was also studied in an individual experiment performed under laboratory conditions over K/Co_4MnAlO_x-industry tablets [4]. It was found that ammonia oxidized to N_2O, NO, NO_2, and probably N_2 (not measured), which means that ammonia can also contribute to the fluctuations of NO_x concentration present in the feed or can also increase N_2O concentration at the reactor outlet. Laboratory and pilot-plant experiments were also performed under different pressures. Pilot-plant experiments were performed at higher pressure, which has a positive effect on the N_2O catalytic decomposition rate [12].

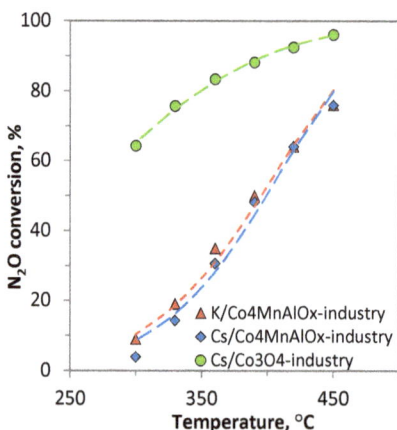

Figure 2. N_2O catalytic decomposition over industrially prepared catalysts (tablets 5 mm × 5 mm) tested in laboratory conditions. Points—experimental data; dashed line—model (Equation (2)). Conditions: VHSV = 3000 m^3 m_{bed}^{-3} h^{-1}, 1000 ppm N_2O + 5 mol% O_2 + 2 mol% H_2O in N_2.

To answer the second question, N_2O decomposition was carried out in kinetic regime on laboratory and commercially prepared catalysts in grain form (Figure 3). First order kinetic constants using an ideal plug flow reactor model were also determined according to Equation (3), both for data in Figure 3a–c (grain catalysts) and for data in Figure 2 (tablet catalysts). Reaction rates (Equation (5)) and internal effectiveness factors (Equation (1)) were determined. A summary of the calculated kinetic parameters is shown in Table 1.

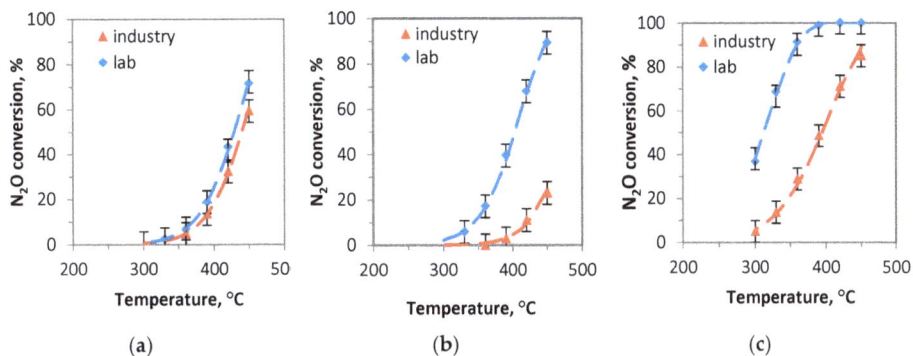

Figure 3. N_2O decomposition over laboratory and commercially prepared catalysts in grain form: (a) K/Co_4MnAlO_x; (b) Cs/Co_4MnAlO_x; (c) Cs/Co_3O_4. Points—experimental data; dashed line—model (Equation (2)). Conditions: 1000 ppm N_2O + 5 mol% O_2 + 2 mol% H_2O in N_2, GHSV = 60,000 l kg^{-1} h^{-1}.

Table 1. Kinetic constants and reaction rates of N_2O catalytic decomposition over laboratory (grain form) and commercially prepared (grain and pellet forms) catalysts and estimated internal effectiveness factors. Conditions: 1000 ppm N_2O + 5 mol% O_2 + 2 mol% H_2O in N_2, 390 °C.

Parameter/Catalyst	K/Co_4MnAlO_x		Cs/Co_4MnAlO_x		Cs/Co_3O_4	
	Lab	Industry	Lab	Industry	Lab	Industry
k_{grain} (m^3 kg^{-1}s^{-1}) × 10^3	3.47	2.42	8.38	0.51	73.15	11.09
$k_{grain}{}^*$ (s^{-1}) × 10^6 [1]	1.51	1.05	3.54	0.22	28.24	4.28
k_{pellet} (m^3 kg^{-1} s^{-1}) × 10^3	-	0.47	-	0.37	-	1.32
r_{grain} (mol kg^{-1}h^{-1}) × 10^2	-	3.37	-	0.75	-	12.12
r_{pellet} (mol kg^{-1} h^{-1}) × 10^2	-	0.70	-	0.55	-	1.90
η (-) [2]	-	0.21	-	0.73	-	0.15

[1] 1st order kinetic constant (s^{-1}) determined according to Equation (4) [2] Calculated as $\eta = r_{grain}/r_{pellet}$ for commercially prepared catalysts

By comparing N_2O conversions, the values of X_A for the laboratory and commercially prepared K/Co_4MnAlO_x catalyst differ only by 12% maximum, at lower temperature the differences were within the margin of experimental error. This implies that the activity of active sites on the K/Co_4MnAlO_x-industry catalyst is nearly the same as on the laboratory prepared sample, K/Co_4MnAlO_x-lab and the reaction rate and kinetic constants of the tablet catalyst is lower mainly due to the inhibiting effect of internal diffusion. The main way to improve N_2O conversion is therefore to increase the internal effectiveness factor, which in this case has a value of 0.21. The possibilities of increasing the internal effectiveness factor are discussed in Chapter 3.

The situation is different with the Cs/Co_4MnAlO_x and Cs/Co_3O_4 catalysts. The N_2O conversion and kinetic constants in kinetic regime (k_{grain}) are significantly lower when comparing the commercial and laboratory prepared samples. In the case of Cs/Co_3O_4, the kinetic constant of the commercially prepared sample is six times lower and in the case of Cs/Co_4MnAlO_x 16 times lower. This implies that the sites on the surface of commercially prepared catalysts are less active than those on the surface of laboratory prepared catalysts. The kinetic constants k_{pellet} are lower than k_{grain} of laboratory prepared samples due to the effect of internal diffusion and another effect, which caused the activity decrease of active sites. The N_2O conversion using commercial tablets could be improved both by increasing the internal effectiveness factor and by increasing the activity of active sites to the same level as laboratory-prepared catalysts. It is important to note that by increasing activity, the internal effectiveness factor will decrease because parameters η and k are related to each other according to

Equation (6)–(8). To determine the cause of different activities of Cs/Co_4MnAlO_x and Cs/Co_3O_4 catalysts, a basic characterization of laboratory and commercially prepared catalysts was performed.

2.2. Physicochemical Properties of Catalysts

Basic physicochemical properties of laboratory and commercially prepared catalysts are shown in Table 2. Analysis of chemical composition shows that the molar ratio of Co:Mn:Al is basically the same for laboratory and commercially prepared Co_4MnAlO_x samples modified by K and Cs and is close to the ratio of 4:1:1, which was the aim during synthesis.

Table 2. Physicochemical properties of laboratory and commercially prepared catalysts.

		K/Co_4MnAlO_x		Cs/Co_4MnAlO_x		Cs/Co_3O_4	
		Industry	Lab	Industry	Lab	Industry	Lab
	Co	45	49.6	45.7	46.0	63.4	68.0
Chemical	Mn	9.3	13.3	8.5	10.6	-	-
composition	Al	5.2	8.9	5.1	n.d.	-	-
(wt%)	K/Cs	1.3	1.8	3.4	3.5	1.0	1.0
	Na	0.1	0.1	0.1	0.5	n.d.	0.04
Co:Mn:Al mol. ratio		4 : 0.7 : 0.7	4 : 1.2 : 1.6	4 : 0.9 : 1.0	4 : 1 : n.d.	-	-
S_{BET} (m²/g)		93	98	68	86	20	13
r (nm) [1]		3.8	12.3	5.3	7.1	10.2	13.5
Alkali metal normalized loading (atoms/nm²) [2]		1.9	2.8	2.1	1.8	2.4	3.5
Phase composition		Spinel, graphite	Spinel	Spinel, graphite	Spinel	Spinel, graphite	Spinel
L_c (nm)		9	7	13	7	18	49
a (nm)		0.8118	0.8110	0.8111	0.8116	0.8084	0.8086
T_{max} (°C) [3]		272; 377	320; 361	284; 390	168; 274; 359	352; 389	370
H_2 (mmol/g) [4]		4.5	4.8	4.2	4.1	14	12.9

[1] Mean pore radius $r = 2V/A$, where V is pore volume and A specific surface area S_{BET} [2] Calculated as K or Cs atoms/nm² [3] Temperature of reduction peak maxima from TPR-H_2 [4] H_2 consumption in 25–450 °C temperature interval from TPR-H_2.

For catalytic activity, the content of alkali promoters is important [2,13,14]. The amount of Cs in laboratory and commercially prepared samples is the same and close to the optimal value of 3.5wt% for Co_4MnAlO_x [1] and 1wt% for Co_3O_4 [3]. In the case of K/Co_4MnAlO_x, the intention was to prepare the commercial catalyst with K content of 1.8wt%, which proved to be the most active in our previous work [2]. Here it was found that the optimal K content (determined by atomic emission spectroscopy, AES) leading to the highest N_2O conversion was 0.9–1.6wt% K in an inert gas and 1.6–2.8wt% K in the presence of O_2 and H_2O. The analysis of K content in a commercially prepared sample showed a lower K content, probably due to inhomogeneity, which is a known disadvantage of the impregnation method. All catalysts contained sodium residuals (0.04–0.50wt%) from the preparation procedure. Since our recent results showed that to affect catalytic activity, significantly higher Na content (>1.15 wt%) is necessary [8], we do not expect residual sodium to influence catalytic activity.

The comparison of the specific surface areas of laboratory and commercially prepared samples (Table 2) shows that, in terms of specific surface area, the transfer of catalyst synthesis to a larger scale was successful. Catalyst synthesis on a larger commercial scale led to a lower mean pore size for all three catalysts, which was caused by the use of high pressure during tabletization [15]. It was found out that on increasing pressure (up to 10 MPa) during tabletization of the catalyst matter (grains of defined size), the diameter of mesopores decreased, while the specific surface area stayed constant up to 8 MPa.

Measured nitrogen adsorption–desorption isotherms of commercially prepared catalysts (Figure 4a) indicate that there is a big difference between the individual samples. While the isotherms of K/Co_4MnAlO_x-industry and Cs/Co_4MnAlO_x-industry correspond to the mesoporous material with generally larger mesopores, the isotherm of Cs/Co_3O_4 corresponds to macroporous material. Comparing K/Co_4MnAlO_x-industry and Cs/Co_4MnAlO_x-industry samples, K/Co_4MnAlO_x-industry possesses smaller mesopores than Cs/Co_4MnAlO_x-industry (Figure 4b) which is at comparable total pore volumes reflected by the higher specific surface area of the pores of K/Co_4MnAlO_x-industry.

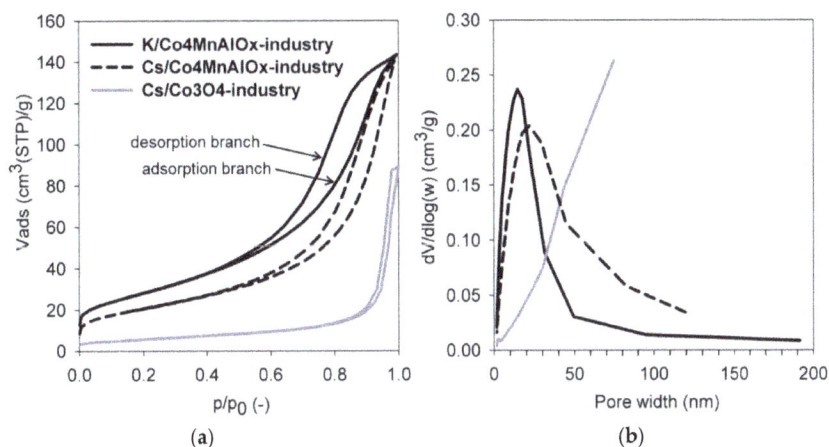

Figure 4. (a) Measured nitrogen adsorption–desorption isotherms and (b) evaluated pore-size distributions of commercially prepared catalysts.

All commercially prepared samples do not possess any micropores, thus, the values of the micropore volume and mesopore surface area are not shown in Table 3. The determined textural properties of the commercially prepared catalysts are summarized in Table 3.

Table 3. Textural properties of commercially prepared catalysts.

Catalyst	S_{BET} (m^2/g)	V_{net} $(cm^3{}_{liq}/g)$	Pore width$_{max}$ (nm)	ρ_c $(kg\ m^{-3})$	ε_p (-)
K/Co_4MnAlO_x-industry	101	0.220	15	2300	0.48
Cs/Co_4MnAlO_x-industry	73	0.217	22	2370	0.50
Cs/Co_3O_4-industry	19	0.138	above 75	2590	0.47

Results of phase composition are summarized in Figure 5 and Table 2. Measured diffraction lines characteristic for a cobalt spinel structure were indexed within the *Fd3m* space group and this structure was confirmed in all investigated samples as the major phase. Graphite was also found as a minor phase in all industrially prepared catalysts, together with the spinel phase, since graphite was added to the catalyst matter for ease of forming. Graphite line was not observed in the catalysts used in the long-term pilot plant testing (not shown here), confirming combustion of graphite during the catalytic test [4]. Small un-labelled diffractions correspond to the Kbeta line of the used X-ray radiation source. All Co_4MnAlO_x samples have slightly higher values of cell parameters (*a*) in comparison to pure Co_3O_4 due to incorporation of Mn and Al ions into the spinel structure. In addition, samples with Co_4MnAlO_x active phase have nanocrystalline structure and higher surface area in comparison to samples with Co_3O_4. Crystallite sizes (expressed as mean coherent domain size L_c) correlate with the determined surface area; the smaller the crystallite size the larger is the surface area. Potassium doped samples K/Co_4MnAlO_x-lab and -industry exhibit approximately similar size for coherent domain and surface area. In the case of cesium doped Cs/Co_4MnAlO_x samples, the laboratory prepared catalyst has a less crystalline structure and higher surface area in comparison with the commercially prepared sample. Laboratory prepared cesium doped cobalt oxide Cs/Co_3O_4 exhibits the highest value of coherent domain size and the lowest specific surface area of all tested samples.

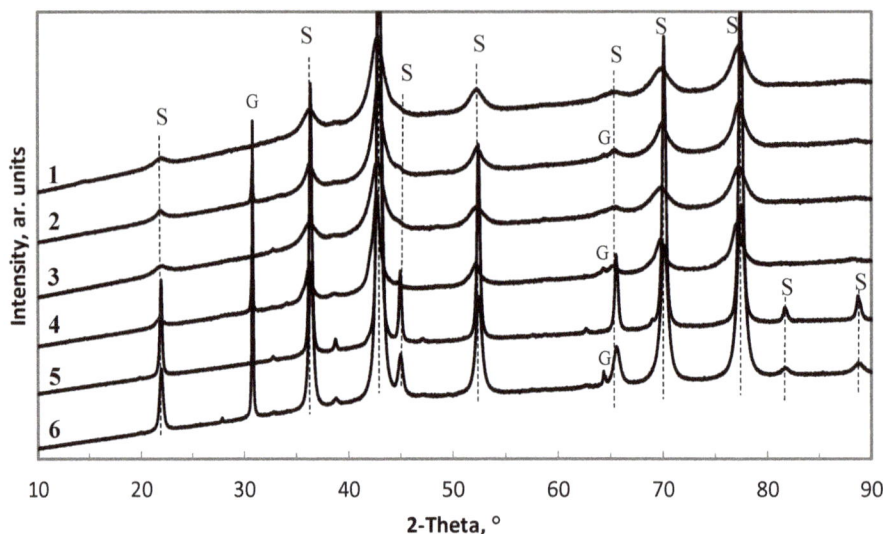

Figure 5. XRD patterns of the laboratory and commercially prepared catalysts: (**1**) K/Co_4MnAlO_x-lab; (**2**) K/Co_4MnAlO_x-industry; (**3**) Cs/Co_4MnAlO_x-lab; (**4**) Cs/Co_4MnAlO_x-industry; (**5**) Cs/Co_3O_4-lab; (**6**) Cs/Co_3O_4-industry. Phase designation: S—spinel, G—graphite.

The reduction patterns of catalysts prepared commercially and in the laboratory are shown in Figure 6. Since the catalytic reaction proceeds up to 450 °C, only species reducible in this temperature region (low temperature region) can contribute to the catalyst activity and for that reason only the low temperature area of TPR-H_2 is shown. For samples containing only cobalt species (besides alkali promoter), the main reduction peak can be ascribed to the reduction of $Co^{III} \rightarrow Co^{II} \rightarrow Co^0$. In the case of Cs/Co_3O_4, both reduction processes overlap (more for the laboratory prepared catalyst), suggesting not only different primary particle size but also different shapes of cobalt spinel nanocrystals [16]. The catalysts containing cobalt, manganese, and aluminum are reduced in two main temperature regions, 200–430 °C and >430 °C. Both reduction peaks consist of overlapping peaks corresponding to the reduction of more species. The low-temperature reduction peak also represents, besides the reduction of cobalt species in a segregated Co_3O_4-like phase, the reduction of Mn^{IV} to Mn^{III} oxides, while $Mn^{III} \rightarrow Mn^{II}$ reduction can take place in both temperature regions [17]. The high temperature peak was attributed to the reduction of Co and Mn ions surrounded by Al ions in the spinel-like phase [8]. The presence of alkali metals in the mixed oxides caused the formation of an easily reducible species (Co and/or Mn) manifested as a low temperature shoulder visible especially in Co_4MnAlO_x containing samples.

The quantitative data (Table 2) confirmed a comparable amount of reducible species (in the temperature range of 25–450 °C) in samples prepared commercially and in laboratory conditions, while samples containing only Co_3O_4 active phase possessed higher amounts of reducible components in comparison with Co_4MnAlO_x containing samples. The TPR profile shapes are similar for therelevant samples prepared commercially and in the laboratory; only small shifts are seen—the biggest changes are visible for Cs/Co_4MnAlO_x where the commercially prepared catalyst exhibits inferior reducibility. Since the reducibility of the cobalt and manganese species is affected by the primary particle size [18], the visible changes are probably connected with different crystallite sizes as was also proved by the XRD results and the inhomogeneity of samples rather than a different chemico-electronic environment.

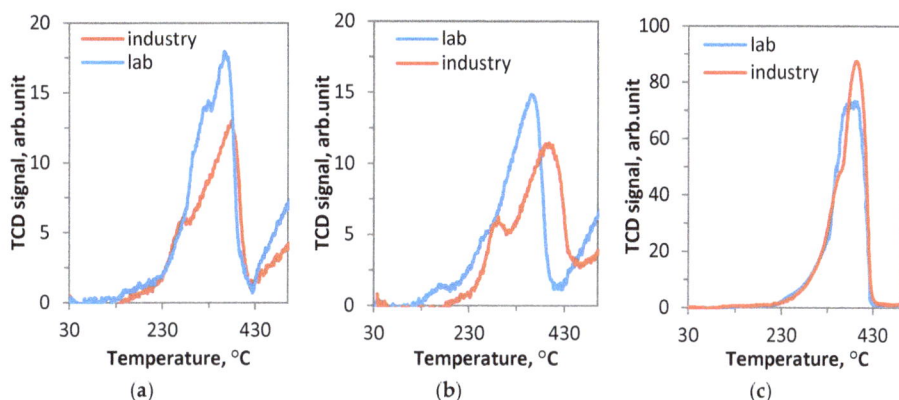

Figure 6. TPR-H$_2$ of the laboratory and commercially prepared catalysts: (**a**) K/Co$_4$MnAlO$_x$; (**b**) Cs/Co$_4$MnAlO$_x$; (**c**) Cs/Co$_3$O$_4$.

Published results imply that normalized loading of alkali metals (atoms/nm^2) on the catalyst surface is an important parameter of cobalt-based catalysts [19,20]. The reason is the electronic nature of the potassium and cesium promotion effect based on the formation of surface dipoles influencing the electron density of cobalt and manganese cations, which was confirmed by the direct correlation between N$_2$O conversion and surface work function catalysts [19]. Figure 7 shows the dependence of N$_2$O conversion on the normalized loading of K and Cs over Co$_4$MnAlO$_x$ and Co$_3$O$_4$, utilizing data on the optimization of K [2] and Cs [1] content in Co$_4$MnAlO$_x$ mixed oxide and Cs content in Co$_3$O$_4$ [3].

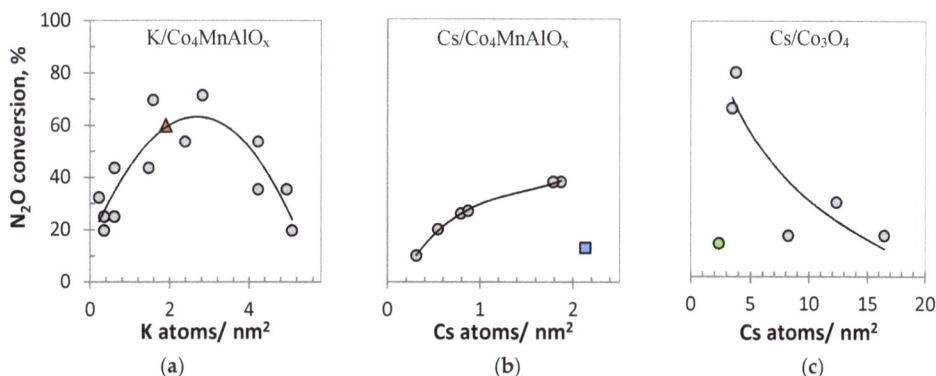

Figure 7. Dependence of N$_2$O conversion on alkali metal loading. Conditions: 1000 ppm N$_2$O + 5 mol% O$_2$ + 2 mol% H$_2$O in N$_2$. ◎ Laboratory prepared catalysts. (**a**) 450 °C, GHSV = 20,000 l kg^{-1} h $^{-1}$, ▲ K/Co$_4$MnAlO$_x$-industry; (**b**) 390 °C, GHSV = 60,000 l kg^{-1} h $^{-1}$, ◻ Cs/Co$_4$MnAlO$_x$-industry; (**c**) 330 °C, GHSV = 60,000 l kg^{-1} h $^{-1}$, ◉ Cs/Co$_3$O$_4$-industry.

It is evident that the conversion of N$_2$O in the presence of K/Co$_4$MnAlO$_x$ reaches an optimum at 1.6–2.8 K atoms/nm^2, while in the case of Cs/Co$_4$MnAl the conversion of N$_2$O increases with higher Cs loading in the observed concentration range and decreases with increasing Cs loading on Co$_3$O$_4$. Loading of K and Cs in samples prepared commercially and in the laboratory are listed in Table 2. In the case of commercially prepared K/Co$_4$MnAlO$_x$, the loading of K is in the optimal range, while in the case of Cs/Co$_4$MnAlO$_x$-industry the Cs loading is higher than in the laboratory prepared sample and the dependence of N$_2$O conversion on Cs loading in this region is unknown. On the contrary, the Cs loading in Cs/Co$_3$O$_4$-industry is lower than 3.5, the value that was determined for Cs/Co$_3$O$_4$-lab. It is important to mention that the reproducibility of the normalized loading during the preparation was difficult. The reason is that even relatively small changes in the specific surface area and/or the alkali metal content caused a change in the normalized loading, which is already significant due to the relatively narrow optimal range of its value.

The characterization results of laboratory and commercially prepared catalysts showed good correlation with catalytic properties in the kinetic regime:

- The K/Co$_4$MnAlO$_x$ catalysts demonstrated very similar characteristics for laboratory and commercially prepared samples. Both samples have the same chemical and phase composition; they differ slightly in K content, which is however still in the optimal range, similar to K normalized loading (atoms/nm^2). The samples have the same specific surface areas and crystallite sizes (within the margin of error), which is manifested in also having similar reducibilities.
- The situation is different in the case of Cs-modified catalysts. Here the characterization results differ more significantly, which is reflected by the differences in conversion and kinetic constant values determined in the kinetic regime (Table 1, Figure 3).
- While the commercially prepared Cs/Co$_4$MnAlO$_x$ has the same chemical and phase composition as the laboratory prepared sample, its crystallite size is 1.8× higher than for the laboratory prepared sample, which leads to both lower specific surface area and worse reducibility. Our previous work proved a direct relationship between the crystallite size and reducibility of Co and Mn in their higher oxidation states (Co^{3+}, Mn^{4+}) in Co$_4$MnAlO$_x$ [17]. Taking into account that the slowest step of N$_2$O decomposition is the desorption of oxygen from the catalyst surface connected to the reduction of active sites, worse reducibility of Co^{3+} and probably even Mn^{4+} (corresponding to the temperature maxima of the 1st reduction peak in TPR-H$_2$) causes a decrease in N$_2$O conversion. This was observed in the case of the commercially prepared Cs/Co$_4$MnAlO$_x$.
- Similarly, the commercially prepared Cs/Co$_3$O$_4$ catalyst has the same chemical and phase composition as the laboratory prepared sample. However, the situation regarding crystallite size is completely reversed compared to Cs/Co$_4$MnAlO$_x$. The commercially prepared Cs/Co$_3$O$_4$ catalyst's crystallite size is 2.7× lower than for the laboratory prepared sample, which corresponds with its higher specific surface area and lower normalized loading of Cs (atoms/nm^2). The reducibility of Co^{3+} and Co^{2+} cannot be determined due to TPR peaks overlapping, but the results indicate different shapes of the cobalt spinel nanocrystals, which affect the catalytic activity [16].

3. Discussion

As determined in Section 2.1, we assume that the k value for the K/Co$_4$MnAlO$_x$ catalyst in the kinetic regime is nearly at its maximum and the increase of N$_2$O conversion on the tablet catalyst can be theoretically achieved mainly by accelerating the diffusion of reaction mixture molecules in the catalyst's pores. In contrast to this, the increase of N$_2$O conversion on the Cs/Co$_4$MnAlO$_x$ and Cs/Co$_3$O$_4$ catalysts is possible, besides accelerating internal diffusion, also by optimization of physico-chemical characteristics (crystallite size, reducibility, Cs loading and shape of Co nanocrystals) as described in Section 2.2.

The aim of this part of the paper is to show the limits of increasing conversion by intensification of mass transfer inside pores by tuning the morphology of the catalysts (pore diameter, porosity, and tortuosity) studied for N$_2$O decomposition.

The measure of reaction rate limitation by internal diffusion is the internal effectiveness factor (Equation (1)). Its dependence on catalyst morphology is represented by the effective diffusion coefficient D_{eff} (Equation (10)), which is influenced by catalyst porosity, tortuosity, and by the ratio of pore diameter and the reacting molecule's dimension. Table 4 shows the calculated values of the overall diffusivity of N_2O in a multicomponent mixture ($\overline{D}$) according to Equation (11) and the contributions of molecular (D_{mol}) and Knudsen (D_{k,N_2O}) diffusivity according to Equations (12)–(13). The results imply that the prevalent transport mechanism inside pores is Knudsen diffusion, which means that the rate of internal diffusion can be theoretically increased by increasing the pore size and increasing the value of D_k for all three catalysts. In addition, the mass transport inside the pores can be improved by increasing the ε_p/q ratio, i.e. increasing the porosity of the tablets and decreasing the pore tortuosity and thus increasing D_{eff}.

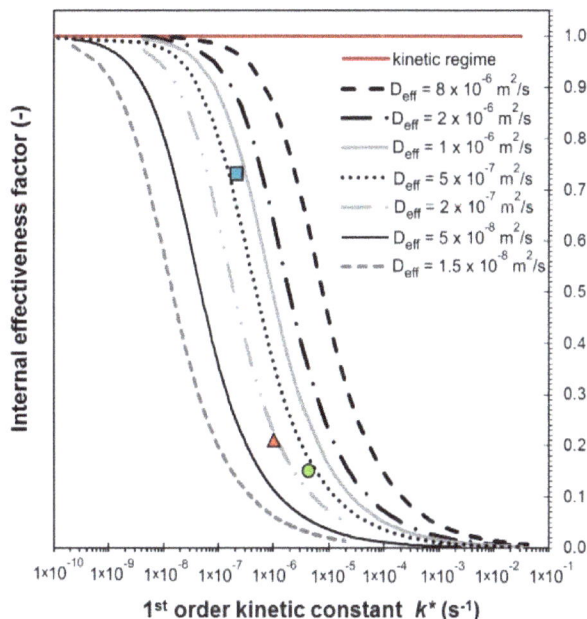

Figure 8. Dependence of internal effectiveness factor on the intrinsic 1st order kinetic constant for different values of effective diffusion coefficients. Calculated for tablets 5.1 mm × 5.1 mm. ▲ K/Co$_4$MnAlO$_x$-industry; ■ Cs/Co$_4$MnAlO$_x$-industry; ○ Cs/Co$_3$O$_4$-industry.

Table 4. Diffusion coefficients and tortuosity determined for commercially prepared catalysts. Parameters: 390 °C, 150 kPa, mean pore diameter, model gas mixture 1000 ppm N_2O + 5 mol% O_2 + 2 mol% H_2O in N_2, catalysts in tablets 5.1 mm × 5.1 mm.

Parameter/Catalyst	K/Co$_4$MnAlO$_x$-Industry	Cs/Co$_4$MnAlO$_x$-Industry	Cs/Co$_3$O$_4$-Industry
D_k (m^2 s^{-1})	$2.82 \cdot 10^{-6}$	$4.14 \cdot 10^{-6}$	$1.41 \cdot 10^{-5}$
D_{mol} (m^2 s^{-1})	$7 \cdot 10^{-3}$	$7 \cdot 10^{-3}$	$7 \cdot 10^{-3}$
$\overline{D}$ (m^2 s^{-1})	$2.82 \cdot 10^{-6}$	$4.14 \cdot 10^{-6}$	$1.41 \cdot 10^{-5}$
D_{eff} (m^2 s^{-1}) [1]	$2 \cdot 10^{-7}$	$7.5 \cdot 10^{-7}$	$3.5 \cdot 10^{-7}$
X_{N_2O} model (%) [2]	50	39	85
X_{N_2O} experiment (%) [3]	50	48	88
q [4] (-)	6.78	2.76	18.92
ε_p/q (-)	0.07	0.18	0.02

[1] Determined from Figure 8 for η evaluated from experimental data mentioned in Table 1 [2] Determined from Figure 9 [3] Experimental data of N_2O decomposition over commercial tablets measured in the laboratory at 390 °C and VHSV = 3000 m^3 m$_{bed}^{-3}$ h^{-1} (Figure 2) [4] Tortuosity calculated from Equation (10) using experimentally determined porosity (Table 3)

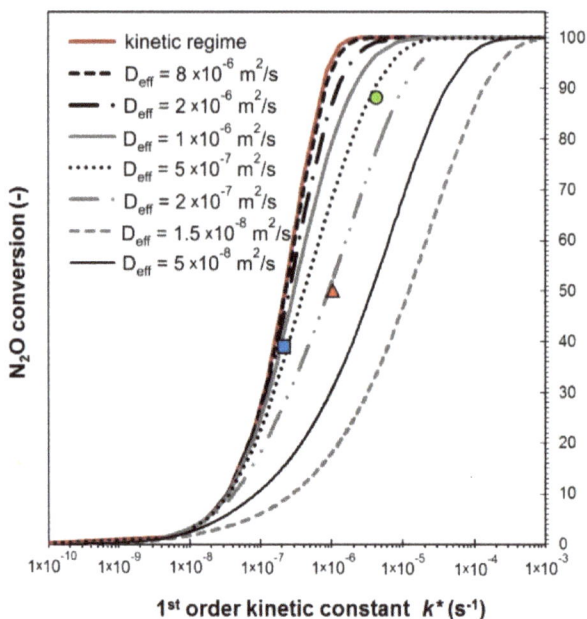

Figure 9. Dependence of theoretically achievable N_2O conversion on the 1st order kinetic constant in kinetic regime. Conversion of N_2O calculated for VHSV = 3000 $m^3 \cdot m_{bed}^{-3} \cdot h^{-1}$ and tablets 5.1 mm × 5.1 mm. ▲ K/Co_4MnAlO_x-industry; ■ Cs/Co_4MnAlO_x-industry; ● Cs/Co_3O_4-industry.

The diagram in Figure 8 shows the dependence of the internal effectiveness factor on the 1st order kinetic constant defined by Equation (4), where η was calculated according to Equations (6)–(8) for different effective diffusion coefficients D_{eff}. Based on the experimentally determined effectiveness factors and intrinsic kinetic constants, the effective diffusion coefficient can be deducted from Figure 8 and then the tortuosity can be calculated using Equation (10). The figure depicts points corresponding to each commercial catalyst. The calculation results for each commercially prepared catalyst are given in Table 4 and the following conclusion can be derived from them:

- The K/Co_4MnAlO_x-industry catalyst has the lowest D_k due to its smallest pores and therefore the lowest overall diffusivity. It also has the second highest tortuosity, which means the second highest ε_p/q ratio (0.07) (the porosity of all catalysts is approximately the same). As a result, it has the lowest effective diffusion coefficient of all three commercial catalysts. Its activity could theoretically be improved via enlargement of pores and lowering their tortuosity, which would lead to an increase of the internal effectiveness factor.
- The Cs/Co_4MnAlO_x-industry has the most favorable morphology for transport because it has the highest effective diffusion coefficient due to the lowest tortuosity and pore size, which is comparable to K/Co_4MnAlO_x-industry. As a result, this catalyst had the highest value of η (0.18) and there is still room to increase its value.

The Cs/Co_3O_4-industry catalyst has the largest pores of all three catalysts, but at the same time the highest tortuosity and thus the least favourable ε_p/q ratio (0.02), resulting in the second highest effective diffusion coefficient. D_{eff} could be further increased by lowering the tortuosity, which would lead to a significantly higher internal effectiveness factor. The diagram in Figure 8 can be used for D_{eff} estimation also for other reactions with 1st order reaction kinetics, when the value of the intrinsic kinetic constant and internal effectiveness factor are known from experiments. The diagram is valid only for shaped catalysts in the form of pellets 5.1 mm × 5.1 mm or for spheres with diameter of 5.1 mm.

To answer the third question of N_2O conversion limits achievable on tablets by tuning their morphological characteristics, the conversions of N_2O were calculated based on the 1st order intrinsic kinetic constant for different effective diffusion coefficients. The pseudo-homogeneous one-dimensional model of an ideal plug flow reactor in an isothermal regime, mentioned in our previous paper [12], was used for the calculation, where the effect of internal and external mass transport was described by the overall effectiveness factor Ω. Considering that the resistance to mass transfer by internal diffusion is prevalent, the overall effectiveness factor was substituted by the internal effectiveness factor, which was incorporated into the 1st order kinetic equation $r = \eta\, k\, c_A$. This model disregards pressure drop.

In Table 4, the calculated N_2O conversions for VHSV = 3000 $m^3.m_{bed}^{-3}.h^{-1}$ and catalyst tablets 5.1 mm $\times$ 5.1 mm are compared with experimental N_2O conversions determined at the same VHSV (Figure 2), and quite good agreement between the X_{N_2O} model and X_{N_2O} experiment was reached. The diagram showing the dependence of theoretically reachable N_2O conversion on the 1st order kinetic constant in the kinetic regime is shown in Figure 9. The dependencies imply that the catalyst with a higher kinetic constant in the kinetic regime after tabletization can demonstrate lower N_2O conversion than a sample which is less active in the kinetic regime, but has a more favorable tablet texture (pore size, porosity, and tortuosity). The figure depicts points corresponding to each commercial catalyst calculated according to reactor model. Their placement implies that the activity of the Cs/Co_4MnAlO_4-industry catalyst (39% N_2O conversion) cannot be further increased by much; since a N_2O conversion of 50% corresponds to kinetic regime. The situation is similar for the Cs/Co_3O_4-industry catalyst, which reaches 90% conversion under the given conditions, while the conversion corresponding to the kinetic regime is 100%. On the other hand, there is a relatively large room for improvement of N_2O conversion on the K/Co_4MnAlO_x-industry sample (50% conversion) since a N_2O conversion of 95% corresponds to the kinetic regime.

4. Materials and Methods

4.1. Preparation of Catalysts at Laboratory Scale

4.1.1. K/Co_4MnAlO_x and Cs/Co_4MnAlO_x

The Co-Mn-Al layered double hydroxide (LDH) precursor with Co:Mn:Al molar ratio of 4:1:1 was prepared by co-precipitation of the corresponding nitrates in $Na_2CO_3/NaOH$ solution at 25 °C and pH 10. The product was washed and dried at 60 °C and then calcined for 4 h at 500 °C in air. The prepared mixed oxide was then crushed and sieved to obtain a fraction with particle size of 0.160–0.315 mm. The next step in preparation was impregnation by the pore filling method using aqueous solutions of KNO_3 or Cs_2CO_3 to obtain a K content of 1.8 wt% and Cs content of 4 wt%, respectively. The dried products were calcined for 4 h at 500 °C in air, sieved to obtain a fraction with particle size of 0.160–0.315 mm and marked as K/Co_4MnAlO_x-lab and Cs/Co_4MnAlO_x-lab.

4.1.2. Cs/Co_3O_4

$Co(NO_3)_2 \cdot 6H_2O$ was added to an NaOH solution at room temperature while stirring. The concentration of the alkaline solution was chosen to maintain the required molar ratio of OH to Co in the reaction mixture. The resulting suspension was maintained under stirring with air bubbling at room temperature for 5 minutes. Subsequently, the solid product was filtered, washed with distilled water, and dried at 30 °C. The dried products were calcined for 4 h at 500 °C in air and sieved to obtain a fraction with a particle size of 0.160–0.315 mm. The next step in preparation was impregnation by the pore filling method using aqueous solutions of Cs_2CO_3 to obtain Cs content of 1wt%. The dried products were calcined for 4 h at 500 °C in air, sieved to obtain a fraction with particle size of 0.160–0.315 mm and marked as Cs/Co_3O_4-lab.

4.2. Preparation of Catalysts on Large Scale

4.2.1. K/Co$_4$MnAlO$_x$ and Cs/Co$_4$MnAlO$_x$

The Co−Mn−Al LDH with Co:Mn:Al molar ratio of 4:1:1 was used as a precursor of Co$_4$MnAlO$_x$ mixed oxide and was prepared by co-precipitation of the corresponding nitrates in an alkaline Na$_2$CO$_3$/NaOH solution at 25 °C and pH 10. Concentrations of 40 and 106 mol/l were used for NaOH and Na$_2$CO$_3$, respectively. The resulting suspension was filtered off, washed with water, dried at 60–105 °C, and calcined for 4 h at 500 °C in air. The resulting product was milled, impregnated with KNO$_3$ or Cs$_2$CO$_3$ to obtain K content of 1.8wt% and Cs content of 4wt%, respectively, calcined (4 h, 500 °C) and shaped into tablets (5 mm × 5 mm). For better shaping, graphite was added. The catalyst was produced by ASTIN Catalysts and Chemicals, Ltd., Litvínov, Czech Republic and marked as K/Co$_4$MnAlO$_x$-industry and Cs/Co$_4$MnAlO$_x$-industry.

4.2.2. Cs/Co$_3$O$_4$-industry

The Co(OH)$_2$–β was used as a precursor of Co$_3$O$_4$ and was prepared by co-precipitation of cobalt nitrates in an alkaline NaOH solution at 25 °C. The resulting suspension was filtered off, washed with water, dried at 60 °C and calcined for 4 h at 500 °C in air. The resulting product was milled, impregnated with Cs$_2$CO$_3$ to obtain Cs content of 1 wt%, dried, calcined (4 h at 500 °C) and shaped into tablets (5 mm × 5 mm). For better shaping, graphite was added. The catalyst was produced by ASTIN Catalysts and Chemicals, Ltd., Litvínov, Czech Republic and marked as Cs/Co$_3$O$_4$.

4.3. Characterization of Catalysts

The chemical composition of samples was determined by atomic absorption spectrometry or atomic emission spectrometry using a SpectrAA880 instrument (Varian, Palo Alto, CA, USA) after dissolving the samples in hydrochloric acid.

Phase composition and microstructural properties were determined using the X-ray powder diffraction (XRD) technique. XRD patterns were obtained using a Rigaku SmartLab diffractometer (Rigaku, Tokio, Japan) with D/teX Ultra 250 detector. The source of X-ray irradiation was Co tube (CoKα, λ$_1$ = 0.178892 nm, λ$_2$ = 0.179278 nm) operated at 40 kV and 40 mA. Incident slits were set up to irradiate a 10 mm × 10 mm area of the sample (automatic divergence slits) constantly. The XRD patterns were collected in a 2θ range 5–90° with a step size of 0.01° and speed 0.5 deg.min^{-1}. The samples were rotated (30 rpm) during the measurement. Lattice parameters were refined using the LeBail method; the sizes of coherent domains were calculated using Scherrer's formula as an average of the five strongest diffractions with hkl symbols (220), (311), (400), (511), and (440) evaluated.

The surface areas of the catalysts were determined by N$_2$ adsorption/desorption at −196 °C using the ASAP 2010 instrument (Micromeritics, Ottawa, Ontario, Canada) and evaluated by the Brunauer–Emmett–Teller (BET) method. Prior to the measurement, both laboratory and commercially prepared samples were crushed to obtain a fraction <0.16 mm and dried at 120 °C for at least 12 h.

In addition, textural properties of the commercially prepared catalysts were determined using the 3Flex physisorption set-up (Micromeritics, Ottawa, Ontario, Canada) and pieces of catalyst tablets were used for measurements. Before the physisorption measurement, the samples were dried at 120 °C for 12 h in vacuum. The nitrogen adsorption–desorption isotherms were measured at 77 K. The specific surface area, S_{BET}, was calculated based on the BET method. The mesopore surface area, S_{meso}, and the micropore volume, V_{micro}, were evaluated by using the t-plot method, applying the Broekhoff-de-Boer standard isotherm. The total pore volume, V_{net}, was evaluated as the nitrogen volume adsorbed at maximum relative pressure ($p/p_0 = 0.99$). The pore-size distribution was evaluated from adsorption data applying the Barrett-Joyner-Halenda (BJH) method, assuming cylindrical pore geometry and using the Broekhoff-de-Boer standard isotherm with Faas correction.

The porous structure (mesopores and macropores) of commercial catalysts was characterized using a mercury porosimeter AutoPore III (Micromeritics, Ottawa, Ontario, Canada) working in the range of 0.1–400 MPa. Prior to the measurement, the samples were dried at 130 °C. Helium pycnometry (AccuPyc1330, Micrometrics, Ottawa, Ontario, Canada) was used to evaluate the sample's true density.

Temperature programmed reduction (H_2-TPR) measurements of the calcined samples (0.025 g) were performed using a system described in detail in [21], with a H_2/N_2 mixture (10 mol% H_2), flow rate 50 ml/min and linear temperature increase of 20 °C/min up to 1000 °C. The change in H_2 concentration was evaluated using an Omnistar 300 mass spectrometer (Pfeiffer Vacuum, Asslar, Germany).

4.4. Catalytic Tests on Laboratory Scale

Catalytic experiments for laboratory prepared catalysts and crushed pellets with particle size 0.165–0.315 mm were performed in a reactor with an internal diameter of 5.5 mm. A catalytic bed contained 0.1–0.3 g of the sample; the total flow rate was 100 ml/min (20 °C, 101,325 Pa) containing 0.1 mol% N_2O, 5 mol% O_2 and 2 mol% water vapor in N_2 leading to GHSV = 20,000–60,000 l kg^{-1}h^{-1} was applied to simulate the real waste gas from a nitric acid plant.

Experiments with commercially prepared pellets were performed in an integral reactor with an internal diameter of 50 mm. The catalytic bed contained 10 ml of the sample, the total flow rate was 500 ml/min (20 °C, 101,325 Pa) containing 0.1 mol% N_2O, 5 mol% O_2 and 2 mol% water vapor in N_2 to simulate the real waste gas from a nitric acid plant. VHSV = 3000 m^3 m$_{bed}$$^{-3}$ h^{-1} was applied.

4.5. Pilot Plant Catalytic Tests

Pilot plant catalytic measurements of N_2O decomposition were performed in a fixed bed stainless steel reactor (0.31 m internal diameter) in the temperature range of 350 to 450 °C and inlet pressure of 0.6 MPa. The reactor was connected to the bypassed tail gas from the nitric acid production plant downstream from the SCR NO_x/NH_3 catalyst. The catalyst tablets (K/Co_4MnAlO_x-industry: 69.1 kg weight, 62.5 cm bed height, 1334 kg m^{-3} bed density; Cs/Co_4MnAlO_x-industry: 74.4 kg weight, 62.5 cm bed height, 1436 kg m^{-3} bed density; Cs/Co_3O_4-industry: 75.9 kg weight, 60 cm bed height, 1527 kg m^{-3} bed density) were placed on a stainless steel grate sieve and a bed of ceramic spheres (diameter of 8 mm) 5 cm in height. On the catalyst layer, again more ceramic spheres (height of 1 cm), a sieve, and a last layer of ceramic spheres (height of 6.5 cm) were placed. The feed to the reactor varied between 300 and 600 kg h^{-1} and contained typically 400–700 ppm of N_2O together with oxygen, water vapor, and a low concentration of NO, NO_2, and NH_3 (0–70 ppm of NO_x, 0–30 ppm of NH_3). The variable composition of gas mixture at the reactor inlet was due to the fact that the inlet gas was the real waste gas from the nitric acid plant downstream from the SCR NO_x/NH_3 unit. Infrared and chemiluminescence online analyzers were used for analysis of the gas at the catalyst bed inlet and outlet: Sick (N_2O), Horiba (NO, NO_2), ABB modul Uras 26 (N_2O), and ABB Limas11 (NO, NO_2, and NH_3). The reactor was equipped with online monitoring of the concentrations of all measured gas components, the temperature in the catalyst bed, and the pressure drop.

4.6. Determination of Internal Effectiveness Factor

The internal effectiveness factor expresses the influence of the internal mass transport limitation on the overall rate of catalytic reaction and is defined by Equation (1).

$$\eta = \frac{Actual\ overall\ rate\ of\ reaction}{Rate\ of\ reaction\ that\ would\ result\ if\ entire\ interior\ was\ exposed\ to\ external\ pellet\ surface\ conditions} \tag{1}$$

Both the experimental determination and the theoretical calculation of η were used. For the determination from experimental data, N_2O conversions at the same reaction conditions had to be used for evaluation of both reaction rates in Equation (1). We assume that the rate of reaction on the catalyst in the form of grains is not affected by internal diffusion and can be used as the reaction rate at external pellet surface conditions. For the evaluation of the actual overall rate of reaction, N_2O conversions over pellets were recalculated to the same GHSV which was used for the testing of grains (GHSV = 60,000 l h^{-1}kg^{-1} = 0.00025 m^{3}s^{-1}kg^{-1}) using 1st order kinetic equation and the material balance of an ideal plug flow reactor:

$$X_{N_2O} = 1 - e^{\frac{-k}{GHSV}} \tag{2}$$

Kinetic constants over catalyst pellets and grains were calculated from N_2O conversions according to Equations (3) and (4):

$$k = \frac{\ln\left(\frac{1}{1-X_{N_2O}}\right)}{\frac{w}{V}} \tag{3}$$

$$k^* = k/\rho_c \tag{4}$$

where k (m^3 s^{-1} kg^{-1}) and k^* (s^{-1}) are 1st order kinetic constants, w (kg) is the weight of catalyst, $\dot{V}$ (m^3 s^{-1}) the total gas flow and ρ_c (kg m^{-3}) is the bulk density of catalyst determined by Hg porosimetry. Then both reaction rates, defined as the amount of N_2O converted per kg of catalysts per hour (mol$_{N2O}$ kg$_{cat}$$^{-1}$ h^{-1}), were calculated (r_{grain} and r_{pellet}, respectively). The ratio of r_{grain}/r_{pellet} provides the internal effectiveness factor of the pellets utilization.

$$r = k \cdot c_{A0} \cdot (1 - X_{N_2O}) \tag{5}$$

where X_{N_2O} is N_2O conversion (-), c_{A0} inlet N_2O concentration (mol m^{-3}).

For the theoretical estimation of the internal effectiveness factor η for catalyst pellets, Equations (6)–(13) were used [22].

$$\eta = \frac{1}{\Phi_{gen}} \cdot tanh\ \Phi_{gen} \quad \text{for } \Phi gen < 4 \tag{6}$$

$$\eta = \frac{1}{\Phi_{gen}} \quad \text{for } \Phi gen > 4 \tag{7}$$

where the generalized Thiele modulus Φ_{gen} is defined:

$$\Phi_{gen} = \frac{V}{S} \cdot \sqrt{\frac{k \cdot \rho_c}{D_{eff}}} \tag{8}$$

where for cylinder pellet with equal diameter and height ($d_p = h$):

$$\frac{V}{S} = \frac{d_p}{6} \tag{9}$$

The effective diffusion coefficient D_{eff} is dependent on the morphology of porous catalyst:

$$D_{eff} = \frac{\varepsilon_p}{q}\overline{D} \tag{10}$$

The ε_p/q ratio between 0.05–0.1 was published previously [23]. For the determination of the overall diffusivity of N_2O in a multicomponent mixture ($\overline{D}$), the contributions of molecular ($D_{i/j}$) and Knudsen (D_{k,N_2O}) diffusivity was considered together with the stoichiometry of the reaction [12]:

$$\frac{1}{\overline{D}} = \frac{1}{D_{k,N_2O}} + \frac{x_{N_2O} + x_{N_2O}}{D_{\frac{N_2O}{N_2}}} + \frac{0.5 \cdot x_{N_2O} + 0.5 \cdot x_{N_2O}}{D_{\frac{N_2O}{O_2}}} + \sum_{j=1}^{n} \frac{x_{N_2O}}{D_{N_2O/j}} \tag{11}$$

where j is a component of gas mixture. The Knudsen diffusivity D_{k,N_2O} for pore with radius r_0 was determined [23] as:

$$D_{k,N_2O} = 97 \cdot r_0 \cdot \sqrt{\frac{T}{M_{N_2O}}} \tag{12}$$

Binary diffusion coefficients $D_{i/j}$ were calculated from Equation (13) [24]:

$$D_{i/j} = \frac{\frac{10^{-2} \cdot T^{\frac{7}{4}} \cdot \left(M_j^{-1} + M_i^{-1}\right)^{\frac{1}{2}}}{p}}{\left[(\Sigma v)_i^{\frac{1}{3}} + (\Sigma v)_j^{\frac{1}{3}}\right]^2} \tag{13}$$

where $(\Sigma v)_{N_2} = 17.9$, $(\Sigma v)_{N_2O} = 35.9$, $(\Sigma v)_{O_2} = 16.6$, $(\Sigma v)_{H_2O} = 12.7$, $(\Sigma v)_{He} = 2.88$, $(\Sigma v)_{NO} = 11.17$.

5. Conclusions

Three cobalt containing mixed oxides modified with alkali metal promoters (K/Co$_4$MnAlO$_x$, Cs/Co$_4$MnAlO$_x$ and Cs/Co$_3$O$_4$) were prepared commercially in the form of tablets. Their catalytic properties were tested for N$_2$O catalytic decomposition in the laboratory and in a pilot plant reactor connected to the waste gas from a nitric acid production plant and compared to the laboratory prepared catalysts with the same chemical composition.

It was found that the different order of catalytic activities in the pilot-plant experiment compared to the laboratory prepared catalysts was caused by the different conditions of the pilot plant experiment, particularly by the presence of NO$_x$ and NH$_3$.

Comparison of N$_2$O conversions of both laboratory and commercially prepared catalysts in the kinetic regime showed that K/Co$_4$MnAlO$_x$ has active sites with nearly the same activity as the laboratory prepared sample and the kinetic constant of the tablets is lower mainly due to the inhibiting effect of internal diffusion. N$_2$O conversion in such a case can be improved by increasing the internal effectiveness factor, i.e. by increasing D_{eff}, which can be achieved by increasing the pore size and also by increasing porosity and decreasing tortuosity.

Conversely, commercially prepared Cs/Co$_4$MnAlO$_x$ and Cs/Co$_3$O$_4$ catalysts possess active sites which are less active than those of laboratory prepared catalysts. N$_2$O conversions on these tablets were lower partly due to the inhibiting effect of internal diffusion and also due to observed differences in reducibility, primary particle sizes, and probably also promoter normalized loading (atoms/nm^2) compared to laboratory prepared catalysts of the same chemical composition. N$_2$O conversion on such tablets can be increased mainly by optimizing the preparation procedure and increasing the activity of the active sites (by increasing k).

The shown dependencies answer the question from the title of the paper: It is possible that a catalyst with a great performance in the kinetic regime can be less effective when prepared commercially in the form of tablets. The major reason for such changes in catalyst activity order after scale up are morphological characteristics (limiting mass transfer rates) and for that reason these can sometimes be even more important than the catalyst composition itself.

Author Contributions: Conceptualization, L.O.; methodology, L.O.; investigation, A.K.; data curation, A.K., L.M., K.P., and D.F.; writing—original draft preparation, L.O.; writing—review and editing, K.P. and L.M.

Funding: This research was funded by ERDF OP RDE project No. CZ.02.1.01/0.0/0.0/15_019/0000853 and by OP RDI project No. CZ.1.05/2.1.00/19.0389.

Acknowledgments: We thank Alexandr Martaus from the Institute of Environmental Technology VŠB-Technical University of Ostrava for the XRD measurements.

Conflicts of Interest: The authors declare no conflict of interest.

Symbols

a	lattice parameter (nm)
c_A	concentration of component A (N_2O) (mol.m^{-3})
c_{A0}	initial concentration of component A (N_2O) (mol.m^{-3})
$\overline{D}$	overall diffusivity (m$_2$.s^{-1})
D_{eff}	effective diffusion coefficient (m^2.s^{-1})
D_{ij}	binary diffusion coefficient of molecular diffusivity (m^2.s^{-1})
D_k	Knudsen diffusivity (m^2.s^{-1})
d_p	equal catalyst particle diameter (m)
k	kinetic constant, 1st order rate law (m^{3}s^{-1}kg^{-1})
k^*	kinetic constant, 1st order rate law (s^{-1})
L_c	size of coherent domains
M_i	molar weight of compound i (g.mol^{-1})
q	tortuosity (-)
r	reaction rate per unit weight of catalyst (mol.kg^{-1}.h^{-1})
r_o	catalyst pore radius (m)
S_{BET}	specific surface area (m^2 g^{-1})
V	volume of the catalyst bed (m^3)
$\dot{V}$	total volumetric flow (m^3 s^{-1})
V_{net}	total pore volume (cm$^3_{liq}$ g^{-1})
v_i	molar volume of component i (m^3.kmol^{-1})
w	catalyst weight (g)
X_A	conversion of component A (N_2O) (-)
x_{N2O}	molar fraction of N_2O entering the reactor (-)
Greek symbols:	
ε_p	porosity of catalyst particle (-)
Φ_{gen}	generalized Thiele modul (-)
η	internal effectiveness factor (-)
ρ_c	bulk density of the catalyst (kg.m^{-3})

References

1. Chromčáková, Ž.; Obalová, L.; Kustrowski, P.; Drozdek, M.; Karásková, K.; Jirátová, K.; Kovanda, F. Optimization of Cs content in Co–Mn–Al mixed oxide as catalyst for N_2O decomposition. *Res. Chem. Intermed.* **2015**, *41*, 9319–9332. [CrossRef]

2. Obalová, L.; Karásková, K.; Jirátová, K.; Kovanda, F. Effect of potassium in calcined Co–Mn–Al layered double hydroxide on the catalytic decomposition of N_2O. *Appl. Catal. B Environ.* **2009**, *90*, 132–140. [CrossRef]

3. Lucie Obalová, S.M. Cesium doped Co_3O_4 spinel for N_2O decomposition. In Proceedings of the 5th International Conference on Chemical Technology, Mikulov, Czech Republic, 10–12 April 2017.

4. Pacultová, K.; Karásková, K.; Kovanda, F.; Jirátová, K.; Šrámek, J.; Kustrowski, P.; Kotarba, A.; Chromčáková, Ž.; Kočí, K.; Obalová, L. K-Doped Co–Mn–Al Mixed Oxide Catalyst for N_2O Abatement from Nitric Acid Plant Waste Gases: Pilot Plant Studies. *Ind. Eng. Chem. Res.* **2016**, *55*, 7076–7084. [CrossRef]

5. Pacultová, K.; Chromčáková, Ž.; Obalová, L. Cs doped cobalt based deN_2O catalyst-Pilot plant results. *Waste Forum* **2017**, *2*, 54–63.

6. Inger, M.; Kowalik, P.; Saramok, M.; Wilk, M.; Stelmachowski, P.; Maniak, G.; Granger, P.; Kotarba, A.; Sojka, Z. Laboratory and pilot scale synthesis, characterization and reactivity of multicomponent cobalt spinel catalyst for low temperature removal of N_2O from nitric acid plant tail gases. *Catal. Today* **2011**, *176*, 365–368. [CrossRef]

7. Inger, M.; Wilk, M.; Saramok, M.; Grzybek, G.; Grodzka, A.; Stelmachowski, P.; Makowski, W.; Kotarba, A.; Sojka, Z. Cobalt Spinel Catalyst for N_2O Abatement in the Pilot Plant Operation–Long-Term Activity and Stability in Tail Gases. *Ind. Eng. Chem. Res.* **2014**, *53*, 10335–10342. [CrossRef]

8. Obalová, L.; Karásková, K.; Wach, A.; Kustrowski, P.; Mamulová-Kutláková, K.; Michalik, S.; Jirátová, K. Alkali metals as promoters in Co–Mn–Al mixed oxide for N_2O decomposition. *Appl. Catal. A Gen.* **2013**, *462–463*, 227–235.

9. Karásková, K.; Obalová, L.; Jirátová, K.; Kovanda, F. Effect of promoters in Co–Mn–Al mixed oxide catalyst on N_2O decomposition. *Chem. Eng. J.* **2010**, *160*, 480–487. [CrossRef]

10. Yamada, K.; Pophal, C.; Segawa, K. Selective catalytic reduction of N_2O by C_3H_6 over Fe-ZSM-5. *Microporous Mesoporous Mater.* **1998**, *21*, 549–555. [CrossRef]

11. Chmielarz, L.; Kuśtrowski, P.; Rafalska-Łasocha, A.; Dziembaj, R. Selective oxidation of ammonia to nitrogen on transition metal containing mixed metal oxides. *Appl. Catal. B Environ.* **2005**, *58*, 235–244. [CrossRef]

12. Obalová, L.; Jirátová, K.; Karásková, K.; Chromčáková, Ž. N_2O catalytic decomposition – From laboratory experiment to industry reactor. *Catal. Today* **2012**, *191*, 116–120. [CrossRef]

13. Pasha, N.; Lingaiah, N.; Babu, N.S.; Reddy, P.S.S.; Prasad, P.S.S. Studies on cesium doped cobalt oxide catalysts for direct N_2O decomposition in the presence of oxygen and steam. *Catal. Commun.* **2008**, *10*, 132–136. [CrossRef]

14. Stelmachowski, P.; Maniak, G.; Kotarba, A.; Sojka, Z. Strong electronic promotion of Co_3O_4 towards N_2O decomposition by surface alkali dopants. *Catal. Commun.* **2009**, *10*, 1062–1065. [CrossRef]

15. Galejová, K.; Obalová, L.; Jiratova, K.; Pacultová, K.; Kovanda, F. N_2O catalytic decomposition—Effect of pelleting pressure on activity of Co-Mn-Al mixed oxide catalysts. *Chem. Pap.* **2009**, *63*, 172–179. [CrossRef]

16. Gudyka, S.; Grzybek, G.; Gryboś, J.; Indyka, P.; Leszczyński, B.; Kotarba, A.; Sojka, Z. Enhancing the deN_2O activity of the supported Co_3O_4 | α-Al_2O_3 catalyst by glycerol-assisted shape engineering of the active phase at the nanoscale. *Appl. Catal. B Environ.* **2017**, *201*, 339–347. [CrossRef]

17. Klyushina, A.; Pacultová, K.; Karásková, K.; Jirátová, K.; Ritz, M.; Fridrichová, D.; Volodarskaja, A.; Obalová, L. Effect of preparation method on catalytic properties of Co-Mn-Al mixed oxides for N_2O decomposition. *J. Mol. Catal. A Chem.* **2016**, *425*, 237–247. [CrossRef]

18. Kaczmarczyk, J.; Zasada, F.; Janas, J.; Indyka, P.; Piskorz, W.; Kotarba, A.; Sojka, Z. Thermodynamic Stability, Redox Properties, and Reactivity of Mn_3O_4, Fe_3O_4, and Co_3O_4 Model Catalysts for N_2O Decomposition: Resolving the Origins of Steady Turnover. *ACS Catal.* **2016**, *6*, 1235–1246. [CrossRef]

19. Obalová, L.; Maniak, G.; Karásková, K.; Kovanda, F.; Kotarba, A. Electronic nature of potassium promotion effect in Co–Mn–Al mixed oxide on the catalytic decomposition of N_2O. *Catal. Commun.* **2011**, *12*, 1055–1058. [CrossRef]

20. Maniak, G.; Stelmachowski, P.; Zasada, F.; Piskorz, W.; Kotarba, A.; Sojka, Z. Guidelines for optimization of catalytic activity of 3d transition metal oxide catalysts in N_2O decomposition by potassium promotion. *Catal. Today* **2011**, *176*, 369–372. [CrossRef]

21. Jirátová, K.; Mikulová, J.; Klempa, J.; Grygar, T.; Bastl, Z.; Kovanda, F. Modification of Co–Mn–Al mixed oxide with potassium and its effect on deep oxidation of VOC. *Appl. Catal. A Gen.* **2009**, *361*, 106–116. [CrossRef]

22. Fogler, H.S. *Elements of Chemical Reaction Engineering*, 3rd ed.; Prentice Hall PTR: Upper Saddle River, NJ, USA, 1999.

23. Kaptejn, F.; Moulijn, J.A. *Handbook of Heterogeneous Catalysis*; Ertl, G., Knözinger, H., Weitkamp, J., Eds.; Wiley: Weinheim, Germany, 1996.

24. Fuller, E.N.; Schettler, P.D.; Giddings, J.C. New method for prediction of binary gas-phase diffusion coefficients. *Ind. Eng. Chem.* **1966**, *58*, 18–27. [CrossRef]

MDPI

St. Alban-Anlage 66

4052 Basel

Switzerland

Tel. +41 61 683 77 34

Fax +41 61 302 89 18

www.mdpi.com

Catalysts Editorial Office

E-mail: catalysts@mdpi.com

www.mdpi.com/journal/catalysts

www.ingramcontent.com/pod-product-compliance
Lightning Source LLC
Chambersburg PA
CBHW051846210326
41597CB00033B/5796